CHEMICAL ENGINEERING

Solutions to the Problems in Chemical Engineering Volumes 2 and 3

Related Butterworth-Heinemann Titles in the Chemical Engineering Series by
J. M. COULSON & J. F. RICHARDSON

Chemical Engineering, Volume 1, Sixth edition
Fluid Flow, Heat Transfer and Mass Transfer
(with J. R. Backhurst and J. H. Harker)

Chemical Engineering, Volume 3, Third edition
Chemical and Biochemical Reaction Engineering, and Control
(edited by J. F. Richardson and D. G. Peacock)

Chemical Engineering, Volume 6, Third edition
Chemical Engineering Design
(R. K. Sinnott)

Chemical Engineering, Solutions to Problems in Volume 1
(J. R. Backhurst, J. H. Harker and J. F. Richardson)

Chemical Engineering, Solutions to Problems in Volume 2
(J. R. Backhurst, J. H. Harker and J. F. Richardson)

Coulson & Richardson's

CHEMICAL ENGINEERING

J. M. COULSON and J. F. RICHARDSON

Solutions to the Problems in Chemical Engineering Volume 2 (5th edition) and Volume 3 (3rd edition)

By

J. R. BACKHURST and J. H. HARKER

University of Newcastle upon Tyne

With

J. F. RICHARDSON

University of Wales Swansea

OXFORD   AMSTERDAM   BOSTON   LONDON   NEW YORK   PARIS
SAN DIEGO   SAN FRANCISCO   SINGAPORE   SYDNEY   TOKYO

Butterworth-Heinemann
An imprint of Elsevier Science
Linacre House, Jordan Hill, Oxford OX2 8DP
225 Wildwood Avenue, Woburn, MA 01801-2041

First published 2002
Transferred to digital printing 2004
Copyright © 2002, J.F. Richardson and J.H. Harker. All rights reserved

The right of J.F. Richardson and J.H. Harker to be identified as the authors of this work
has been asserted in accordance with the Copyright, Designs
and Patents Act 1988

No part of this publication may be
reproduced in any material form (including
photocopying or storing in any medium by electronic
means and whether or not transiently or incidentally
to some other use of this publication) without the
written permission of the copyright holder except
in accordance with the provisions of the Copyright,
Designs and Patents Act 1988 or under the terms of a
licence issued by the Copyright Licensing Agency Ltd,
90 Tottenham Court Road, London, England W1T 4LP.
Applications for the copyright holder's written permission
to reproduce any part of this publication should be
addressed to the publishers

British Library Cataloguing in Publication Data
A catalogue record for this book is available from the British Library

Library of Congress Cataloguing in Publication Data
A catalogue record for this book is available from the Library of Congress

ISBN 0 7506 5639 5

For information on all Butterworth-Heinemann publications visit our website at www.bh.com

Contents

Preface	vii
Preface to the Second Edition of Volume 5	ix
Preface to the First Edition of Volume 5	xi
Factors for Conversion of SI units	xiii

Solutions to Problems in Volume 2

2-1	Particulate solids	1
2-2	Particle size reduction and enlargement	8
2-3	Motion of particles in a fluid	14
2-4	Flow of fluids through granular beds and packed columns	34
2-5	Sedimentation	39
2-6	Fluidisation	44
2-7	Liquid filtration	59
2-8	Membrane separation processes	76
2-9	Centrifugal separations	79
2-10	Leaching	83
2-11	Distillation	98
2-12	Absorption of gases	150
2-13	Liquid–liquid extraction	171
2-14	Evaporation	181
2-15	Crystallisation	216
2-16	Drying	222
2-17	Adsorption	231
2-18	Ion exchange	234
2-19	Chromatographic separations	235

Solutions to Problems in Volume 3

3-1	Reactor design — general principles	237
3-2	Flow characteristics of reactors — flow modelling	262
3-3	Gas–solid reactions and reactors	265
3-4	Gas–liquid and gas–liquid–solid reactors	271

3-5	Biochemical reaction engineering	285
3-7	Process control	294

(*Note: The equations quoted in Sections* 2.1–2.19 *appear in Volume* 2 *and those in Sections* 3.1–3.7 *appear in Volume* 3. *As far as possible, the nomenclature used in this volume is the same as that used in Volumes* 2 *and* 3 *to which reference may be made.*)

Preface

Each of the volumes of the Chemical Engineering Series includes numerical examples to illustrate the application of the theory presented in the text. In addition, at the end of each volume, there is a selection of problems which the reader is invited to solve in order to consolidate his (or her) understanding of the principles and to gain a better appreciation of the order of magnitude of the quantities involved.

Many readers who do not have ready access to assistance have expressed the desire for solutions manuals to be available. This book, which is a successor to the old Volume 5, is an attempt to satisfy this demand as far as the problems in Volumes 2 and 3 are concerned. It should be appreciated that most engineering problems do not have unique solutions, and they can also often be solved using a variety of different approaches. If therefore the reader arrives at a different answer from that in the book, it does not necessarily mean that it is wrong.

This edition of the Solutions Manual which relates to the fifth edition of Volume 2 and to the third edition of Volume 3 incorporates many new problems. There may therefore be some mismatch with earlier editions and, as the volumes are being continually revised, they can easily get out-of-step with each other.

None of the authors claims to be infallible, and it is inevitable that errors will occur from time to time. These will become apparent to readers who use the book. We have been very grateful in the past to those who have pointed out mistakes which have then been corrected in later editions. It is hoped that the present generation of readers will prove to be equally helpful!

J. F. R.

Preface to the Second Edition of Volume 5

It is always a great joy to be invited to prepare a second edition of any book and on two counts. Firstly, it indicates that the volume is proving useful and fulfilling a need, which is always gratifying and secondly, it offers an opportunity of making whatever corrections are necessary and also adding new material where appropriate. With regard to corrections, we are, as ever, grateful in the extreme to those of our readers who have written to us pointing out, mercifully minor errors and offering, albeit a few of what may be termed 'more elegant solutions'. It is important that a volume such as this is as accurate as possible and we are very grateful indeed for all the contributions we have received which, please be assured, have been incorporated in the preparation of this new edition.

With regard to new material, this new edition is now in line with the latest edition, that is the Fourth, of Volume 2 which includes new sections, formerly in Volume 3 with, of course, the associated problems. The sections are: 17, Adsorption; 18, Ion Exchange; 19, Chromatographic Separations and 20, Membrane Separation Processes and we are more than grateful to Professor Richardson's colleagues at Swansea, J. H. Bowen, J. R. Conder and W. R. Bowen, for an enormous amount of very hard work in preparing the solutions to these problems. A further and very substantial addition to this edition of Volume 5 is the inclusion of solutions to the problems which appear in *Chemical Engineering, Volume 3—Chemical & Biochemical Reactors & Process Control* and again, we are greatly indebted to the authors as follows:

3.1 Reactor Design—J. C. Lee
3.2 Flow Characteristics of Reactors—J. C. Lee
3.3 Gas–Solid Reactions and Reactors—W. J. Thomas and J. C. Lee
3.4 Gas–Liquid and Gas–Liquid–Solid Reactors—J. C. Lee
3.5 Biological Reaction Engineering—M. G. Jones and R. L. Lovitt
3.6 Process Control—A. P. Wardle

and also of course, to Professor Richardson himself, who, with a drive and enthusiasm which seems to be getting ever more vigorous as the years proceed, has not only arranged for the preparation of this material and overseen our efforts with his usual meticulous efficiency, but also continues very much in master-minding this whole series. We often reflect on the time when, in preparing 150 solutions for the original edition of Volume 4, the worthy Professor pointed out that we had only 147 correct, though rather reluctantly agreed that we might still just merit first class honours! Whatever, we always have and we are sure that we always will owe him an enormous debt of gratitude.

We must also offer thanks to our seemingly ever-changing publishers for their drive, efficiency and encouragement and especially to the present staff at Butterworth-Heinemann for not inconsiderable efforts in locating the manuscript for the present edition which was apparently lost somewhere in all the changes and chances of the past months.

We offer a final thought as to the future where there has been a suggestion that the titles Volume 4 and Volume 5 may find themselves hijacked for new textural volumes, coupled with a proposal that the solutions offered here hitherto may just find a new resting place on the Internet. Whatever, we will continue with our efforts in ensuring that more and more solutions find their way into the text in Volumes 1 and 2 and, holding to the view expressed in the Preface to the First Edition of Volume 4 that '... worked examples are essential to a proper understanding of the methods of treatment given in the various texts', that the rest of the solutions are accessible to the widest group of students and practising engineers as possible.

Newcastle upon Tyne, 1997　　　　　　　　　　　　　　　　　　　　　　　J. R. BACKHURST
　　J. H. HARKER

(Note: Some of the chapter numbers quoted here have been amended in the later editions of the various volumes.)

Preface to the First Edition of Volume 5

IN THE preface to the first edition of *Chemical Engineering*, Volume 4, we quoted the following paragraph written by Coulson and Richardson in their preface to the first edition of *Chemical Engineering*, Volume 1:

> 'We have introduced into each chapter a number of worked examples which we believe are essential to a proper understanding of the methods of treatment given in the text. It is very desirable for a student to understand a worked example before tackling fresh practical problems himself. Chemical Engineering problems require a numerical answer, and it is essential to become familiar with the different techniques so that the answer is obtained by systematic methods rather than by intuition.'

It is with these aims in mind that we have prepared Volume 5, which gives our solutions to the problems in the third edition of *Chemical Engineering*, Volume 2. The material is grouped in sections corresponding to the chapters in that volume and the present book is complementary in that extensive reference has been made to the equations and sources of data in Volume 2 at all stages. The book has been written concurrently with the revision of Volume 2 and SI units have been used.

In many ways these problems are more taxing and certainly longer than those in Volume 4, which gives the solutions to problems in Volume 1, and yet they have considerable merit in that they are concerned with real fluids and, more importantly, with industrial equipment and conditions. For this reason we hope that our efforts will be of interest to the professional engineer in industry as well as to the student, who must surely take some delight in the number of tutorial and examination questions which are attempted here.

We are again delighted to acknowledge the help we have received from Professors Coulson and Richardson in so many ways. The former has the enviable gift of providing the minimum of data on which to frame a simple key question, which illustrates the crux of the problem perfectly, whilst the latter has in a very gentle and yet thorough way corrected our mercifully few mistakes and checked the entire work. Our colleagues at the University of Newcastle upon Tyne have again helped us, in many cases unwittingly, and for this we are grateful.

Newcastle upon Tyne, 1978

J. R. BACKHURST
J. H. HARKER

Factors for conversion of SI units

mass		*pressure*	
1 lb	0.454 kg	1 lbf/in^2	6.895 kN/m^2
1 ton	1016 kg	1 atm	101.3 kN/m^2
		1 bar	100 kN/m
length		1 ft water	2.99 kN/m^2
1 in	25.4 mm	1 in water	2.49 N/m^2
1 ft	0.305 m	1 in Hg	3.39 kN/m^2
1 mile	1.609 km	1 mm Hg	133 N/m^2
time		*viscosity*	
1 min	60 s	1 P	0.1 N s/m^2
1 h	3.6 ks	1 lb/ft h	0.414 mN s/m^2
1 day	86.4 ks	1 stoke	10^{-4} m^2/s
1 year	31.5 Ms	1 ft^2/h	0.258 cm^2/s
area		*mass flow*	
1 in^2	645.2 mm^2	1 lb/h	0.126 g/s
1 ft^2	0.093 m^2	1 ton/h	0.282 kg/s
		1 lb/h ft^2	1.356 g/s m^2
volume		*thermal*	
1 in^3	16,387.1 mm^3	1 Btu/h ft^2	3.155 W/m^2
1 ft^3	0.0283 m^3	1 Btu/h ft^2 °F	5.678 W/m^2 K
1 UK gal	4546 cm^3	1 Btu/lb	2.326 kJ/kg
1 US gal	3786 cm^3	1 Btu/lb °F	4.187 kJ/kg K
		1 Btu/h ft °F	1.731 W/m K
force		*energy*	
1 pdl	0.138 N	1 kWh	3.6 MJ
1 lb	4.45 N	1 therm	106.5 MJ
1 dyne	10^{-5} N		
energy		*calorific value*	
1 ft lb	1.36 J	1 Btu/ft^3	37.26 kJ/m^3
1 cal	4.187 J	1 Btu/lb	2.326 kJ/kg
1 erg	10^{-7} J		
1 Btu	1.055 kJ		
power		*density*	
1 h.p.	745 W	1 lb/ft^3	16.02 kg/m^3
1 Btu/h	0.293 W		

SECTION 2-1

Particulate Solids

PROBLEM 1.1

The size analysis of a powdered material on a mass basis is represented by a straight line from 0 per cent at 1 μm particle size to 100 per cent by mass at 101 μm particle size. Calculate the surface mean diameter of the particles constituting the system.

Solution

See Volume 2, Example 1.1.

PROBLEM 1.2

The equations giving the number distribution curve for a powdered material are $dn/dd = d$ for the size range 0–10 μm, and $dn/dd = 100,000/d^4$ for the size range 10–100 μm where d is in μm. Sketch the number, surface and mass distribution curves and calculate the surface mean diameter for the powder. Explain briefly how the data for the construction of these curves may be obtained experimentally.

Solution

See Volume 2, Example 1.2.

PROBLEM 1.3

The fineness characteristic of a powder on a cumulative basis is represented by a straight line from the origin to 100 per cent undersize at a particle size of 50 μm. If the powder is initially dispersed uniformly in a column of liquid, calculate the proportion by mass which remains in suspension in the time from commencement of settling to that at which a 40 μm particle falls the total height of the column. It may be assumed that Stokes' law is applicable to the settling of the particles over the whole size range.

Solution

For settling in the Stokes' law region, the velocity is proportional to the diameter squared and hence the time taken for a 40 μm particle to fall a height h m is:

$$t = h/40^2 k$$

where k a constant.

During this time, a particle of diameter d μm has fallen a distance equal to:

$$kd^2 h/40^2 k = hd^2/40^2$$

The proportion of particles of size d which are still in suspension is:

$$= 1 - (d^2/40^2)$$

and the fraction by mass of particles which are still in suspension is:

$$= \int_0^{40} [1 - (d^2/40^2)] dw$$

Since $dw/dd = 1/50$, the mass fraction is:

$$= (1/50) \int_0^{40} [1 - (d^2/40^2)] dd$$

$$= (1/50)[d - (d^3/4800)]_0^{40}$$

$$= 0.533 \text{ or } \underline{\underline{53.3 \text{ per cent}}} \text{ of the particles remain in suspension.}$$

PROBLEM 1.4

In a mixture of quartz of density 2650 kg/m³ and galena of density 7500 kg/m³, the sizes of the particles range from 0.0052 to 0.025 mm.

On separation in a hydraulic classifier under free settling conditions, three fractions are obtained, one consisting of quartz only, one a mixture of quartz and galena, and one of galena only. What are the ranges of sizes of particles of the two substances in the original mixture?

Solution

Use is made of equation 3.24, Stokes' law, which may be written as:

$$u_0 = kd^2(\rho_s - \rho),$$

where $k (= g/18\mu)$ is a constant.

For large galena: $u_0 = k(25 \times 10^{-6})^2(7500 - 1000) = 4.06 \times 10^{-6} k$ m/s

For small galena: $u_0 = k(5.2 \times 10^{-6})^2(7500 - 1000) = 0.176 \times 10^{-6} k$ m/s

For large quartz: $u_0 = k(25 \times 10^{-6})^2(2650 - 1000) = 1.03 \times 10^{-6} k$ m/s

For small quartz: $u_0 = k(5.2 \times 10^{-6})^2(2650 - 1000) = 0.046 \times 10^{-6} k$ m/s

If the time of settling was such that particles with a velocity equal to $1.03 \times 10^{-6}k$ m/s settled, then the bottom product would contain quartz. This is not so and hence the maximum size of galena particles still in suspension is given by:

$$1.03 \times 10^{-6}k = kd^2(7500 - 1000) \quad \text{or} \quad d = 0.0000126 \text{ m} \quad \text{or} \quad 0.0126 \text{ mm}.$$

Similarly if the time of settling was such that particles with a velocity equal to $0.176 \times 10^{-6}k$ m/s did not start to settle, then the top product would contain galena. This is not the case and hence the minimum size of quartz in suspension is given by:

$$0.176 \times 10^{-6}k = kd^2(2650 - 1000) \quad \text{or} \quad d = 0.0000103 \text{ m} \quad \text{or} \quad 0.0103 \text{ mm}.$$

It may therefore be concluded that, assuming streamline conditions, the fraction of material in suspension, that is containing quartz *and* galena, is made up of particles of sizes in the range <u>0.0103–0.0126 mm</u>

PROBLEM 1.5

A mixture of quartz and galena of a size range from 0.015 mm to 0.065 mm is to be separated into two pure fractions using a hindered settling process. What is the minimum apparent density of the fluid that will give this separation? How will the viscosity of the bed affect the minimum required density?

The density of galena is 7500 kg/m³ and the density of quartz is 2650 kg/m³.

Solution

See Volume 2, Example 1.4.

PROBLEM 1.6

The size distribution of a dust as measured by a microscope is as follows. Convert these data to obtain the distribution on a mass basis, and calculate the specific surface, assuming spherical particles of density 2650 kg/m³.

Size range (μm)	Number of particles in range (−)
0–2	2000
2–4	600
4–8	140
8–12	40
12–16	15
16–20	5
20–24	2

Solution

From equation 1.4, the mass fraction of particles of size d_1 is given by:

$$x_1 = n_1 k_1 d_1^3 \rho_s,$$

where k_1 is a constant, n_1 is the number of particles of size d_1, and ρ_s is the density of the particles = 2650 kg/m^3.

$\Sigma x_1 = 1$ and hence the mass fraction is:

$$x_1 = n_1 k_1 d_1^3 \rho_s / \Sigma n k d^3 \rho_s.$$

In this case:

d	n	$kd^3 n\rho_s$	x
1	200	5,300,000k	0.011
3	600	42,930,000k	0.090
6	140	80,136,000k	0.168
10	40	106,000,000k	0.222
14	15	109,074,000k	0.229
18	5	77,274,000k	0.162
22	2	56,434,400k	0.118
		$\Sigma = 477,148,400k$	$\Sigma = 1.0$

The surface mean diameter is given by equation 1.14:

$$d_s = \Sigma(n_1 d_1^3)/\Sigma(n_1 d_1^2)$$

and hence:

d	n	nd^2	nd^3
1	2000	2000	2000
3	600	5400	16,200
6	140	5040	30,240
10	40	4000	40,000
14	15	2940	41,160
18	5	1620	29,160
22	2	968	21,296
		$\Sigma = 21,968$	$\Sigma = 180,056$

Thus: $d_s = (180,056/21,968) = 8.20$ μm

This is the size of a particle with the same specific surface as the mixture.

The volume of a particle 8.20 μm in diameter = $(\pi/6)8.20^3 = 288.7$ μm^3.

The surface area of a particle 8.20 μm in diameter = $(\pi \times 8.20^2) = 211.2$ μm^2

and hence: the specific surface = (211.2/288.7)

$$= 0.731 \text{ μm}^2/\text{μm}^3 \text{ or } \underline{0.731 \times 10^6 \text{ m}^2/\text{m}^3}$$

PROBLEM 1.7

The performance of a solids mixer was assessed by calculating the variance occurring in the mass fraction of a component amongst a selection of samples withdrawn from the mixture. The quality was tested at intervals of 30 s and the data obtained are:

mixing time (s)	30	60	90	120	150
sample variance (—)	0.025	0.006	0.015	0.018	0.019

If the component analysed represents 20 per cent of the mixture by mass and each of the samples removed contains approximately 100 particles, comment on the quality of the mixture produced and present the data in graphical form showing the variation of mixing index with time.

Solution

See Volume 2, Example 1.3.

PROBLEM 1.8

The size distribution by mass of the dust carried in a gas, together with the efficiency of collection over each size range is as follows:

Size range, (μm)	0–5	5–10	10–20	20–40	40–80	80–160
Mass (per cent)	10	15	35	20	10	10
Efficiency (per cent)	20	40	80	90	95	100

Calculate the overall efficiency of the collector and the percentage by mass of the emitted dust that is smaller than 20 μm in diameter. If the dust burden is 18 g/m^3 at entry and the gas flow is 0.3 m^3/s, calculate the mass flow of dust emitted.

Solution

See Volume 2, Example 1.6.

PROBLEM 1.9

The collection efficiency of a cyclone is 45 per cent over the size range 0–5 μm, 80 per cent over the size range 5–10 μm, and 96 per cent for particles exceeding 10 μm.

Calculate the efficiency of collection for a dust with a mass distribution of 50 per cent 0–5 μm, 30 per cent 5–10 μm and 20 per cent above 10 μm.

Solution

See Volume 2, Example 1.5.

PROBLEM 1.10

A sample of dust from the air in a factory is collected on a glass slide. If dust on the slide was deposited from one cubic centimetre of air, estimate the mass of dust in g/m^3 of air in the factory, given the number of particles in the various size ranges to be as follows:

Size range (μm)	0–1	1–2	2–4	4–6	6–10	10–14
Number of particles (—)	2000	1000	500	200	100	40

It may be assumed that the density of the dust is 2600 kg/m^3, and an appropriate allowance should be made for particle shape.

Solution

If the particles are spherical, the particle diameter is d m and the density $\rho = 2600$ kg/m^3, then the volume of 1 particle $= (\pi/6)d^3$ m^3, the mass of 1 particle $= 2600(\pi/6)d^3$ kg and the following table may be produced:

Size (μm)	0–1	1–2	2–4	4–6
Number of particles (—)	2000	1000	500	200
Mean diameter (μm)	0.5	1.5	3.0	5.0
(m)	0.5×10^{-6}	1.5×10^{-6}	3.0×10^{-6}	5.0×10^{-6}
Volume (m^3)	6.54×10^{-20}	3.38×10^{-18}	1.41×10^{-17}	6.54×10^{-17}
Mass of one particle (kg)	1.70×10^{-16}	8.78×10^{-15}	3.68×10^{-14}	1.70×10^{-13}
Mass of one particles in size range (kg)	3.40×10^{-13}	8.78×10^{-12}	1.83×10^{-11}	3.40×10^{-11}

Size (μm)	6–10	10–14
Number of particles (—)	100	40
Mean diameter (μm)	8.0	12.0
(m)	8.0×10^{-6}	12.0×10^{-6}
Volume (m^3)	2.68×10^{-16}	9.05×10^{-16}
Mass of one particle (kg)	6.97×10^{-13}	2.35×10^{-12}
Mass of one particles in size range (kg)	6.97×10^{-11}	9.41×10^{-11}

Total mass of particles $= 2.50 \times 10^{-10}$ kg.

As this mass is obtained from 1 cm^3 of air, the required dust concentration is given by:

$$(2.50 \times 10^{-10}) \times 10^3 \times 10^6 = \underline{\underline{0.25 \text{ g/m}^3}}$$

PROBLEM 1.11

A cyclone separator 0.3 m in diameter and 1.2 m long, has a circular inlet 75 mm in diameter and an outlet of the same size. If the gas enters at a velocity of 1.5 m/s, at what particle size will the theoretical cut occur?

The viscosity of air is 0.018 mN s/m^2, the density of air is 1.3 kg/m^3 and the density of the particles is 2700 kg/m^3.

Solution

See Volume 2, Example 1.7.

SECTION 2-2

Particle Size Reduction and Enlargement

PROBLEM 2.1

A material is crushed in a Blake jaw crusher such that the average size of particle is reduced from 50 mm to 10 mm, with the consumption of energy of 13.0 kW/(kg/s). What will be the consumption of energy needed to crush the same material of average size 75 mm to average size of 25 mm:

(a) assuming Rittinger's Law applies,
(b) assuming Kick's Law applies?

Which of these results would be regarded as being more reliable and why?

Solution

See Volume 2, Example 2.1.

PROBLEM 2.2

A crusher was used to crush a material with a compressive strength of 22.5 MN/m². The size of the feed was *minus* 50 mm, *plus* 40 mm and the power required was 13.0 kW/(kg/s). The screen analysis of the product was:

Size of aperture (mm)		Amount of product (per cent)
through	6.0	all
on	4.0	26
on	2.0	18
on	0.75	23
on	0.50	8
on	0.25	17
on	0.125	3
through	0.125	5

What power would be required to crush 1 kg/s of a material of compressive strength 45 MN/m² from a feed of *minus* 45 mm, *plus* 40 mm to a product of 0.50 mm average size?

Solution

A dimension representing the mean size of the product is required. Using Bond's method of taking the size of opening through which 80 per cent of the material will pass, a value of just over 4.00 mm is indicated by the data. Alternatively, calculations may be made as follows:

Size of aperture (mm)	Mean d_1 (mm)	n_1	nd_1	nd_1^2	nd_1^3	nd_1^4
6.00						
	5.00	0.26	1.3	6.5	32.5	162.5
4.00						
	3.00	0.18	0.54	1.62	4.86	14.58
2.00						
	1.375	0.23	0.316	0.435	0.598	0.822
0.75						
	0.67	0.08	0.0536	0.0359	0.0241	0.0161
0.50						
	0.37	0.17	0.0629	0.0233	0.0086	0.00319
0.25						
	0.1875	0.03	0.0056	0.00105	0.00020	0.000037
0.125						
	0.125	0.05	0.00625	0.00078	0.000098	0.000012
	Totals:		2.284	8.616	37.991	177.92

From equation 1.11, the mass mean diameter is:

$$d_v = \Sigma n_1 d_1^4 / \Sigma n_1 d_1^3$$
$$= (177.92/37.991) = 4.683 \text{ mm.}$$

From equation 1.14, the surface mean diameter is:

$$d_s = \Sigma n_1 d_1^3 / \Sigma n_1 d_1^2$$
$$= (37.991/8.616) = 4.409 \text{ mm.}$$

From equation 1.18, the length mean diameter is:

$$d_l = \Sigma n_1 d_1^2 / \Sigma n_1 d_1$$
$$= (8.616/2.284) = 3.772 \text{ mm.}$$

From equation 1.19, the mean length diameter is:

$$d_1' = \Sigma n_1 d_1 / \Sigma n_1$$
$$= (2.284/1.0) = 2.284 \text{ mm.}$$

In the present case, which is concerned with power consumption per unit mass, the mass mean diameter is probably of the greatest relevance. For the purposes of calculation a mean value of 4.0 mm will be used, which agrees with the value obtained by Bond's method.

For coarse crushing, Kick's law may be used as follows:

Case 1:
mean diameter of feed = 45 mm, mean diameter of product = 4 mm,
energy consumption = 13.0 kJ/kg, compressive strength = 22.5 N/m²
In equation 2.4:
$$13.0 = K_K \times 22.5 \ln(45/4)$$
and:
$$K_K = (13.0/54.4) = 0.239 \text{ kW/(kg/s) (MN/m}^2)$$

Case 2:
mean diameter of feed = 42.5 mm, mean diameter of product = 0.50 mm
compressive strength = 45 MN/m²

Thus: $E = 0.239 \times 45 \ln(42.5/0.50) = (0.239 \times 199.9) = 47.8$ kJ/kg

or, for a feed of 1 kg/s, the energy required = <u>47.8 kW</u>.

PROBLEM 2.3

A crusher reducing limestone of crushing strength 70 MN/m² from 6 mm diameter average size to 0.1 mm diameter average size, requires 9 kW. The same machine is used to crush dolomite at the same output from 6 mm diameter average size to a product consisting of 20 per cent with an average diameter of 0.25 mm, 60 per cent with an average diameter of 0.125 mm and a balance having an average diameter of 0.085 mm. Estimate the power required, assuming that the crushing strength of the dolomite is 100 MN/m² and that crushing follows Rittinger's Law.

Solution

The mass mean diameter of the crushed dolomite may be calculated thus:

n_1	d_1	$n_1 d_1^3$	$n_1 d_1^4$
0.20	0.250	0.003125	0.00078
0.60	0.125	0.001172	0.000146
0.20	0.085	0.000123	0.000011
Totals:		0.00442	0.000937

and from equation 1.11:

$$d_v = \Sigma n_1 d_1^4 / \Sigma n_1 d_1^3 = (0.000937/0.00442) = 0.212 \text{ mm}.$$

For Case 1:
$$E = 9.0 \text{ kW}, f_c = 70.0 \text{ MN/m}^2, L_1 = 6.0 \text{ mm, and } L_2 = 0.1 \text{ mm}$$

and in equation 2.3:
$$9.0 = K_R \times 70.0[(1/0.1) - (1/6.0)]$$

or:
$$K_R = 0.013 \text{ kW mm/(MN/m}^2)$$

For Case 2:
$$f_c = 100.0 \text{ MN/m}^2, \quad L_1 = 6.0 \text{ mm} \quad \text{and} \quad L_2 = 0.212 \text{ mm}$$

Hence:
$$E = 0.013 \times 100.0[(1/0.212) - (1/6.0)]$$
$$= \underline{5.9 \text{ kW}}$$

PROBLEM 2.4

If crushing rolls 1 m diameter are set so that the crushing surfaces are 12.5 mm apart and the angle of nip is 31°, what is the maximum size of particle which should be fed to the rolls?

If the actual capacity of the machine is 12 per cent of the theoretical, calculate the throughput in kg/s when running at 2.0 Hz if the working face of the rolls is 0.4 m long and the feed density is 2500 kg/m³.

Solution

See Volume 2, Example 2.2.

PROBLEM 2.5

A crushing mill reduces limestone from a mean particle size of 45 mm to the following product:

Size (mm)	Amount of product (per cent)
12.5	0.5
7.5	7.5
5.0	45.0
2.5	19.0
1.5	16.0
0.75	8.0
0.40	3.0
0.20	1.0

It requires 21 kJ/kg of material crushed. Calculate the power required to crush the same material at the same rate, from a feed having a mean size of 25 mm to a product with a mean size of 1 mm.

Solution

The mean size of the product may be obtained thus:

n_1	d_1	$n_1 d_1^3$	$n_1 d_1^4$
0.5	12.5	3906	48,828
7.5	7.5	3164	23,731
45.0	5.0	5625	28,125
19.0	2.5	296.9	742.2
16.0	1.5	54.0	81.0
8.0	0.75	3.375	2.531
3.0	0.40	0.192	0.0768
1.0	0.20	0.008	0.0016
Totals:		13,049	101,510

and from equation 1.11, the mass mean diameter is:

$$d_v = \Sigma n_1 d_1^4 / \Sigma n_1 d_1^3$$
$$= (101,510/13,049) = 7.78 \text{ mm}$$

Kick's law is used as the present case may be regarded as coarse crushing.

Case 1:

$$E = 21 \text{ kJ/kg}, L_1 = 45 \text{ mm and } L_2 = 7.8 \text{ mm}.$$

In equation 2.4:

$$21 = K_K f_c \ln(45/7.8)$$

and:

$$K_K f_c = 11.98 \text{ kJ/kg}$$

Case 2:

$$L_1 = 25 \text{ mm and } L_2 = 1.0 \text{ mm}.$$

Thus:

$$E = 11.98 \ln(25/1.0)$$
$$= \underline{\underline{38.6 \text{ kJ/kg}}}$$

PROBLEM 2.6

A ball-mill 1.2 m in diameter is run at 0.8 Hz and it is found that the mill is not working satisfactorily. Should any modification in the condition of operation be suggested?

Solution

See Volume 2, Example 2.3.

PROBLEM 2.7

Power of 3 kW is supplied to a machine crushing material at the rate of 0.3 kg/s from 12.5 mm cubes to a product having the following sizes: 80 per cent 3.175 mm, 10 per cent 2.5 mm and 10 per cent 2.25 mm. What power should be supplied to this machine to crush 0.3 kg/s of the same material from 7.5 mm cube to 2.0 mm cube?

Solution

The mass mean diameter is calculated thus:

n_1	d_1	$n_1 d_1^3$	$n_1 d_1^4$
0.8	3.175	25.605	81.295
0.1	2.5	1.563	3.906
0.1	2.25	1.139	2.563

and from equation 1.11:

$$d_v = \Sigma n_1 d_1^4 / \Sigma n_1 d_1^3$$
$$= (87.763/28.307) = 3.1 \text{ mm}$$

(Using Bond's approach, the mean diameter is clearly 3.175 mm.)

For the size ranges involved, the crushing may be considered as intermediate and Bond's law will be used.

Case 1:
$$E = (3/0.3) = 10 \text{ kW/(kg/s)}, \quad L_1 = 12.5 \text{ mm} \quad \text{and} \quad L_2 = 3.1 \text{ mm}.$$

Thus in equation 2.5:

$$q = (L_1/L_2) = 4.03 \text{ and } E = 2C\sqrt{(1/L_2)}(1 - 1/q^{0.5})$$

or:
$$10 = 2C\sqrt{(1/3.1)}(1 - 1/4.03^{0.5})$$
$$= (2C \times 0.568 \times 0.502).$$

Thus:
$$C = 17.54 \text{ kW mm}^{0.5}/(\text{kg/s})$$

Case 2:
$$L_1 = 7.5 \text{ mm}, \quad L_2 = 2.0 \text{ mm} \quad \text{and} \quad q = (7.5/2.0) = 3.75.$$

Hence:
$$E = 2 \times 17.54(1/2.0)(1 - 1/3.75^{0.5})$$
$$= (35.08 \times 0.707 \times 0.484)$$
$$= 12.0 \text{ kJ/kg}$$

For a feed of 0.3 kg/s, the power required $= (12.0 \times 0.3) = \underline{\underline{3.6 \text{ kW}}}$

SECTION 2-3

Motion of Particles in a Fluid

PROBLEM 3.1

A finely ground mixture of galena and limestone in the proportion of 1 to 4 by mass, is subjected to elutriation by a current of water flowing upwards at 5 mm/s. Assuming that the size distribution for each material is the same, and is as follows, estimate the percentage of galena in the material carried away and in the material left behind. The absolute viscosity of water is 1 mN s/m^2 and Stokes' equation should be used.

Diameter (μm)	20	30	40	50	60	70	80	100
Undersize (per cent mass)	15	28	48	54	64	72	78	88

The density of galena is 7500 kg/m^3 and the density of limestone is 2700 kg/m^3.

Solution

See Volume 2, Example 3.2.

PROBLEM 3.2

Calculate the terminal velocity of a steel ball, 2 mm diameter and of density 7870 kg/m^3 in an oil of density 900 kg/m^3 and viscosity 50 mN s/m^2.

Solution

For a sphere:

$$(R'_0/\rho u_0^2) Re_0'^2 = (2d^3/3\mu^2)\rho(\rho_s - \rho)g \qquad \text{(equation 3.34)}$$

$$= (2 \times 0.002^3/3 \times 0.05^2) 900 (7870 - 900) 9.81$$

$$= 131.3$$

$$\log_{10} 131.3 = 2.118$$

From Table 3.4: $\log_{10} Re_0' = 0.833$

or: $Re_0' = 6.80$

Thus: $u_0 = (6.80 \times 0.05)/(900 \times 0.002) = \underline{0.189 \text{ m/s}}$

PROBLEM 3.3

What is the terminal velocity of a spherical steel particle of 0.40 mm diameter, settling in an oil of density 820 kg/m^3 and viscosity 10 mN s/m^2? The density of steel is 7870 kg/m^3.

Solution

See Volume 2, Example 3.1.

PROBLEM 3.4

What are the settling velocities of mica plates, 1 mm thick and ranging in area from 6 to 600 mm^2, in an oil of density 820 kg/m^3 and viscosity 10 mN s/m^2? The density of mica is 3000 kg/m^3.

Solution

	Smallest particles	Largest particles
A'	6×10^{-6} m^2	6×10^{-4} m^2
d_p	$\sqrt{[(4 \times 6 \times 10^{-6})/\pi]} = 2.76 \times 10^{-3}$ m	$\sqrt{[(4 \times 6 \times 10^{-4})/\pi]} = 2.76 \times 10^{-2}$ m
d_p^3	2.103×10^{-8} m^3	2.103×10^{-5} m^3
volume	6×10^{-9} m^3	6×10^{-7} m^3
k'	0.285	0.0285

$$(R_0'/\rho u^2) Re_0'^2 = (4k'/\mu^2 \pi)(\rho_s - \rho) \rho d_p^3 g \quad \text{(equation 3.52)}$$

$$= [(4 \times 0.285)/(\pi \times 0.01^2)](3000 - 820)(820 \times 2.103 \times 10^{-8} \times 9.81)$$

$$= 1340 \text{ for smallest particle and } 134{,}000 \text{ for largest particle}$$

	Smallest particles	Largest particles
$\log_{10}(R_0'/\rho u^2) Re_0'^2$	3.127	5.127
$\log_{10} Re_0'$	1.581	2.857 (from Table 3.4)
Correction from Table 3.6	−0.038	−0.300 (estimated)
Corrected $\log_{10} Re_0'$	1.543	2.557
Re_0'	34.9	361
u	0.154 m/s	0.159 m/s

Thus it is seen that all the mica particles settle at approximately the same velocity.

PROBLEM 3.5

A material of density 2500 kg/m^3 is fed to a size separation plant where the separating fluid is water which rises with a velocity of 1.2 m/s. The upward vertical component of the velocity of the particles is 6 m/s. How far will an approximately spherical particle, 6 mm diameter, rise relative to the walls of the plant before it comes to rest in the fluid?

Solution

See Volume 2, Example 3.4.

PROBLEM 3.6

A spherical glass particle is allowed to settle freely in water. If the particle starts initially from rest and if the value of the Reynolds number with respect to the particle is 0.1 when it has attained its terminal falling velocity, calculate:

(a) the distance travelled before the particle reaches 90 per cent of its terminal falling velocity,
(b) the time elapsed when the acceleration of the particle is one hundredth of its initial value.

Solution

When $Re' < 0.2$, the terminal velocity is given by equation 3.24:

$$u_0 = (d^2 g/18\mu)(\rho_s - \rho)$$

Taking the densities of glass and water as 2750 and 1000 kg/m³, respectively, and the viscosity of water as 0.001 Ns/m², then:

$$u_0 = [(9.81 d^2)/(18 \times 0.001)](2750 - 1000) = 9.54 \times 10^5 d^2 \text{ m/s}$$

The Reynolds number, $Re' = 0.1$ and substituting for u_0:

$$d(9.54 \times 10^5 d^2)(1000/0.001) = 0.1$$

or: $d = 4.76 \times 10^{-5}$ m

$a = 18\mu/d^2 \rho_s = (18 \times 0.001)/[4.76 \times 10^{-5})^2 \times 2750]$ (equation 3.89)

$= 2889$ s^{-1}

and: $b = [1 - (\rho/\rho_s)]g = [1 - (1000/2750)]9.81 = 6.24$ m/s² (equation 3.90)

In equation 3.88:

$$y = \frac{b}{a}t + \frac{v}{a} - \frac{b}{a^2} + \left(\frac{b}{a^2} - \frac{v}{a}\right)e^{-at}$$

In this case $v = 0$ and differentiating gives:

$$\dot{y} = (b/a)(1 - e^{-at})$$

or, since $(b/a) = u_0$, the terminal velocity:

$$\dot{y} = u_0(1 - e^{-at})$$

When $\dot{y} = 0.9u_0$, then: $\quad 0.9 = (1 - e^{-2889t})$

or: $\quad\quad\quad\quad\quad\quad 2889t = 2.303 \quad\text{and}\quad t = 8.0 \times 10^{-4}$ s

Thus in equation 3.88:

$y = (6.24 \times 8.0 \times 10^{-4})/2889 - (6.24/2889^2) + (6.24/2889^2)\exp(-2889 \times 8.0 \times 10^{-4})$

$\quad = (1.73 \times 10^{-6}) - (7.52 \times 10^{-7}) + (7.513 \times 10^{-8})$

$\quad = 1.053 \times 10^{-6}$ m $\quad$ or $\quad$ <u>1.05 mm</u>

From equation 3.86:

$$\ddot{y} = b - a\dot{y}$$

At the start of the fall, $\dot{y} = 0$ and the initial acceleration, $\ddot{y} = b$.

When $\ddot{y} = 0.01b$, then:

$$0.01b = b - a\dot{y}$$

or: $\quad\quad\quad\quad\quad\quad \dot{y} = (0.89 \times 6.24)/2889 = 0.00214$ m/s

Thus: $\quad\quad\quad\quad\quad 0.00214 = (6.24/2889)(1 - e^{-2889t})$

or: $\quad\quad\quad\quad\quad\quad 2889t = 4.605$

and: $\quad\quad\quad\quad\quad\quad \underline{t = 0.0016 \text{ s}}$

PROBLEM 3.7

In a hydraulic jig, a mixture of two solids is separated into its components by subjecting an aqueous slurry of the material to a pulsating motion, and allowing the particles to settle for a series of short time intervals such that their terminal falling velocities are not attained. Materials of densities 1800 and 2500 kg/m^3 whose particle size ranges from 0.3 mm to 3 mm diameter are to be separated. It may be assumed that the particles are approximately spherical and that Stokes' Law is applicable. Calculate approximately the maximum time interval for which the particles may be allowed to settle so that no particle of the less dense material falls a greater distance than any particle of the denser material. The viscosity of water is 1 mN s/m^2.

Solution

For Stokes' law to apply, $Re' < 0.2$ and equation 3.88 may be used:

$$y = \frac{b}{a}t + \frac{v}{a} - \frac{b}{a^2} + \left(\frac{b}{a^2} - \frac{v}{a}\right)e^{-at}$$

or, assuming the initial velocity $v = 0$:

$$y = \frac{b}{a}t - \frac{b}{a^2} + \frac{b}{a^2}e^{-at}$$

where: $b = [1 - (\rho/\rho_s)]g$ and $a = 18\mu/d^2\rho_s$. (equations 3.89 and 3.90)

For small particles of the dense material:

$$b = [1 - (1000/2500)]9.81 = 5.89 \text{ m/s}^2$$
$$a = (18 \times 0.001)/[(0.3 \times 10^{-3})^2 2500] = 80 \text{ s}^{-1}$$

For large particles of the light material:

$$b = [1 - (1000/1800)]9.81 = 4.36 \text{ m/s}^2$$
$$a = (18 \times 0.001)/[(3 \times 10^{-3})^2 1800] = 1.11 \text{ s}^{-1}$$

In order that these particles should fall the same distance, from equation 3.88:

$$(5.89/80)t - (5.89/80^2)(1 - e^{-80t}) = (4.36/1.11)t - (4.36/1.11^2)(1 - e^{-1.11t})$$

Thus: $\quad 3.8504t + 3.5316\, e^{-1.11t} - 0.00092\, e^{-80t} = 3.5307$

and, solving by trial and error:

$$\underline{\underline{t = 0.01 \text{ s}}}$$

PROBLEM 3.8

Two spheres of equal terminal falling velocity settle in water starting from rest at the same horizontal level. How far apart vertically will the particles be when they have both reached 99 per cent of their terminal falling velocities? It may be assumed that Stokes' law is valid and this assumption should be checked.

The diameter of one sphere is 40 μm and its density is 1500 kg/m³ and the density of the second sphere is 3000 kg/m³. The density and viscosity of water are 1000 kg/m³ and 1 mN s/m² respectively.

Solution

Assuming Stokes' law is valid, the terminal velocity is given by equation 3.24 as:

$$u_0 = (d^2 g/18\mu)(\rho_s - \rho)$$

For particle 1:

$$u_0 = \{[(40 \times 10^{-6})^2 \times 9.81]/(18 \times 1 \times 10^{-3})\}(1500 - 1000)$$
$$= 4.36 \times 10^{-4} \text{ m/s}$$

Since particle 2 has the same terminal velocity:

$$4.36 \times 10^{-4} = [(d_2^2 \times 9.81)/(18 \times 1 \times 10^{-3})](3000 - 1000)$$

From which: $\quad d_2 = (2 \times 10^{-5})$ m $\quad$ or $\quad 20$ μm

From equation 3.83: $\quad a = 18\mu/d^2 \rho_s$

For particle 1: $\quad a_1 = (18 \times 1 \times 10^{-3})/((40 \times 10^{-6})^2 \times 1500) = 7.5 \times 10^3$ s^{-1}

and for particle 2: $\quad a_2 = (18 \times 1 \times 10^{-3})/((20 \times 10^{-6})^2 \times 3000) = 1.5 \times 10^4$ s^{-1}

From equation 3.90: $\quad b = (1 - \rho/\rho_s)g$

For particle 1: $\quad b_1 = (1 - 1000/1500)9.81 = 3.27$ m/s^2

and for particle 2: $\quad b_2 = (1 - 1000/3000)9.81 = 6.54$ m/s^2

The initial velocity of both particles, $v = 0$, and from equation 3.88:

$$y = \frac{b}{a}t - \frac{b}{a^2} + \frac{b}{a^2}e^{-at}$$

Differentiating:

$$\dot{y} = (b/a)(1 - e^{-at})$$

or, from equation 3.24:

$$\dot{y} = u_t(1 - e^{-at})$$

When $\dot{y} = u_0$, the terminal velocity, it is not possible to solve for t and hence $\dot{y}$ will be taken as $0.99 u_0$.

For particle 1:

$$(0.99 \times 4.36 \times 10^{-4}) = (4.36 \times 10^{-4})[1 - \exp(-7.5 \times 10^3 t)]$$

and: $\quad t = 6.14 \times 10^{-4}$ s

The distance travelled in this time is given by equation 3.88:

$$y = (3.27/7.5 \times 10^3)6.14 \times 10^{-4} - [3.27/(7.5 \times 10^3)^2]$$
$$[1 - \exp(-7.5 \times 10^3 \times 6.14 \times 10^{-4})] = 2.10 \times 10^{-7} \text{ m}$$

For particle 2:

$$(0.99 \times 4.36 \times 10^{-4}) = (4.36 \times 10^{-4})[1 - \exp(-1.5 \times 10^4 t)]$$

and: $\quad t = 3.07 \times 10^{-4}$ s

Thus: $\quad y = ((6.54/1.5 \times 10^4)3.07 \times 10^{-4}) - [6.54/(1.5 \times 10^4)^2]$
$$[1 - \exp(-1.5 \times 10^4 \times 3.07 \times 10^{-4})] = 1.03 \times 10^{-7} \text{ m}$$

Particle 2 reaches 99 per cent of its terminal velocity after 3.07×10^{-4} s and it then travels at 4.36×10^{-4} m/s for a further $(6.14 \times 10^{-4} - 3.07 \times 10^{-4}) = 3.07 \times 10^{-4}$ s during which time it travels a further $(3.07 \times 10^{-4} \times 4.36 \times 10^{-4}) = 1.338 \times 10^{-7}$ m.

Thus the total distance moved by particle 1 = 2.10×10^{-7} m
and the total distance moved by particle 2 = $(1.03 \times 10^{-7} + 1.338 \times 10^{-7})$
$$= 2.368 \times 10^{-7} \text{ m.}$$

The distance apart when both particles have attained their terminal velocities is:
$$(2.368 \times 10^{-7} - 2.10 \times 10^{-7}) = \underline{2.68 \times 10^{-8} \text{ m}}$$

For Stokes' law to be valid, Re' must be less than 0.2.

For particle 1, $Re = (40 \times 10^{-6} \times 4.36 \times 10^{-4} \times 1500)/(1 \times 10^{-3}) = 0.026$
and for particle 2, $Re = (20 \times 10^{-6} \times 4.36 \times 10^{-4} \times 3000)/(1 \times 10^{-3}) = 0.026$
and Stokes' law applies.

PROBLEM 3.9

The size distribution of a powder is measured by sedimentation in a vessel having the sampling point 180 mm below the liquid surface. If the viscosity of the liquid is 1.2 mN s/m^2, and the densities of the powder and liquid are 2650 and 1000 kg/m^3 respectively, determine the time which must elapse before any sample will exclude particles larger than 20 μm.

If Stokes' law applies when the Reynolds number is less than 0.2, what is the approximate maximum size of particle to which Stokes' Law may be applied under these conditions?

Solution

The problem involves determining the time taken for a 20 μm particle to fall below the sampling point, that is 180 mm. Assuming that Stokes' law is applicable, equation 3.88 may be used, taking the initial velocity as $v = 0$.

Thus: $y = (bt/a) - (b/a^2)(1 - e^{-at})$

where: $b = g(1 - \rho/\rho_s) = 9.81[1 - (1000/2650)] = 6.108$ m/s^2

and: $a = 18\mu/d^2\rho_s = (18 \times 1.2 \times 10^{-3})/[(20 \times 10^{-6})^2 \times 2650] = 20{,}377$ s^{-1}

In this case: $y = 180$ mm or 0.180 m

Thus: $0.180 = (6.108/20{,}377)t - (6.108/20{,}377^2)(1 - e^{-20{,}377t})$
$$= 0.0003t + (1.4071 \times 10^{-8} e^{-20{,}377t})$$

Ignoring the exponential term as being negligible, then:
$$t = (0.180/0.0003) = \underline{600 \text{ s}}$$

The velocity is given by differentiating equation 3.88 giving:
$$\dot{y} = (b/a)(1 - e^{-at})$$

When $t = 600$ s:

$$\dot{y} = [(6.108d^2 \times 2650)/(18 \times 0.0012)]\{1 - \exp[-(18 \times 0.0012 \times 600)/d^2 \times 2650]\}$$
$$= 7.49 \times 10^5 d^2 [1 - \exp(-4.89 \times 10^{-3} d^{-2})]$$

For $Re' = 0.2$, then

$$d(7.49 \times 10^5 d^2)[1 - \exp(-4.89 \times 10^{-3} d^{-2})] \times 2650/0.0012 = 0.2$$

or:
$$1.65 \times 10^{12} d^3 [1 - \exp(-4.89 \times 10^{-3} d^{-2})] = 0.2$$

When d is small, the exponential term may be neglected and:

$$d^3 = 1.212 \times 10^{-13}$$

or:
$$d = 5.46 \times 10^{-5} \text{ m} \quad \text{or} \quad \underline{\underline{54.6 \text{ } \mu\text{m}}}$$

PROBLEM 3.10

Calculate the distance a spherical particle of lead shot of diameter 0.1 mm settles in a glycerol/water mixture before it reaches 99 per cent of its terminal falling velocity.

The density of lead is 11,400 kg/m³ and the density of liquid is 1000 kg/m³. The viscosity of liquid is 10 mN s/m².

It may be assumed that the resistance force may be calculated from Stokes' Law and is equal to $3\pi \mu d u$, where u is the velocity of the particle relative to the liquid.

Solution

The terminal velocity, when Stokes' law applies, is given by:

$$\frac{1}{6}\pi d^3 (\rho_s - \rho) g = 3\pi \mu d u_0$$

or: $\quad u_0 = \dfrac{d^2 g}{18\mu}(\rho_s - \rho) = \dfrac{d^2 \rho_s}{18\mu} g(1 - \rho/\rho_s) = (b/a)$ (equations 3.24, 3.89 and 3.90)

where: $\quad b = g(1 - \rho/\rho_s) = 9.81[1 - (1000/11,400)] = 8.95 \text{ m/s}^2$

and: $\quad a = 18\mu/d^2 \rho_s = (18 \times 10 \times 10^{-3})/[(0.1 \times 10^{-3})^2 11,400] = 1579 \text{ s}^{-1}$

Thus: $\quad u_0 = (8.95/1579) = 5.67 \times 10^{-3}$ m/s

When 99 per cent of this velocity is attained, then:

$$\dot{y} = (0.99 \times 5.67 \times 10^{-3}) = 5.61 \times 10^{-3} \text{ m/s}$$

Assuming that the initial velocity v is zero, then equation 3.88 may be differentiated to give:

$$\dot{y} = (b/a)(1 - e^{-at})$$

Thus: $\quad (5.61 \times 10^{-3}) = (5.67 \times 10^{-3})(1 - e^{-1579t})$ and $t = 0.0029$ s

Substituting in equation 3.88:

$$y = (b/a)t - (b/a^2)(1 - e^{-at})$$
$$= (5.67 \times 10^{-3} \times 0.0029) - (5.67 \times 10^{-3}/1579)(1 - e^{-1579 \times 0.0029})$$
$$= (1.644 \times 10^{-5}) - (3.59 \times 10^{-6} \times 9.89 \times 10^{-1})$$
$$= 1.29 \times 10^{-5} \text{ m} \quad \text{or} \quad \underline{\underline{0.013 \text{ mm}}}$$

PROBLEM 3.11

What is the mass of a sphere of material of density 7500 kg/m³ whose terminal velocity in a large deep tank of water is 0.6 m/s?

Solution

$$\frac{R_0'}{\rho u_0^2} Re_0'^{-1} = \frac{2\mu g}{3\rho^2 u_0^3}(\rho_s - \rho) \qquad \text{(equation 3.41)}$$

Taking the density and viscosity of water as 1000 kg/m³ and 0.001 N s/m² respectively, then:

$$(R_0'/\rho u_0^2)/Re_0' = [(2 \times 0.001 \times 9.81)/(3 \times 1000^2 \times 0.6^3)](7500 - 1000)$$
$$= 0.000197$$

Thus: $\quad \log_{10}(R_0'/\rho u_0^2)/Re_0' = \overline{4}.296$

From Table 3.5, $\quad \log_{10} Re_0' = 3.068$

$$Re_0' = 1169.5$$

and: $\quad d = (1169.5 \times 0.001)/(0.6 \times 1000)$
$$= 0.00195 \text{ m} \quad \text{or} \quad 1.95 \text{ mm.}$$

The mass of the sphere $= \pi d^3 \rho_s/6$
$$= (\pi \times 0.00195^3 \times 7500)/6$$
$$= 2.908 \times 10^{-5} \text{ kg} \quad \text{or} \quad \underline{\underline{0.029 \text{ g}}}$$

PROBLEM 3.12

Two ores, of densities 3700 and 9800 kg/m³ are to be separated in water by a hydraulic classification method. If the particles are all of approximately the same shape and each is sufficiently large for the drag force to be proportional to the square of its velocity in the fluid, calculate the maximum ratio of sizes which can be completely separated if the particles attain their terminal falling velocities. Explain why a wider range of sizes can be separated if the time of settling is so small that the particles do not reach their terminal velocities.

An explicit expression should be obtained for the distance through which a particle will settle in a given time if it starts from rest and if the resistance force is proportional to the square of the velocity. The acceleration period should be taken into account.

Solution

If the total drag force is proportional to the square of the velocity, then when the terminal velocity u_0 is attained:
$$F = k_1 u_0^2 d_m^2$$
since the area is proportional to d_p^2
and the accelerating force $= (\rho_s - \rho) g k_2 d_p^3$ where k_2 is a constant depending on the shape of the particle and d_p is a mean projected area.
When the terminal velocity is reached, then:
$$k_1 u_0^2 d_p^2 = (\rho_s - \rho) g k_2 d_p^3$$
and:
$$u_0 = [(\rho_s - \rho) g k_3 d_p]^{0.5}$$

In order to achieve complete separation, the terminal velocity of the smallest particle (diameter d_1) of the dense material must exceed that of the largest particle (diameter d_2) of the light material. For equal terminal falling velocities:
$$[(9800 - 1000)9.81 k_3 d_1]^{0.5} = [(3700 - 1000)9.81 k_3 d_2]^{0.5}$$
and:
$$(d_2/d_1) = (8800/2700) = \underline{\underline{3.26}}$$

which is the maximum range for which complete separation can be achieved if the particles settle at their terminal velocities.

If the particles are allowed to settle in a suspension for only very short periods, they will not attain their terminal falling velocities and a better degree of separation may be obtained. All particles will have an initial acceleration $g(1 - \rho/\rho_s)$ because no fluid frictional force is exerted at zero particle velocity. Thus the initial acceleration is a function of density only, and is unaffected by both size and shape. A very small particle of the denser material will therefore always commence settling at a higher rate than a large particle of the less dense material. Theoretically, therefore, it should be possible to effect complete separation irrespective of the size range, provided that the periods of settling are sufficiently short. In practice, the required periods will often be so short that it is impossible to make use of this principle alone. As the time of settling increases some of the larger particles of the less dense material will catch up and then overtake the smaller particles of the denser material.

If the total drag force is proportional to the velocity squared, that is to $\dot{y}^2$, then the equation of motion for a particle moving downwards under the influence of gravity may be written as:
$$m\ddot{y} = mg(1 - \rho/\rho_s) - k_1 \dot{y}^2$$
Thus:
$$\ddot{y} = g(1 - \rho/\rho_s) - (k_1/m)\dot{y}^2$$
or:
$$\ddot{y} = b - c\dot{y}^2$$

where $b = g(1 - \rho/\rho_s)$, $c = k_1/m$, and k_1 is a proportionality constant.

Thus: $$d\dot{y}/(b - c\dot{y}^2) = dt$$

or: $$d\dot{y}/(f^2 - \dot{y}^2) = c\,dt$$

where $f = (b/c)^{0.5}$.

Integrating: $$(1/2f)\ln[(f + \dot{y})/(f - \dot{y})] = ct + k_4$$

When $t = 0$, then: $\dot{y} = 0$ and $k_4 = 0$

Thus: $$(1/2f)\ln[(f + \dot{y})/(f - \dot{y})] = ct$$
$$(f + \dot{y})/(f - \dot{y}) = e^{2fct}$$
$$f - \dot{y} = 2f/(1 + e^{2fct})$$
$$y = ft - 2f \int_0^t dt/(1 - e^{2fct})$$
$$y = ft - (1/c)\ln[e^{2fct}/(1 + e^{2fct})] + k_5$$

when $t = 0$, then: $y = 0$ and $k_5 = (1/c)\ln 0.5$

Thus: $$\underline{\underline{y = ft - (1/c)\ln(0.5 e^{2fct})/(1 + e^{2fct})}}$$

where $f = (b/c)^{0.5}$, $b = g(1 - \rho/\rho_s)$, and $c = k_1/m$.

PROBLEM 3.13

Salt, of density 2350 kg/m³, is charged to the top of a reactor containing a 3 m depth of aqueous liquid of density 1100 kg/m³ and of viscosity 2 mN s/m² and the crystals must dissolve completely before reaching the bottom. If the rate of dissolution of the crystals is given by:

$$-\frac{dd}{dt} = 3 \times 10^{-6} + 2 \times 10^{-4} u$$

where d is the size of the crystal (m) at time t (s) and u is its velocity in the fluid (m/s), calculate the maximum size of crystal which should be charged. The inertia of the particles may be neglected and the resistance force may be taken as that given by Stokes' Law ($3\pi\mu du$) where d is the equivalent spherical diameter of the particle.

Solution

See Volume 2, Example 3.5.

PROBLEM 3.14

A balloon of mass 7 g is charged with hydrogen to a pressure of 104 kN/m². The balloon is released from ground level and, as it rises, hydrogen escapes in order to maintain a constant differential pressure of 2.7 kN/m², under which condition the diameter of the balloon is 0.3 m. If conditions are assumed to remain isothermal at 273 K as the balloon rises, what is the ultimate height reached and how long does it take to rise through the first 3000 m?

It may be assumed that the value of the Reynolds number with respect to the balloon exceeds 500 throughout and that the resistance coefficient is constant at 0.22. The inertia of the balloon may be neglected and at any moment, it may be assumed that it is rising at its equilibrium velocity.

Solution

Volume of balloon $= (4/3)\pi(0.15)^3 = 0.0142$ m³.

Mass of balloon $= 7$ g or 0.007 kg.

The upthrust $=$ (weight of air at a pressure of P N/m²)

  $-$ (weight of hydrogen at a pressure of $(P + 2700)$ N/m²).

The density of air ρ_a at 101,300 N/m² and 273 K $= (28.9/22.4) = 1.29$ kg/m³, where the mean molecular mass of air is taken as 28.9 kg/kmol.

The net upthrust force W on the balloon is given by:

$$W = 9.81\{0.0142[(\rho_a P/101,300) - \rho_a(2/28.9)(P + 2700)/101,300] - 0.007\}$$

$$= 0.139[0.0000127P - 0.000000881(P + 2700)] - 0.0687$$

$$= (0.00000164P - 0.0690) \text{ N} \tag{i}$$

The balloon will stop using when $W = 0$, that is when:

$$P = (0.0690/0.00000164) = 42,092 \text{ N/m}^2.$$

From equation 2.43 in Volume 1, the variation of pressure with height is given by:

$$g \, dz + v \, dP = 0$$

For isothermal conditions:

$$v = (1/\rho_a)(101,300/P) \text{ m}^3$$

Thus: $\quad dz + [101,300/(9.81 \times 1.29P)] \, dP = 0$

and, on integration: $\quad (z_2 - z_1) = 8005 \ln(101,300/P)$

When $P = 42,092$ N/m², $(z_2 - z_1) = 8005 \ln(101,300/42,092) = \underline{7030 \text{ m}}$

The resistance force per unit projected area R on the balloon is given by:

$$(R/\rho_a u^2) = 0.22$$

or: $\quad R = 0.22\rho_a(P/101,300)(\pi \times 0.3^2/4)(dz/dt)^2$ N/m²

$$= 1.98 \times 10^{-7} P(dz/dt)^2$$

This must be equal to the net upthrust force W, given by equation (i),

or: $\quad 0.00000164 P - 0.0690 = (1.98 \times 10^{-7} P)(dz/dt)^2$

and: $\quad (dz/dt)^2 = (8.28 - 3.49 \times 10^5)/P$

But: $\quad z = 8005 \ln(101,300/P)$

Therefore: $\quad (dz/dt)^2 = 8.28 - [(3.49 \times 10^5) e^{z/8005}]/101,300$

and: $\quad (dz/dt) = 1.89(2.41 - e^{1.25 \times 10^{-4} z})^{0.5}$

The time taken to rise 3000 m is therefore given by:

$$t = (1/1.89) \int_0^{3000} dz/(2.41 - e^{1.25 \times 10^{-4} z})^{0.5}$$

Writing the integral as: $\quad I = \int_0^{3000} dz/(a - e^{bz})^{0.5}$

and putting: $\quad (a - e^{bz}) = x^2$

then: $\quad dz = 2x \, dx/[b(a - x^2)]$

and: $\quad I = (-2/b) \int dx/(a - x^2)$

$$= (-2/b)[1/2(\sqrt{a})] \left[\ln \frac{\sqrt{a} - \sqrt{(a - e^{bz})}}{\sqrt{a} + \sqrt{(a - e^{bz})}} \right]_0^{3000}$$

$$= [1/(b\sqrt{a})] \ln \frac{[\sqrt{a} - \sqrt{(a - e^{3000b})}][\sqrt{a} + \sqrt{(a - 1)}]}{[\sqrt{a} + \sqrt{(a - e^{3000b})}][\sqrt{a} - \sqrt{(a - 1)}]}$$

Substituting:
$$a = 2.41 \quad \text{and} \quad b = 1.25 \times 10^{-4}$$

then: $\quad I = 5161 \ln[(1.55 - 0.977)/(1.55 + 0.977)][(1.55 - 1.19)/(1.55 + 1.19)]$

$\quad\quad = 2816$

Thus: $\quad t = [2816(1/1.89)] = \underline{1490 \text{ s}} \text{ (25 min)}$

PROBLEM 3.15

A mixture of quartz and galena of densities 3700 and 9800 kg/m^3 respectively with a size range of 0.3 to 1 mm is to be separated by a sedimentation process. If Stokes' Law is applicable, what is the minimum density required for the liquid if the particles all settle at their terminal velocities?

A separating system using water as the liquid is considered in which the particles were to be allowed to settle for a series of short time intervals so that the smallest particle of galena settled a larger distance than the largest particle of quartz. What is the approximate maximum permissible settling period?

According to Stokes' Law, the resistance force F acting on a particle of diameter d, settling at a velocity u in a fluid of viscosity μ is given by:

$$F = 3\pi \mu\, du$$

The viscosity of water is 1 mN s/m^2.

Solution

For particles settling in the Stokes' law region, equation 3.32 applies:

$$d_g/d_A = [(\rho_A - \rho)/(\rho_B - \rho)]^{0.5}$$

For separation it is necessary that a large particle of the less dense material does not overtake a small particle of the dense material,

or: $\qquad (1/0.3) = [(9800 - \rho)/(3700 - \rho)]^{0.5}$ and $\rho = 3097$ kg/m^3

Assuming Stokes' law is valid, the distance travelled including the period of acceleration is given by equation 3.88:

$$y = (b/a)t + (v/a) - (b/a^2) - [(b/a^2) - (v/a)]e^{-at}$$

When the initial velocity $v = 0$, then:

$$y = (b/a)t + (b/a^2)(e^{-at} - 1)$$

where: $\qquad a = 18\mu/d^2 \rho_s$ $\qquad$ (equation 3.89)

and: $\qquad b = g(1 - \rho/\rho_s)$ $\qquad$ (equation 3.90)

For a small particle of galena

$$b = 9.81[1 - (1000/9800)] = 8.81 \text{ m/s}^2$$

$$a = (18 \times 1 \times 10^{-3})/[(0.3 \times 10^{-3})^2 \times 9800] = 20.4 \text{ s}^{-1}$$

For a large particle of quartz

$$b = 9.81[1 - (1000/3700)] = 7.15 \text{ m/s}^2$$

$$a = (18 \times 1 \times 10^{-3})/[(1 \times 10^{-3})^2 \times 3700] = 4.86 \text{ s}^{-1}$$

In order to achieve separation, these particles must travel at least the same distance in time t.

Thus: $\qquad (8.81/20.4)t + (8.81/20.4^2)(e^{-20.4t} - 1)$

$$= (7.15/4.86)t + (7.15/4.86^2)(e^{-4.86t} - 1)$$

or: $\qquad (0.0212\, e^{-20.4t} - 0.303\, e^{-4.86t}) = 1.039t - 0.282$

and solving by trial and error: $\qquad t = 0.05$ s

PROBLEM 3.16

A glass sphere, of diameter 6 mm and density 2600 kg/m³, falls through a layer of oil of density 900 kg/m³ into water. If the oil layer is sufficiently deep for the particle to have reached its free falling velocity in the oil, how far will it have penetrated into the water before its velocity is only 1 per cent above its free falling velocity in water? It may be assumed that the force on the particle is given by Newton's law and that the particle drag coefficient $R'/\rho u^2 = 0.22$.

Solution

The settling velocity in water is given by equation 3.25, assuming Newton's law, or:

$$u_0^2 = 3dg(\rho_s - \rho)/\rho$$

For a solid density of 2600 kg/m³ and a particle diameter of $(6/1000) = 0.006$ m,

then: $u_0^2 = (3 \times 0.006 \times 9.81)(2600 - 1000)/1000$ and $u_0 = 0.529$ m/s

The Reynolds number may now be checked taking the viscosity of water as 0.001 Ns/m².

Thus: $Re' = (0.529 \times 0.006 \times 1000)/0.001 = 3174$

which is very much in excess of 500, which is the minimum value for Newton's law to be applicable.

The settling velocity in an oil of density 900 kg/m³ is also given by equation 3.25 as:

$$u_0^2 = (3 \times 0.006 \times 9.81)(2600 - 900)/900 \quad \text{and} \quad u_0 = 0.577 \text{ m/s.}$$

Using the nomenclature of Chapter 3 in Volume 2, a force balance on the particle in water gives:

$$m\ddot{y} = mg(1 - \rho/\rho_s) - A\rho\dot{y}^2(R'/\rho u^2)$$

Substituting $R'/\rho u^2 = 0.22$, then:

$$\ddot{y} = g(1 - \rho/\rho_s) - 0.22(A\rho/m)\dot{y}^2$$
$$= 9.81(1 - (1000/2600)) - 0.22((\pi/4)d^2\rho\dot{y}^2)/((\pi/6)d^3\rho_s)$$
$$= 6.03 - (0.33\dot{y}^2(1000/2600)/0.006) = 6.03 - 21.4\dot{y}^2$$

or from equation 3.97:

$$\ddot{y} = b - c\dot{y}^2$$

Following the reasoning in Volume 2, Section 3.6.3, for downward motion, then:

$$y = ft + (1/c)\ln(1/2f)[f + v + (f - v)e^{-2fct}] \quad \text{(equation 3.101)}$$

where $f = (b/c)^{0.5}$.

Thus:
$$\dot{y} = f + (1/c)\{1/[(f+v) + (f-v)e^{-2fct}][(f-v)e^{-2fct}(-2fc)]\}$$
$$= f[1 - \{c/[1 + (f+v)e^{2fct}/(f-v)]\}]$$
$$= f\left(1 - \frac{c}{1 + (f+v)e^{2fct}/(f-v)}\right)$$

When $t = \infty$:
$$y = f = (b/c)^{0.5}$$
$$= (6.03/21.4)^{0.5} = 0.529 \text{ m/s, as before.}$$

The initial velocity, $v = 0.577$ m/s.

Thus:
$$(f+v)/(f-v) = (0.529 + 0.577)/(0.529 - 0.577) = -23.04$$
$$2fc = (2 \times 0.529 \times 21.4) = 22.6$$

When $(\dot{y}/f) = 1.01$, then:
$$1.01 = 1 - 2/(1 - 23.04\, e^{22.6t})$$

and:
$$e^{22.6t} = 8.72 \quad \text{and} \quad \underline{t = 0.096 \text{ s}}$$

In equation 3.101:
$$y = (0.529 \times 0.0958) + (1/21.4)\ln(1/(2 \times 0.529))(0.529 + 0.577)$$
$$+ (0.529 - 0.577)\exp[-(22.6 \times 0.0958)]$$

and: $y = 0.048$ m or $\underline{48 \text{ mm}}$

PROBLEM 3.17

Two spherical particles, one of density 3000 kg/m³ and diameter 20 μm, and the other of density 2000 kg/m³ and diameter 30 μm start settling from rest at the same horizontal level in a liquid of density 900 kg/m³ and of viscosity 3 mN s/m². After what period of settling will the particles be again at the same horizontal level? It may be assumed that Stokes' Law is applicable, and the effect of added mass of the liquid moved with each sphere may be ignored.

Solution

For motion of a sphere in the Stokes' law region equation 3.88 is valid:
$$y = (b/a)t + (v/a) - (b/a^2) + [(b/a^2) - (v/a)]e^{-at}$$

When the initial velocity, $v = 0$, then:
$$y = (b/a)t - (b/a^2)(1 - e^{-at}) \tag{i}$$

From equation 3.89, $a = 18\mu/(d^2\rho_s)$
and hence, for particle 1:
$$a_1 = (18 \times 3 \times 10^{-3})/[(20 \times 10^{-6})^2 \times 3000] = 45{,}000$$
and for particle 2:
$$a_2 = (18 \times 3 \times 10^{-3})/[(30 \times 10^{-6})^2 \times 2000] = 30{,}000$$

Similarly:
$$b = g(1 - (\rho/\rho_s)) \qquad \text{(equation 3.30)}$$

For particle 1:
$$b_1 = 9.81[1 - (900/3000)] = 6.867$$

and for particle 2:
$$b_2 = 8.81[1 - (900/2000)] = 5.395$$

Substituting for a_1, a_2, b_1, b_2 in equation (i), then:
$$y_1 = (6.867/45{,}000)t - (6.867/45{,}000^2)(1 - e^{-45{,}000t}) \qquad \text{(ii)}$$
$$y_2 = (5.395/30{,}000)t - (5.395/30{,}000^2)(1 - e^{-30{,}000t}) \qquad \text{(iii)}$$

Putting $y_1 = y_2$, that is equating (ii) and (iii), then:
$$t = 0.0002203(1 - e^{-30000t}) - 0.0001247(1 - e^{-45000t})$$

and solving by trial and error:
$$\underline{\underline{t = 7.81 \times 10^{-5}\text{s}}}$$

PROBLEM 3.18

A binary suspension consists of equal masses of spherical particles of the same shape and density whose free falling velocities in the liquid are 1 mm/s and 2 mm/s, respectively. The system is initially well mixed and the total volumetric concentration of solids is 0.2. As sedimentation proceeds, a sharp interface forms between the clear liquid and suspension consisting only of small particles, and a second interface separates the suspension of fines from the mixed suspension. Using a suitable model for the behaviour of the system, estimate the falling rates of the two interfaces. It may be assumed that the sedimentation velocity u_c in a concentrated suspension of voidage e is related to the free falling velocity u_0 of the particles by:
$$u_c/u_0 = e^{2.3}$$

Solution

In the mixture, the relative velocities of the particles, u_P are given by:
for the large particles:
$$u_{PL} = u_{0L} e^{n-1} \qquad \text{(from equation 5.108)}$$

and for the small particles:
$$u_{PS} = u_{0S}e^{n-1}$$

If the upward fluid velocity is u_F m/s, then the sedimentation velocities are:
for the large particles:
$$u_{cL} = u_{0L}e^{n-1} - u_F$$
and for the small particles:
$$u_{cS} = u_{0S}e^{n-1} - u_F$$

Combining these equations and noting that the concentrations of large and small particles are equal then:
$$u_F e = u_{cL}(1-e)/2 + u_{cS}(1-e)/2$$
$$= (u_{0L}e^{n-1} - u_F)(1-e)/2 + (u_{0S}e^{n-1} - u_F)(1-e)/2$$

Thus: $\quad u_F = (e^{n-1}(1-e)/2)(u_{0L} + u_{0S})$

and: $\quad u_{cL} = u_{0L}e^{n-1} - (e^{n-1}(1-e)/2)(u_{0L} + u_{0S})$
$$= e^{n-1}[u_{0L}(1+e)/2 - u_{0S}(1-e)/2] \qquad (i)$$

Similarly:
$$u_{cS} = e^{n-1}[u_{0S}(1+e)/2 - u_{0L}(1-e)/2] \qquad (ii)$$

If, in the upper zone, the settling velocity of the fine particles and the voidage are u_x and e_x respectively,

then: $\quad (u_x/u_{0S}) = e_x^n \qquad (iii)$

The rate at which solids are entering the upper, single-size zone is $(u_{cL} - u_{cS})(1-e)/2$, per unit area, and the rate at which the zone is growing $= (u_{cL} - u_S)$

Thus: $\quad (1 - e_x) = (u_{cL} - u_{cS})(1-e)/2(u_{cL} - u_x) \qquad (iv)$

In equation (i):
$$u_{cL} = (1 - 0.2)^{2.3-1}(2(1+1-0.2)/2 - 1(1-(1-0.2))/2) = \underline{0.733} \text{ mm/s}$$

and: $u_{cS} = (1 - 0.2)^{2.3-1}(1(1+1-0.2)/2 - 2(1-(1-0.2))/2) = \underline{0.523}$ mm/s

In equation (iii):
$$(u_x/1) = e_x^{2.3} \qquad (v)$$

and in equation (iv):
$$(1 - e_x) = (0.733 - 0.523)(1 - 0.8)/2(0.733 - u_x)$$
$$= 0.021/(0.733 - u_x) \qquad (vi)$$

By solving equations (v) and (vi) simultaneously and, by assuming values of e_x in the range 0.7–0.9, it is found that $e_x = 0.82$, at which $u_x = \underline{0.634 \text{ mm/s}}$

This is the settling rate of the upper interface. The settling rate of the lower interface is, as before:

$$u_{cL} = \underline{\underline{0.733 \text{ mm/s}}}$$

PROBLEM 3.19

What will be the terminal falling velocity of a glass sphere 1 mm in diameter in water if the density of glass is 2500 kg/m³?

Solution

For a sphere, $\quad (R'_0/\rho u_0^2) Re_0'^2 = (2d^3/3\mu^2)\rho(\rho_s - \rho)g \quad$ (equation 3.34)

Noting that: $\quad d = 1 \text{ mm}$ or 0.001 m

$\mu = 1 \text{ mNs/m}^2 = 0.001 \text{ Ns/m}^2$, per water

and: $\quad \rho = 1000 \text{ kg/m}^3$ per water

then: $\quad (R'_0/\rho u_0^2) Re_0'^2 = [(2 \times 0.001^3)/(3 \times 0.001^2)]1000(2500 - 1000)9.81$

$= 9810$

$\log_{10} 9810 = 3.992$

From Table 3.4: $\quad \log_{10} Re'_0 = 2.16$

or: $\quad Re'_0 = 144.5$

Thus: $\quad u_0 = (1445 \times 0.001)/(1000 \times 0.001)$

$= \underline{\underline{0.145 \text{ m/s}}}$

PROBLEM 3.20

What is the mass of a sphere of density 7500 kg/m³ which has a terminal falling velocity of 0.7 m/s in a large tank of water?

Solution

For a sphere diameter d, the volume $= \pi d^3/6 = 0.524 d^3$ m³

The mass of the sphere is then:

$$m = 0.524 d^3 \times 7500 = 3926 d^3 \text{ kg}$$

or: $\quad d = 0.0639 \, m^{0.3}$ m

From equation 3.34:

$$R'_0/\rho u_0^2 = (2dg/3\rho u_0^2)(\rho_s - \rho)$$

and: $(R'_0/\rho u_0^2)Re'^{-1} = [(2dg/3\rho u_0^2)(\rho_s - \rho)](\mu/du_0\rho)$
$$= [2g(\rho_s - \rho)\mu]/(3\rho^2 u_0^3)$$
$$= [(2 \times 9.81)(7500 - 1000) \times 1 \times 10^{-3}]/(3 \times 1000^2 \times 0.7^3)$$
$$= 1.24 \times 10^{-4}$$

From Figure 3.6:
$$Re' = 1800$$
and: $d = 1800\mu/(u_0\rho)$
$$= (1800 \times 1 \times 10^{-3})/(0.7 \times 1000)$$
$$= 2.57 \times 10^{-3} \text{ m} \quad \text{or} \quad 2.6 \text{ mm}$$

The mass of the sphere is then:
$$m = 3926(2.57 \times 10^{-3})^3$$
$$= 6.6 \times 10^{-5} \text{kg} \quad \text{or} \quad \underline{\underline{0.066 \text{ g}}}$$

As Re is in the Newton's law region, it is more accurate to use:
$$R'/\rho u_0^2 = 0.22 \qquad \text{(equation 3.18)}$$
or: $[2dg(\rho_s - \rho)]/3\rho u_0^2 = 0.22$

from which: $d = \rho u_0^2/[3g(\rho_s - \rho)]$
$$= (1000 \times 0.7^2)/[(3 \times 9.81)(7500 - 1000)]$$
$$= 2.56 \times 10^{-3} \text{ m} \quad \text{or} \quad 2.6 \text{ mm, as before.}$$

SECTION 2-4

Flow of Fluids Through Granular Beds and Packed Columns

PROBLEM 4.1

In a contact sulphuric acid plant the secondary converter is a tray type converter, 2.3 m in diameter with the catalyst arranged in three layers, each 0.45 m thick. The catalyst is in the form of cylindrical pellets 9.5 mm in diameter and 9.5 mm long. The void fraction is 0.35. The gas enters the converter at 675 K and leaves at 720 K. Its inlet composition is:

$$SO_3\ 6.6,\ SO_2\ 1.7,\ O_2\ 10.0,\ N_2\ 81.7 \text{ mole per cent}$$

and its exit composition is:

$$SO_3\ 8.2,\ SO_2\ 0.2,\ O_2\ 9.3,\ N_2\ 82.3 \text{ mole per cent}$$

The gas flowrate is 0.68 kg/m²s. Calculate the pressure drop through the converter. The viscosity of the gas is 0.032 mN s/m².

Solution

From the Carman equation:

$$\frac{R}{\rho u_1^2} = \frac{e^3}{S(1-e)} \frac{(-\Delta P)}{l} \frac{1}{\rho u_c^2} \quad \text{(equation 4.15)}$$

$$\frac{R}{\rho u_1^2} = 5/Re_1 + 0.4/Re_1^{0.1} \quad \text{(equation 4.16)}$$

and:

$$Re_1 = \frac{G'}{S(1-e)\mu} \quad \text{(equation 4.13)}$$

$$S = 6/d = 6/(9.5 \times 10^{-3}) = 631 \text{ m}^2/\text{m}^3$$

Hence:

$$Re_1 = 0.68/(631 \times 0.65 \times 0.032 \times 10^{-3}) = 51.8$$

and:

$$\frac{R}{\rho u^2} = \left(\frac{5}{51.8}\right) + \left(\frac{0.4}{(51.8)^{0.1}}\right) = 0.366$$

From equation 4.15:

$$-\Delta P = 0.366 \times 631 \times 0.65 \times (3 \times 0.45) \times 0.569 \times (1.20)^2/(0.35)^3$$

$$= 3.87 \times 10^3 \text{ N/m}^2 \text{ or } \underline{\underline{3.9 \text{ kN/m}^2}}$$

PROBLEM 4.2

Two heat-sensitive organic liquids of an average molecular mass of 155 kg/kmol are to be separated by vacuum distillation in a 100 mm diameter column packed with 6 mm stoneware Raschig rings. The number of theoretical plates required is 16 and it has been found that the HETP is 150 mm. If the product rate is 5 g/s at a reflux ratio of 8, calculate the pressure in the condenser so that the temperature in the still does not exceed 395 K (equivalent to a pressure of 8 kN/m^2). It may be assumed that $a = 800$ m^2/m^3, $\mu = 0.02$ mN s/m^2, $e = 0.72$ and that the temperature changes and the correction for liquid flow may be neglected.

Solution

See Volume 2, Example 4.1.

PROBLEM 4.3

A column 0.6 m diameter and 4 m high is, packed with 25 mm ceramic Raschig rings and used in a gas absorption process carried out at 101.3 kN/m^2 and 293 K. If the liquid and gas properties approximate to those of water and air respectively and their flowrates are 2.5 and 0.6 kg/m^{2}s, what is the pressure drop across the column? In making calculations, Carman's method should be used. By how much may the liquid flow rate be increased before the column floods?

Solution

Carman's correlation for flow through randomly packed beds is given by:

$$R_1/\rho u_1^2 = 5/Re_1 + 1.0/Re_1^{0.1} \quad \text{(equation 4.19)}$$

where:
$$R/\rho u^2 = \left(\frac{e^3}{S(1-e)}\right)\left(\frac{(-\Delta P)}{l}\right)\left(\frac{1}{\rho u_c^2}\right) \quad \text{(equation 4.15)}$$

and:
$$Re_1 = \frac{G'}{S(1-e)\mu} \quad \text{(equation 4.13)}$$

Using the data given, then:

$$\rho_{air} = \left(\frac{29}{22.4}\right)\left(\frac{273}{293}\right) = 1.21 \text{ kg/m}^3$$

$$G' = 0.6 \text{ kg/m}^2 \text{ s}$$

and:
$$u = (0.6/1.21) = 0.496 \text{ m/s}$$

From Table 4.3 for 25 mm Raschig rings:

$$S = 190 \text{ m}^2/\text{m}^3 \text{ and } e = 0.71$$

Thus: $Re_1 = 0.6/(190 \times 0.29 \times 0.018 \times 10^{-3}) = 605$

$$\frac{R}{\rho u^2} = \left(\frac{(0.71)^3}{190 \times 0.29}\right)\left(\frac{(-\Delta P)}{4}\right)\left(\frac{1}{1.21 \times (0.496)^2}\right)$$

$$= 5.90 \times 10^{-3}(-\Delta P)$$

Hence: $5.90 \times 10^{-3}(-\Delta P) = (5/605) + (1.0/(605)^{0.1}) = 0.535$

and: $-\Delta P = 90.7 \text{ N/m}^2 \text{ or } 0.091 \text{ kN/m}^2$

The pressure drop of the wet, drained packing is given by:

$$-\Delta P_w = 90.7[1 + 3.30/25] = 102.5 \text{ N/m}^2 \qquad \text{(equation 4.46)}$$

To take account fully of the liquid flow, reference 56 in Chapter 4 of Volume 2 provides a correction factor which depends on the liquid flowrate and the Raschig size. This factor acts as a multiplier for the dry pressure drop which in this example, is equal to 1.3, giving the pressure drop in this problem as:

$$(1.3 \times 90.7) = 118 \text{ N/m}^2 \text{ or } \underline{0.118 \text{ kN/m}^2}$$

Figure 4.18 may be used to calculate the liquid flowrate which would cause the column to flood. At a value of the ordinate of 0.048, the flooding line gives:

$$\frac{L'}{G'}\sqrt{\left(\frac{\rho V}{\rho L}\right)} = 2.5$$

from which: $L' = \underline{4.3 \text{ kg/m}^2\text{s}}$

PROBLEM 4.4

A packed column, 1.2 m in diameter and 9 m tall, is packed with 25 mm Raschig rings, and used for the vacuum distillation of a mixture of isomers of molecular mass 155 kg/kmol. The mean temperature is 373 K, the pressure at the top of the column is maintained at 0.13 kN/m^2 and the still pressure is 1.3–3.3 kN/m^2. Obtain an expression for the pressure drop on the assumption that this is not appreciably affected by the liquid flow and may be calculated using a modified form of Carman's equation. Show that, over the range of operating pressures used, the pressure drop is approximately directly proportional to the mass rate of flow rate of vapour, and calculate the pressure drop at a vapour rate of 0.125 kg/m^2. The specific surface of packing, $S = 190$ m^2/m^3, the mean voidage of bed, $e = 0.71$, the viscosity of vapour, $\mu = 0.018$ mN s/m^2 and the molecular volume $= 22.4$ m^3/kmol.

Solution

The proof that the pressure drop is approximately proportional to the mass flow rate of vapour is given in Problem 4.5. Using the data specified in this problem:

$$Re_1 = G/S(1-e)\mu$$
$$= 0.125/(190 \times 0.29 \times 0.018 \times 10^{-3}) = 126.0$$

The modified Carman's equation states that:

$$R/\rho u^2 = 5/Re_1 + 1/Re_1^{0.1} \quad \text{(equation 4.19)}$$
$$= (5/126.0) + (1/(126.0)^{0.1}) = 0.656$$

As in Problem 4.2:

$$\frac{R}{\rho u^2} = \frac{e^3}{S(1-e)} \frac{(-dP)}{dl} \frac{1}{\rho u^2} \quad \text{(equation 4.15)}$$

$$= \frac{e^3}{S(1-e)} \frac{(-dP)}{dl} \frac{\rho}{G'^2}$$

Thus:
$$-\int \rho \, dP = \frac{R}{\rho u^2} \frac{S(1-e)}{e^3} G'^2 l$$

$$= (0.656 \times 190 \times 0.29 \times 9 G'^2)/(0.71)^3 = 909 G'^2 \text{ kg/m}^3$$

$\rho/P = \rho_s/P_s$ where subscript s refers to the still.

$$\rho_s = \left(\frac{155}{22.4}\right)\left(\frac{273}{373}\right)\left(\frac{P_s}{101.3 \times 10^3}\right) = 5 \times 10^{-5} P_s \text{ kg/m}^3$$

and:
$$\rho_s/P_s = 5 \times 10^{-5}, \text{ and } \rho = 5 \times 10^{-5} P$$

$$-\int_{P_c}^{P_s} \rho \, dP = 2.5 \times 10^{-5} (P_s^2 - P_c^2)$$

$(P_s - P_c) = -\Delta P$, and if $-\Delta P \simeq P_s$, then $(P_s^2 - P_c^2) \simeq (-\Delta P)^2$

Thus:
$$-\int_{P_c}^{P_s} \rho \, dP = 2.5 \times 10^{-5} (-\Delta P)^2 = 909 G'^2 \text{ or } \underline{-\Delta P \propto G'}$$

If $G' = 0.125$, $-\Delta P = [909 \times (0.125)^2/2.5 \times 10^{-5}]^{0.5} = 754 \text{ N/m}^2$ or $\underline{\underline{0.754 \text{ kN/m}^2}}$

PROBLEM 4.5

A packed column, 1.22 m in diameter and 9 m high, and packed with 25 mm Raschig rings, is used for the vacuum distillation of a mixture of isomers of molecular mass 155 kg/kmol. The mean temperature is 373 K, the pressure at the top of the column is maintained at 0.13 kN/m², and the still pressure is 1.3 kN/m². Obtain an expression for the pressure drop on the assumption that this is not appreciably affected by the liquid flow and may be calculated using the modified form of Carman's equation.

Show that, over the range of operating pressures used, the pressure drop is approximately directly proportional to the mass rate of flow of vapour, and calculate approximately the flow of vapour. The specific surface of the packing is 190 m^2/m^3, the mean voidage of the bed is 0.71, the viscosity of the vapour is 0.018 mN s/m^2 and the kilogramme molecular volume is 22.4 m^3/kmol.

Solution

The modified form of Carman's equation states that:

$$R/\rho u^2 = 5/Re_1 + (1/Re_1)^{0.1} \qquad \text{(equation 4.19)}$$

where:
$$Re_1 = G'/S(1-e)\mu$$

In this case:

$$Re_1 = G'/[190(1-0.71) \times 0.018 \times 10^{-3}] = 1008 G' \quad \text{where } G' \text{ is in kg/m}^2\text{s}$$

Thus $(R/\rho u^2) = (5/1008 G') + (1/1008 G')^{0.1} = (0.005/G') + (0.501/G'^{0.1})$

$$(R/\rho u^2) = [e^3/S(1-e)][(-dP)/dl](\rho/G'^2)$$

so that, as in Problem 7.4:

$$-\int \rho dP = (R/\rho u^2)[S(1-e)/e^3] G'^2 l$$

$$= [(0.005/G') + (0.501/G'^{0.1})][\{190 \times 0.29 \times 9\}/0.71^3] G'^2$$

$$= 6.93 G' + 694 G'^{1.9}$$

As before:

$$-\int \rho dP = 2.5 \times 10^{-5} (-\Delta P)^2$$

Thus:
$$(-\Delta P)^2 = 2.8 \times 10^5 G' + 2.8 \times 10^7 G'^{1.9}$$

Neglecting the first term:

$$\underline{\underline{-\Delta P = 5.30 \times 10^3 G'^{0.95} \text{ N/m}^2}}$$

and, when $-\Delta P = (1300 - 130) = 1170$ N/m^2, then:

$$\underline{\underline{G' = 0.018 \text{ kg/m}^2\text{s}}}$$

SECTION 2-5

Sedimentation

PROBLEM 5.1

A slurry containing 5 kg of water/kg of solids is to be thickened to a sludge containing 1.5 kg of water/kg of solids in a continuous operation. Laboratory tests using five different concentrations of the slurry yielded the following results:

concentration Y (kg water/kg solid)	5.0	4.2	3.7	3.1	2.5
rate of sedimentation u_c (mm/s)	0.17	0.10	0.08	0.06	0.042

Calculate the minimum area of a thickener to effect the separation of 0.6 kg/s of solids.

Solution

Basis: 1 kg of solids:

1.5 kg water is carried away in underflow so that $U = 1$

Concentration Y (kg rate/kg solids)	Water to overflow $(Y - U)$	Sedimentation rate u_c (mm/s)	$(Y - U)/u_c$ (s/mm)
5.0	3.5	0.17	20.56
4.2	2.7	0.10	27.0
3.7	2.2	0.08	27.5
3.1	1.6	0.06	26.67
2.5	1.0	0.042	23.81

The maximum value of $(Y - U)/u_c = 27.5$ s/mm or 27,500 s/m.

$$A = \frac{Q(Y - U)}{u_c} \frac{C\rho_s}{\rho} \quad \text{(equation 5.54)}$$

$C\rho_s = 0.6$ kg/s and $\rho = 1000$ kg/m^3

Hence: $A = (27{,}500 \times 0.6)/1000 = \underline{\underline{16.5 \text{ m}^2}}$

PROBLEM 5.2

A slurry containing 5 kg of water/kg of solids is to be thickened to a sludge containing 1.5 kg of water/kg of solids in a continuous operation.

Laboratory tests using five different concentrations of the slurry yielded the following data:

concentration (kg water/kg solid)	5.0	4.2	3.7	3.1	2.5
rate of sedimentation (mm/s)	0.20	0.12	0.094	0.070	0.052

Calculate the minimum area of a thickener to effect the separation of 1.33 kg/s of solids.

Solution

See Volume 2, Example 5.1.

PROBLEM 5.3

When a suspension of uniform coarse particles settles under the action of gravity, the relation between the sedimentation velocity u_c and the fractional volumetric concentration C is given by:

$$\frac{u_c}{u_0} = (1-C)^n,$$

where $n = 2.3$ and u_0 is the free falling velocity of the particles. Draw the curve of solids flux ψ against concentration and determine the value of C at which ψ is a maximum and where the curve has a point of inflexion. What is implied about the settling characteristics of such a suspension from the Kynch theory? Comment on the validity of the Kynch theory for such a suspension.

Solution

The given equation is: $\quad u_c/u_0 = (1-C)^{2.3}$

The flux is the mass rate of sedimentation per unit area and is given by:

$$\psi = u_c C = u_0 C(1-C)^{2.3} \text{ m/s} \qquad \text{(from equation 5.31)}$$

A plot of ψ as a function of C is shown in Figure 5a.

To find the maximum flux, this equation may be differentiated to give:

$$\frac{d\psi}{dC} = u_0[(1-C)^{2.3} - 2.3C(1-C)^{1.3}]$$

$$= u_0(1-C)^{1.3}(1-3.3C)$$

For a maximum, $d\psi/dC = 0$ and $\quad \underline{C = 0.30}$

At the point of inflexion:

$$d^2\psi/dC^2 = 0$$

Thus:
$$\frac{d^2\psi}{dC^2} = u_0[-3.3(1-C)^{1.3} - 1.3(1-C)^{0.3}(1-3.3C)]$$

$$= u_0(1-C)^{0.3}(7.6C - 4.6)$$

When $d^2\psi/dC^2 = 0$, $\underline{\underline{C = 0.61}}$

The maximum flux and the point of inflexion are shown in Figure 5a. The Kynch theory is discussed fully in Section 5.2.3.

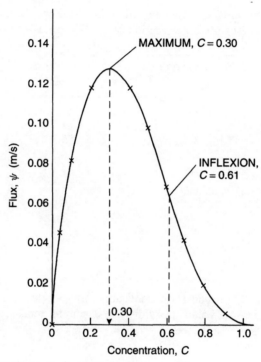

Figure 5a. Flux-concentration curve for suspension when $n = 2.3$

PROBLEM 5.4

For the sedimentation of a suspension of uniform fine particles in a liquid, the relation between observed sedimentation velocity u_c and fractional volumetric concentration C is given by:

$$\frac{u_c}{u_0} = (1-C)^{4.6}$$

where u_0 is the free falling velocity of an individual particle. Calculate the concentration at which the rate of deposition of particles per unit area is a maximum and determine this maximum flux for 0.1 mm spheres of glass of density 2600 kg/m^3 settling in water of density 1000 kg/m^3 and viscosity 1 mN s/m^2.

It may be assumed that the resistance force F on an isolated sphere is given by Stokes' Law.

Solution

See Volume 2, Example 5.3.

PROBLEM 5.5

Calculate the minimum area and diameter of a thickener with a circular basin to treat 0.1 m^3/s of a slurry of a solids concentration of 150 kg/m^3. The results of batch settling tests are:

Solids concentration (kg/m^3)	Settling velocity (μm/s)
100	148
200	91
300	55.33
400	33.25
500	21.40
600	14.50
700	10.29
800	7.38
900	5.56
1000	4.20
1100	3.27

A value of 1290 kg/m^3 for underflow concentration was selected from a retention time test. Estimate the underflow volumetric flow rate assuming total separation of all solids and that a clear overflow is obtained.

Solution

The settling rate of the solids, G' kg/m^{2}s, is calculated as $G' = u_s c$ where u_c is the settling velocity (m/s) and c the concentration of solids (kg/m^3) and the data are plotted in Figure 5b. From the point $u = 0$ and $c = 1290$ kg/m^3, a line is drawn which is tangential to the curve. This intercepts the axis at $G' = 0.0154$ kg/m^{2}s.

The area of the thickener is then:

$$A = (0.1 \times 150)0.0154 = \underline{974 \text{ m}^2}$$

and the diameter is:

$$d = [(4 \times 974)/\pi]^{0.5} = 35.2 \text{ m}$$

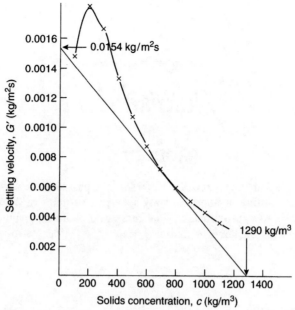

Figure 5b. Construction for Problem 5.5

The volumetric flow rate of underflow, obtained from a mass balance, is:

$$= [(0.1 \times 150)/1290] = \underline{\underline{0.0116 \text{ m}^3/\text{s}}}$$

SECTION 2-6

Fluidisation

PROBLEM 6.1

Oil, of density 900 kg/m³ and viscosity 3 mN s/m², is passed vertically upwards through a bed of catalyst consisting of approximately spherical particles of diameter 0.1 mm and density 2600 kg/m³. At approximately what mass rate of flow per unit area of bed will (a) fluidisation, and (b) transport of particles occur?

Solution

See Volume 2, Example 6.2.

PROBLEM 6.2

Calculate the minimum velocity at which spherical particles of density 1600 kg/m³ and of diameter 1.5 mm will be fluidised by water in a tube of diameter 10 mm on the assumption that the Carman-Kozeny equation is applicable. Discuss the uncertainties in this calculation. Repeat the calculation using the Ergun equation and explain the differences in the results obtained.

Solution

The Carman-Kozeny equation takes the form:

$$u_{mf} = 0.0055[e^3/(1-e)][d^2(\rho_s - \rho)g/\mu] \qquad \text{(equation 6.4)}$$

As a wall effect applies in this problem, use is made of equation 4.23 to determine the correction factor, f_w where:

$$f_w = (1 + 0.5 S_c/S)^2$$

where: S_c = surface area of the container/volume of bed

$$= (\pi \times 0.01 \times 1)/[(\pi/4)(0.01^2 \times 1)] = 400 \text{ m}^2/\text{m}^3$$

$S = 6/d$ for a spherical particle

$$= [6/(1.5 \times 10^{-3})] = 4000 \text{ m}^2/\text{m}^3$$

Thus: $f_w = [1 + 0.5(400/4000)]^2 = 1.10$

The uncertainty in this problem lies in the chosen value of the voidage e. If e is taken as 0.45 then:

$$u_{mf} = 0.0055[0.45^3/(1-0.45)][(1.5 \times 10^{-3})^2(1600-1000) \times 9.81]/(1 \times 10^{-3})$$
$$= 0.0120 \text{ m/s}$$

Allowing for the wall effect:

$$u_{mf} = (0.0120 \times 1.10) = \underline{\underline{0.0133 \text{ m/s}}}$$

By definition:

Galileo number, $Ga = d^3\rho(\rho_s - \rho)g/\mu^2$
$$= (1.5 \times 10^{-3})^3 \times 1000(1600-1000) \times 9.81/(1 \times 10^{-3})^2$$
$$= 1.99 \times 10^4$$

Assuming a value of 0.45 for e_{mf}, equation 6.14 gives:

$$Re'_{mf} = 23.6\{\sqrt{[1 + (9.39 \times 10^{-5})(1.99 \times 10^4)]} - 1\} = 16.4$$

and from equation 6.15:

$$u_{mf} = [(1 \times 10^{-3}) \times 16.4]/(1.5 \times 10^{-3} \times 1000)$$
$$= \underline{\underline{0.00995 \text{ m/s}}}$$

As noted in Section 6.1.3 of Volume 2, the Carman-Kozeny equation applies only to conditions of laminar flow and hence to low values of the Reynolds number for flow in the bed. In practice, this restricts its application to fine particles. Approaches based on both the Carman-Kozeny and the Ergun equations are very sensitive to the value of the voidage and it seems likely that both equations overpredict the pressure drop for fluidised systems.

PROBLEM 6.3

In a fluidised bed, *iso*-octane vapour is adsorbed from an air stream onto the surface of alumina microspheres. The mole fraction of *iso*-octane in the inlet gas is 1.442×10^{-2} and the mole fraction in the outlet gas is found to vary with time as follows:

Time from start (s)	Mole fraction in outlet gas ($\times 10^2$)
250	0.223
500	0.601
750	0.857
1000	1.062
1250	1.207
1500	1.287
1750	1.338
2000	1.373

Show that the results may be interpreted on the assumptions that the solids are completely mixed, that the gas leaves in equilibrium with the solids and that the adsorption isotherm is linear over the range considered. If the flowrate of gas is 0.679×10^{-6} kmol/s and the mass of solids in the bed is 4.66 g, calculate the slope of the adsorption isotherm. What evidence do the results provide concerning the flow pattern of the gas?

Solution

See Volume 2, Example 6.4.

PROBLEM 6.4

Cold particles of glass ballotini are fluidised with heated air in a bed in which a constant flow of particles is maintained in a horizontal direction. When steady conditions have been reached, the temperatures recorded by a bare thermocouple immersed in the bed are:

Distance above bed support (mm)	Temperature (K)
0	339.5
0.64	337.7
1.27	335.0
1.91	333.6
2.54	333.3
3.81	333.2

Calculate the coefficient for heat transfer between the gas and the particles, and the corresponding values of the particle Reynolds and Nusselt numbers. Comment on the results and on any assumptions made. The gas flowrate is 0.2 kg/m² s, the specific heat in air is 0.88 kJ/kg K, the viscosity of air is 0.015 mN s/m², the particle diameter is 0.25 mm and the thermal conductivity of air 0.03 is W/mK.

Solution

See Volume 2, Example 6.5.

PROBLEM 6.5

The relation between bed voidage e and fluid velocity u_c for particulate fluidisation of uniform particles which are small compared with the diameter of the containing vessel is given by:

$$\frac{u_c}{u_0} = e^n$$

where u_0 is the free falling velocity.

Discuss the variation of the index n with flow conditions, indicating why this is independent of the Reynolds number Re with respect to the particle at very low and very high values of Re. When are appreciable deviations from this relation observed with liquid fluidised systems?

For particles of glass ballotini with free falling velocities of 10 and 20 mm/s the index n has a value of 2.39. If a mixture of equal volumes of the two particles is fluidised, what is the relation between the voidage and fluid velocity if it is assumed that complete segregation is obtained?

Solution

The variation of the index n with flow conditions is fully discussed in Chapters 5 and 6 of Volume 2. The ratio u_c/u_0 is in general, dependent on the Reynolds number, voidage, and the ratio of particle diameter to that of the containing vessel. At low velocities, that is when $Re < 0.2$, the drag force is attributable entirely to skin friction, and at high velocities when $Re > 500$ skin friction becomes negligible and in these regions the ratio u_c/u_0 is independent of Re. For intermediate regions, the data given in Table 5.1 apply.

Considering unit volume of each particle say 1 m³, then:

Voidage of large particles = e_1, volume of liquid = $e_1/(1-e_1)$.

Voidage of small particles = e_2, volume of liquid = $e_2/(1-e_2)$.

Total volume of solids = 2 m³.

Total volume of liquid = $e_1/(1-e_1) + e_2/(1-e_2)$.

Total volume of system = $2 + e_1/(1-e_1) + e_2/(1-e_2)$ m³

Thus:
$$\text{voidage} = \frac{e_1/(1-e_1) + e_2/(1-e_2)}{2 + e_1/(1-e_1) + e_2/(1-e_2)}$$

$$= \frac{e_1(1-e_2) + e_2(1-e_1)}{2(1-e_1)(1-e_2) + e_1(1-e_2) + e_2(1-e_1)}$$

That is:
$$e = \frac{e_1 + e_2 - 2e_1e_2}{2 - e_1 - e_2}$$

But, since the free falling velocities are in the ratio 1:2, then:

$$e_1 = \left(\frac{u}{u_{01}}\right)^{1/2.4} \quad \text{and} \quad e_2 = \left(\frac{u}{u_{01}/2}\right)^{1/2.4}$$

Thus:
$$e_2 = e_1 2^{1/2.4}$$

at:
$$e = \frac{e_1 + e_1 \times 2^{1/2.4} - 2^{3.4/2.4} \times e_1^2}{2 - e_1 - 2^{1/2.4} \times e_1}$$

$$e = \frac{(u/20)^{1/2.4}(1 + 2^{1/2.4}) - 2^{3.4/2.4}(u/20)^{1/1.2}}{2 - (1 + 2^{1/2.4})(u/20)^{1/2.4}}$$

$$e = \frac{3u^{0.42} - u^{0.83}}{9 - 3u^{0.42}} \quad \text{(with } u \text{ in mm/s)}$$

and:
$$9e = 3eu^{0.42} = 3u^{0.42} - u^{0.84}$$

or:
$$u^{0.84} - 3(1+e)u^{0.42} + 9e = 0$$

and:
$$\underline{u^{0.42} = 1.5(1+e) + [2.24(1+e)^2 - 9e]}$$

This relationship is plotted in Figure 6a.

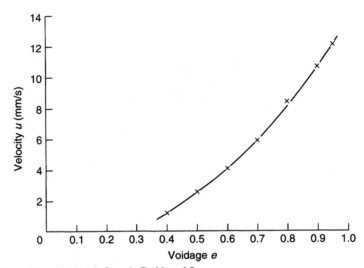

Figure 6a. Plot of the relationship for u in Problem 6.5

PROBLEM 6.6

Obtain a relationship for the ratio of the terminal falling velocity of a particle to the minimum fluidising velocity for a bed of similar particles. It may be assumed that Stokes' Law and the Carman-Kozeny equation are applicable. What is the value of the ratio if the bed voidage at the minimum fluidising velocity is 0.4?

Solution

In a fluidised bed the total frictional force must be equal to the effective weight of bed. Thus:
$$-\Delta P = (1-e)(\rho_s - \rho)lg \qquad \text{(equation 6.1)}$$

Substituting equation 6.1 into equation 4.9, and putting $K'' = 5$, gives:
$$u_{mf} = 0.0055 \frac{e^3}{1-e} \frac{d^2(\rho_s - \rho)g}{\mu} \qquad \text{(equation 6.4)}$$

48

Hence:
$$\frac{u_0}{u_{mf}} = \left(\frac{d^2 g(\rho_s - \rho)}{18\mu \times 0.0055}\right)\left(\frac{(1-e)}{e^3}\right)\left(\frac{\mu}{d^2(\rho_s - \rho)g}\right)$$
$$= \frac{(1-e)}{(18 \times 0.0055 e^3)} = 10.1(1-e)/e^3$$

If $e = 0.4$, then: $u_0/u_{mf} = \underline{94.7}$.

The use of the Carman–Kozeny equation is discussed in Section 4.2.3 of Chapter 4. It is interesting to note that if $e = 0.48$, which was the value taken in Problem 6.1, then $u_0/u_f = 47.5$, which agrees with the solution to that problem.

PROBLEM 6.7

A packed bed consisting of uniform spherical particles of diameter 3 mm and density 4200 kg/m^3, is fluidised by means of a liquid of viscosity 1 mN s/m^2 and density 1100 kg/m^3. Using Ergun's equation for the pressure drop through a bed height l and voidage e as a function of superficial velocity, calculate the minimum fluidising velocity in terms of the settling velocity of the particles in the bed.

State clearly any assumptions made and indicate how closely the results might be confirmed by an experiment.

Ergun's equation:
$$\frac{-\Delta P}{l} = 150\frac{(1-e)^2}{e^3}\frac{\mu u}{d^2} + 1.75\frac{(1-e)}{e^3}\frac{\rho u^2}{d}$$

Solution

The pressure drop through a fluidised bed of height l is:
$$-\Delta P/l = (1-e)(\rho_s - \rho)g \qquad \text{(equation 6.1)}$$

Writing $u = u_{mf}$, the minimum fluidising velocity in Ergun's equation and substituting for $-\Delta P/H$ gives:
$$(1-e)(\rho_s - \rho)g = \frac{150(1-e)^2/\mu u_{mf}}{e^3 d^2} + \frac{1.75(1-e)\rho u_{mf}^2}{e^3 d} \qquad \text{(equation 6.7)}$$

or:
$$(\rho_s - \rho)g = \frac{150(1-e)\mu u_{mf}}{e^3 d^2} + \frac{1.75\rho u_{mf}^2}{e^3 d}$$

If $d = 3 \times 10^{-3}$ m, $\rho_s = 4200$ kg/m^3, $\rho = 1100$ kg/m^3, $\mu = 1 \times 10^{-3}$ Ns/m^2, and if e is taken as 0.48 as in Problem 6.1, these values may be substituted to give:
$$3.04 = 7.84 u_{mf} + 580 u_{mf}^2$$

or:
$$u_{mf} = \underline{0.066 \text{ m/s}} \text{ neglecting the negative root.}$$

If Stokes' law applies, then:

$$u_0 = d^2 g(\rho_s - \rho)/18\mu \qquad \text{(equation 3.24)}$$
$$= (3 \times 10^{-3})(9.81 \times 3100/18 \times 10^{-3}) = 15.21 \text{ m/s}$$

Thus: $\quad Re = (3 \times 10^{-3} \times 15.21 \times 4200/10^{-3}) = 1.92 \times 10^5$

which is outside the range of Stokes' law. A Reynolds number of this order lies in the region (c) of Figure 3.4 where:

$$u_0^2 = 3dg(\rho_s - \rho)/\rho \qquad \text{(equation 3.25)}$$

Thus: $\quad u_0^2 = (3 \times 3 \times 10^{-3} \times 9.81 \times 3100)/1100$

and: $\quad u_0 = 0.5$ m/s

A check on the value of the Reynolds number gives:

$$Re = (3 \times 10^{-3} \times 0.5 \times 4200)/10^{-3} = 6.3 \times 10^3$$

which is within the limits of region (c).

Hence: $\quad u_0/u_{mf} = (0.5/0.066) = \underline{\underline{7.5}}$

Empirical relationships for the minimum fluidising velocity are presented as a function of Reynolds number and this problem illustrates the importance of using the equations applicable to the particle Reynolds number in question.

PROBLEM 6.8

Ballotini particles, 0.25 mm in diameter, are fluidised by hot air flowing at the rate of 0.2 kg/m² cross-section of bed to give a bed of voidage 0.5 and a cross-flow of particles is maintained to remove the heat. Under steady state conditions, a small bare thermocouple immersed in the bed gives the following data:

Distance above bed support (mm)	Temperature (°C)	(K)
0	66.3	339.5
0.625	64.5	337.7
1.25	61.8	335.0
1.875	60.4	333.6
2.5	60.1	333.3
3.75	60.0	333.2

Assuming plug flow of the gas and complete mixing of the solids, calculate the coefficient for heat transfer between the particles and the gas. The specific heat capacity of air is 0.85 kJ/kg K.

A fluidised bed of total volume 0.1 m³ containing the same particles is maintained at an approximately uniform temperature of 425 K by external heating, and a dilute aqueous solution at 375 K is fed to the bed at the rate of 0.1 kg/s so that the water is completely evaporated at atmospheric pressure. If the heat transfer coefficient is the same as that previously determined, what volumetric fraction of the bed is effectively carrying out the evaporation? The latent heat of vaporisation of water is 2.6 MJ/kg.

Solution

See Volume 2, Example 6.6.

PROBLEM 6.9

An electrically heated element of surface area 12 cm² is completely immersed in a fluidised bed. The resistance of the element is measured as a function of the voltage applied to it giving the following data:

Potential (V)	1	2	3	4	5	6
Resistance (ohms)	15.47	15.63	15.91	16.32	16.83	17.48

The relation between resistance R_w and temperature T_w is:

$$\frac{R_w}{R_0} = 0.004 T_w - 0.092$$

where R_0, is the resistance of the wire at 273 K is 14 ohms and T_w is in K. Estimate the bed temperature and the value of the heat transfer coefficient between the surface and the bed.

Solution

The heat generation rate by electrical heating $= V^2/R$

The rate of heat dissipation $= hA(T_w - T_B)$

where T_w and T_B are the wire and bed temperatures respectively.

At equilibrium, $V^2/R_w = hA(T_w - T_B)$

But: $\quad R_w/R_0 = 0.004 T_w - 0.092$

so that: $\quad T_w = 250(R_w/R_0) + 23$

Thus: $$V^2 = \frac{250 h A \bar{R}_w R_w}{R_0} - hA\bar{R}_w(T_B - 23)$$

where $\bar{R}_w$ is a mean value of R_w noting that the mean cannot be used inside the bracket in the equation for T_w.

Thus a plot of V^2 against R_w should yield a line of slope $= 250\, hA\bar{R}_w/R_0$. This is shown in Figure 6b from which the value of the slope is 17.4.

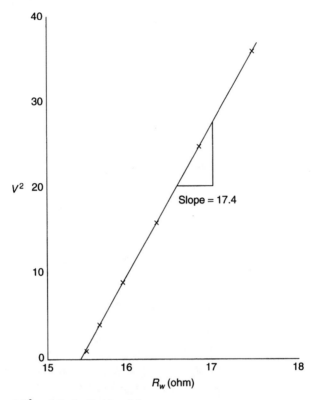

Figure 6b. A plot of V^2 and R_w for Problem 6.9

Hence: $h = (17.4 \times 14)/(250 \times 12 \times 10^{-4} \times 16.5) = \underline{49.2\ \text{W/m}^2\text{K}}$

The bed temperature is found by the intercept at $V^2 = 0$ that is when $R_w = 15.4$ ohm

Thus: $T_B = 250(15.4/14) + 23 = \underline{298\ \text{K}}$.

PROBLEM 6.10

(a) Explain why the sedimentation velocity of uniform coarse particles in a suspension decreases as the concentration is increased. Identify and, where possible, quantify the various factors involved.

(b) Discuss the similarities and differences in the hydrodynamics of a sedimenting suspension of uniform particles and of an evenly fluidised bed of the same particles in the liquid.

(c) A liquid fluidised bed consists of equal volumes of spherical particles 0.5 mm and 1.0 mm in diameter. The bed is fluidised and complete segregation of the two species occurs. When the liquid flow is stopped the particles settle to form a segregated two-layer bed. The liquid flow is then started again. When the velocity is such that the larger particles are at their incipient fluidisation point what will be the approximate voidage of the fluidised bed composed of the smaller particles?

It may be assumed that the drag force F of the fluid on the particles under the free falling conditions is given by Stokes' law and that the relation between the fluidisation velocity u_c and voidage, e, for particles of terminal velocity, u_0, is given by:

$$u_c/u_0 = e^{4.8}$$

For Stokes' law, the force F on the particles is given by $F = 3\pi\mu d u_0$, where d is the particle diameter and μ is the viscosity of the liquid.

Solution

Parts (a) and (b) of this Problem are considered in Volume 2, Chapter 6. Attention is now concentrated on part (c).

A force balance gives:

$$3\pi\mu d u_c = (\pi/6)d^3(\rho_s - \rho)g$$

Thus:
$$u_c = (d^2 g/18\mu)(\rho_s - \rho)g$$

That is:
$$u_c = k_i d^2 \tag{i}$$

For larger particles and assuming that the voidage at $u_{mf} = 0.45$,

then:
$$u_c/u_{0L} = 0.45^n \tag{ii}$$

For smaller particles:

$$u_c/u_{0S} = e^n$$

u_c has the same k_i both large and small particles and from equation (i):

$$u_{0S} = u_{0L}/4$$

Thus:
$$4u_c/u_{0L} = e^n \tag{iii}$$

Dividing equation (iii) by equation (ii) gives:

$$4 = (e/0.45)^n = (e/0.45)^{4.8}$$

Thus:
$$1.334 = e/0.45$$

and:
$$e = \underline{0.60}$$

PROBLEM 6.11

The relation between the concentration of a suspension and its sedimentation velocity is of the same form as that between velocity and concentration in a fluidised bed. Explain this in terms of the hydrodynamics of the two systems. A suspension of uniform spherical particles in a liquid is allowed to settle and, when the sedimentation velocity is measured as a function of concentration, the following results are obtained:

Fractional volumetric concentration (C)	Sedimentation velocity (u_c m/s)
0.35	1.10
0.25	2.19
0.15	3.99
0.05	6.82

Estimate the terminal falling velocity u_0 of the particles at infinite dilution. On the assumption that Stokes' law is applicable, calculate the particle diameter d.

The particle density, $\rho_s = 2600$ kg/m^3, the liquid density, $\rho = 1000$ kg/m^3, and the liquid viscosity, $\mu = 0.1$ Ns/m^2.

What will be the minimum fluidising velocity of the system? Stokes' law states that the force on a spherical particle $= 3\pi \mu d u_0$.

Solution

The relation between u_c and C is:

$$u_c/u_0 = (1 - C)^n \tag{i}$$

u_0 and n may be obtained by plotting $\log_{10} u_c$ and $\log_{10}(1 - C)$. This will give a straight line of slope n and u_0 is given by the intercept which corresponds to $c = 0$ at $\ln(1 - c) = 0$.

Alternatively on algebraic solution may be sought as follows:

C	$1 - c$	$\log_{10}(1 - C)$	u_c	$\log_{10} u_c$
0.35	0.65	-0.187	1.10	0.0414
0.25	0.75	-0.125	2.19	0.8404
0.15	0.85	-0.0706	3.99	0.6010
0.05	0.95	-0.0222	6.82	0.8338

Taking logarithms of equation (i) gives:

$$\log_{10} u_c = \log_{10} u_0 + n \log_{10}(1 - c) \tag{ii}$$

Inserting values from the table into equation (ii) gives:

$$0.0414 = \log_{10} u_c + n(-0.187) \tag{iii}$$

$$0.8404 = \log_{10} u_c + n(-0.125) \tag{iv}$$

$$0.6010 = \log_{10} u_c + n(-0.0706) \tag{v}$$

$$0.8338 = \log_{10} u_c + n(-0.0222) \tag{vi}$$

Any two of the equations (iii)–(vi) may be used to evaluate $\log_{10} u_0$ and n.

Substituting (iii) from (v):

$$0.5396 = 0.116n \quad \text{and} \quad n = 4.8$$

Substituting (iv) from (vi):

$$0.4934 = 0.103n \quad \text{and} \quad n = 4.8 \text{ which is consistent.}$$

Substituting in equation (v):

$$\log_{10} u_0 = 0.6010 + (0.0706 \times 4.8) = 0.94$$

Substituting in equation (vi):

$$\log_{10} u_0 = 0.8338 + (0.0222 \times 4.8) = 0.94 \text{ which is consistent}$$

and hence: $u_0 = \underline{\underline{8.72 \text{ mm/s}}}$

A force balance gives:

$$3\pi \mu d u_0 = (\pi/6) d^3 (\rho_s - \rho) g$$

and:
$$d = [(18 \mu u_0/(g(\rho_s - \rho)))]^{0.5}$$
$$= [(18 \times 0.1 \times 8.72 \times 10^{-3})/(9.81(2600 - 1000))]^{0.5}$$
$$= 0.001 \text{ m} \quad \text{or} \quad \underline{\underline{1 \text{ mm}}}$$

Assuming that $e_{mf} = 0.45$, then:

$$u_{mf} = u_0 e_{mf}$$
$$= 8.72 \times 0.45^{4.8}$$
$$= \underline{\underline{0.19 \text{ mm/s}}}$$

PROBLEM 6.12

A mixture of two sizes of glass spheres of diameters 0.75 and 1.5 mm is fluidised by a liquid and complete segregation of the two species of particles occurs, with the smaller particles constituting the upper portion of the bed and the larger particles in the lower portion. When the voidage of the lower bed is 0.6, what will be the voidage of the upper bed?

The liquid velocity is increased until the smaller particles are completely transported from the bed. What is the minimum voidage of the lower bed at which this phenomenon will occur?

It may be assumed that the terminal falling velocities of both particles may be calculated from Stokes' law and that the relationship between the fluidisation velocity u and the bed voidage e is given by:

$$(u_c/u_0) = e^{4.6}$$

Solution

A force balance gives:

$$(\pi/6) d^3 (\rho_s - \rho) g = 3\pi \mu d u_0$$

from which:
$$u_0 = (d^2 g / 18 \mu)(\rho_s - \rho)$$

and hence, for constant viscosity and densities:

$$u_0 = k_1 d^2$$

For large particles of diameter d_L:

$$u_0 = k_1 d_L^2$$

The voidage, 0.6, achieved at velocity u is given by:

$$u/k_1 d_L^2 = 0.6^{4.6}$$

For small particles of diameter d_s, the voidage e at this velocity is given by:

$$u/k_1 d_s^2 = e^{4.6}$$

Dividing:

$$d_s^2/d_L^2 = (0.6/e)^{4.6}$$

Since $d_s/d_L = 0.5$, $(d_s/d_L)^2 = 0.25$ and:

$$0.25 = (0.6/e)^{4.6}$$

from which:

$$e = \underline{\underline{0.81}}$$

For transport of the smaller particles just to occur, the voidage of the upper bed is unity and:

$$0.25 = (e/1)^{4.6}$$

from which, for the large particles:

$$e = \underline{\underline{0.74}}$$

PROBLEM 6.13

(a) Calculate the terminal falling velocities in water of glass particles of diameter 12 mm and density 2500 kg/m³, and of metal particles of diameter 1.5 mm and density 7500 kg/m³.

It may be assumed that the particles are spherical and that, in both cases, the friction factor, $R'/\rho u^2$ is constant at 0.22, where R' is the force on the particle per unit of projected area of the particle, ρ is the fluid density and u the velocity of the particle relative to the fluid.

(b) Why is the sedimentation velocity lower when the particle concentration in the suspension is high? Compare the behaviour of the concentrated suspension of particles settling under gravity in a liquid with that of a fluidised bed of the same particles.

(c) At what water velocity will fluidised beds of the glass and metal particles have the same densities? The relation between the fluidisation velocity u_c terminal velocity u_0 and bed voidage e is given for both particles by:

$$(u_c/u_0) = e^{2.30}$$

Solution

For spheres, a take balance gives:

$$R'(\pi/4)d^2 = (\pi/6)d^3(\rho_s - \rho)g$$

or: $\quad R' = (2d/3)(\rho_s - \rho)g = (R'/\rho u_0^2)\rho u_0^2 = 0.22\rho u_0^2 \approx (2/g)\rho u_0^2$

Thus: $\quad u_0 = [3dg(\rho_s - \rho)/\rho]^{0.5}$

For the metal particles:

$$u_0 = [(3 \times 1.5 \times 10^{-3} \times 6500 \times 9.81)/1000]^{0.5}$$
$$= \underline{\underline{0.536 \text{ m/s}}}$$

For the glass particles:

$$u_0 = [(3 \times 12 \times 10^{-3} \times 1500 \times 9.81)/1000]^{0.5}$$
$$= \underline{\underline{0.727 \text{ m/s}}}.$$

For the fluidised bed:

The density of the suspension $= e\rho + (1-e)\rho_s$

For the metal particles:

$$(u_c/0.536) = e_1^{2.30} \quad \text{(i)}$$

For the glass particles:

$$(u_c/0.727) = e_2^{2.30} \quad \text{(ii)}$$

and from equations (i) and (ii):

$$(e_1/e_2) = (0.727/0.536)^{1/2.30} = 1.142$$

For equal bed densities:

$$e_1\rho(1-e_1)\rho_{s1} = e_2\rho + (1-e_2)\rho_{s2} \quad \text{(iii)}$$

Thus: $\quad (1.142e_1)1000 + (1 - 1.142e_2)7500 = 1000e_2 + (1 - e_2)2500$

from which: $\quad e_3 = 0.844$

and: $\quad u_c = (0.727 \times 0.844^{2.30}) = \underline{\underline{0.492 \text{ m/s}}}$

Substituting in equation (iii):

$$e_1 = 0.964$$

and: $\quad u_c = (0.536 \times 0.964^{2.30}) = \underline{\underline{0.493 \text{ m/s}}}$

PROBLEM 6.14

Glass spheres are fluidised by water at a velocity equal to one half of their terminal falling velocities. Calculate:

(a) the density of the fluidised bed,
(b) the pressure gradient in the bed attributable to the presence of the particles.

The particles are 2 mm in diameter and have a density of 2500 kg/m³. The density and viscosity of water are 1000 kg/m³ and 1 mNs/m² respectively.

Solution

The Galileo number is given by:

$$Ga = d^3 \rho(\rho_s - \rho)g/\mu^2$$
$$= [(2 \times 10^{-3})^3 \times 1000(2500 - 1000) \times 9.81]/(1 \times 10^{-3})^2$$
$$= 117{,}720$$

From equation 5.79:

$$(4.8 - n)/(n - 2.4) = 0.043 Ga^{0.57}$$
$$= 0.043 \times 117{,}720^{0.57} = 33.4$$

and: $\quad n = 2.47$

$$u/u_0 = 0.5 = e^{2.47}$$

and hence: $\quad e = \underline{0.755}$

The bed density is given by:

$$(1 - e)\rho_s + e\rho = (1 - 0.755) \times 2500 + (0.755 \times 1000)$$
$$= \underline{\underline{1367 \text{ kg/m}^3}}$$

The pressure gradient due to the solids is given by:

$$\{[(1 - e)\rho_s + e\rho] - \rho\}g = (1 - e)(\rho_s - \rho)g$$
$$= (1 - 0.755)(2500 - 1000)9.81$$
$$= \underline{\underline{3605 \text{ (N/m}^2)/\text{m}}}$$

SECTION 2-7

Liquid Filtration

PROBLEM 7.1

A slurry, containing 0.2 kg of solid/kg of water, is fed to a rotary drum filter, 0.6 m in diameter and 0.6 m long. The drum rotates at one revolution in 360 s and 20 per cent of the filtering surface is in contact with the slurry at any given instant. If filtrate is produced at the rate of 0.125 kg/s and the cake has a voidage of 0.5, what thickness of cake is formed when filtering at a pressure difference of 65 kN/m^2? The density of the solid is 3000 kg/m^3.

The rotary filter breaks down and the operation has to be carried out temporarily in a plate and frame press with frames 0.3 m square. The press takes 120 s to dismantle and 120 s to reassemble, and, in addition, 120 s is required to remove the cake from each frame. If filtration is to be carried out at the same overall rate as before, with an operating pressure difference of 75 kN/m^2, what is the minimum number of frames that must be used and what is the thickness of each? It may be assumed that the cakes are incompressible and the resistance of the filter media may be neglected.

Solution

See Volume 2, Example 7.6.

PROBLEM 7.2

A slurry containing 100 kg of whiting/m^3 of water, is filtered in a plate and frame press, which takes 900 s to dismantle, clean and re-assemble. If the filter cake is incompressible and has a voidage of 0.4, what is the optimum thickness of cake for a filtration pressure of 1000 kN/m^2? The density of the whiting is 3000 kg/m^3. If the cake is washed at 500 kN/m^2 and the total volume of wash water employed is 25 per cent of that of the filtrate, how is the optimum thickness of cake affected? The resistance of the filter medium may be neglected and the viscosity of water is 1 mN s/m^2. In an experiment, a pressure of 165 kN/m^2 produced a flow of water of 0.02 cm^3/s though a centimetre cube of filter cake.

Solution

See Volume 2, Example 7.2.

PROBLEM 7.3

A plate and frame press gave a total of 8 m³ of filtrate in 1800 s and 11.3 m³ in 3600 s when filtration was stopped. Estimate the washing time if 3 m³ of wash water is used. The resistance of the cloth may be neglected and a constant pressure is used throughout.

Solution

For constant pressure filtration with no cloth resistance:

$$t = \frac{r\mu v}{2A^2(-\Delta P)} V^2 \qquad \text{(equation 7.1)}$$

At $t_1 = 1800$ s, $V_1 = 8$ m³, and when $t_2 = 3600$ s, $V_2 = 11$ m³

Thus:
$$(3600 - 1800) = \frac{r\mu v}{2A^2(-\Delta P)} (11^2 - 8^2)$$

$$\frac{r\mu v}{2A^2(-\Delta P)} = 316$$

Since:
$$\frac{dV}{dt} = \frac{A^2(-\Delta P)}{r\mu v V}$$

$$= \frac{1}{(2 \times 31.6 V)} = \frac{0.0158}{V}$$

The final rate of filtration = $(0.0158/11) = 1.44 \times 10^{-3}$ m³/s.

For thorough washing in a plate and frame filter, the wash water has twice the thickness of cake to penetrate and half the area for flow that is available to the filtrate. Thus the flow of wash water at the same pressure will be one-quarter of the filtration rate.

Hence: rate of washing = $(1.44 \times 10^{-3})/4 = 3.6 \times 10^{-4}$ m³/s

and: time of washing = $3/(3.6 \times 10^{-4}) = \underline{8400 \text{ s}}$ (2.3 h)

PROBLEM 7.4

In the filtration of a sludge, the initial period is effected at a constant rate with the feed pump at full capacity, until the pressure differences reaches 400 kN/m². The pressure is then maintained at this value for a remainder of the filtration. The constant rate operation requires 900 s and one-third of the total filtrate is obtained during this period.

Neglecting the resistance of the filter medium, determine (a) the total filtration time and (b) the filtration cycle with the existing pump for a maximum daily capacity, if the time for removing the cake and reassembling the press is 1200 s. The cake is not washed.

Solution

For a filtration carried out at a constant filtration rate for time t_1 in which time a volume V_1 is collected and followed by a constant pressure period such that the total filtration time is t and the total volume of filtrate is V, then:

$$V^2 - V_1^2 = \frac{2A^2(-\Delta P)}{r\mu v}(t - t_1) \qquad \text{(equation 7.13)}$$

Assuming no cloth resistance, then:

for the constant rate period: $\quad t_1 = \dfrac{r\mu v}{A^2(-\Delta P)} V_1^2 \qquad$ (equation 7.10)

Using the data given: $t_1 = 900$ s, volume $= V_1$

Thus: $\quad \dfrac{r\mu v}{A^2(-\Delta P)} = \dfrac{900}{V_1^2}$

(a) *For the constant pressure period*: $V = 3V_1$ and $(t - t_1) = t_p$

Thus: $\quad 8V_1^2 = \dfrac{2V_1^2}{900} t_p$

$$t_p = 3600 \text{ s}$$

Thus: total filtration time $= (900 + 3600) = \underline{\underline{4500 \text{ s}}}$

and: total cycle time $= (4500 + 1200) = \underline{\underline{5700 \text{ s}}}$

(b) *For the constant rate period*:

$$t_1 = \frac{r\mu v}{A^2(-\Delta P)} V_1^2 = \frac{V_1^2}{K}$$

For the constant pressure period:

$$t - t_1 = \frac{r\mu v}{2A^2(-\Delta P)}(V^2 - V_1^2) = \frac{V^2 - V_1^2}{2K}$$

Total filtration time, $\quad t = \dfrac{1}{K}\left(V_1^2 + \dfrac{V^2 - V_1^2}{2}\right) = \dfrac{(V^2 + V_1^2)}{2K}$

Rate of filtration $\quad = \dfrac{V}{t + t_d} \quad$ where t_d is the downtime

$$= \frac{2KV}{V^2 + V_1^2 + 2Kt_d}$$

For the rate to be a maximum,

$$\frac{d(\text{rate})}{dV} = 0 \quad \text{or} \quad V_1^2 - V^2 + 2Kt_d = 0$$

Thus: $$t_d = \frac{1}{2K}(V^2 - V_1^2) = (t - t_1)$$

But: $$t_d = 1200 = (t - 900) \quad \text{and} \quad t = 2100 \text{ s}$$

Thus: total cycle time $= (2100 + 1200) = \underline{\underline{3300 \text{ s}}}$

PROBLEM 7.5

A rotary filter, operating at 0.03 Hz, filters at the rate of 0.0075 m³/s. Operating under the same vacuum and neglecting the resistance of the filter cloth, at what speed must the filter be operated to give a filtration rate of 0.0160 m³/s?

Solution

For constant pressure filtration in a rotary filter:

$$V^2 = \frac{2A^2(-\Delta P)t}{r\mu v}$$

or: $\qquad V^2 \propto t \propto 1/N \qquad$ (equation 7.11)

where N is the speed of rotation.

As $V \propto 1/N^{0.5}$ and the rate of filtration is V/t, then:

$$V/t \propto (1/N^{0.5})(1/t) \propto (N/N^{0.5}) \propto N^{0.5}$$

Thus: $\qquad (V/t)_1/(V/t)_2 = N_1^{0.5}/N_2^{0.5}$

$$0.0075/0.0160 = 0.03^{0.5}/N_2^{0.5}$$

and: $\qquad N_2 = \underline{\underline{0.136 \text{ Hz}}} \ (7.2 \text{ rpm})$

PROBLEM 7.6

A slurry is filtered in a plate and frame press containing 12 frames, each 0.3 m square and 25 mm thick. During the first 180 s, the filtration pressure is slowly raised to the final value of 400 kN/m² and, during this period, the rate of filtration is maintained constant. After the initial period, filtration is carried out at constant pressure and the cakes are completely formed in a further 900 s. The cakes are then washed with a pressure difference of 275 kN/m² for 600 s, using *thorough washing*. What is the volume of filtrate collected per cycle and how much wash water is used?

A sample of the slurry was tested, using a vacuum leaf filter of 0.05 m² filtering surface and a vacuum equivalent to a pressure difference of 71.3 kN/m². The volume of filtrate collected in the first 300 s was 250 cm³ and, after a further 300 s, an additional 150 cm³ was collected. It may be assumed that cake is incompressible and the cloth resistance is the same in the leaf as in the filter press.

Solution

See Volume 2, Example 7.1.

PROBLEM 7.7

A sludge is filtered in a plate and frame press fitted with 25 mm frames. For the first 600 s the slurry pump runs at maximum capacity. During this period the pressure difference rises to 500 kN/m² and a quarter of the total filtrate is obtained. The filtration takes a further 3600 s to a complete at constant pressure and 900 s is required for emptying and resetting the press.

It is found that, if the cloths are precoated with filter aid to a depth of 1.6 mm, the cloth resistance is reduced to 25 per cent of its former value. What will be the increase in the overall throughput of the press if the precoat can be applied in 180 s?

Solution

See Volume 2, Example 7.7.

PROBLEM 7.8

Filtration is carried out in a plate and frame filter press, with 20 frames 0.3 m square and 50 mm thick, and the rate of filtration is maintained constant for the first 300 s. During this period, the pressure is raised to 350 kN/m², and one-quarter of the total filtrate per cycle is obtained. At the end of the constant rate period, filtration is continued at a constant pressure of 350 kN/m² for a further 1800 s, after which the frames are full. The total volume of filtrate per cycle is 0.7 m³ and dismantling and refitting of the press takes 500 s. It is decided to use a rotary drum filter, 1.5 m long and 2.2 m in diameter, in place of the filter press. Assuming that the resistance of the cloth is the same in the two plants and that the filter cake is incompressible, calculate the speed of rotation of the drum which will result in the same overall rate of filtration as was obtained with the filter press. The filtration in the rotary filter is carried out at a constant pressure difference of 70 kN/m², and the filter operates with 25 per cent of the drum submerged in the slurry at any instant.

Solution

Data from the plate and frame filter press are used to evaluate the cake and cloth resistance for use with the rotary drum filter.

For the *constant rate period*:

$$V_1^2 + \frac{LA}{v}V_1 = \frac{A^2(-\Delta P)}{r\mu v}t_1 \qquad \text{(equation 7.17)}$$

For the subsequent constant pressure period:

$$(V^2 - V_1^2) + \frac{2LA}{v}(V - V_1) = \frac{2A^2(-\Delta P)}{r\mu v}(t - t_1) \qquad \text{(equation 7.18)}$$

From the data given:

$$t_1 = 300 \text{ s}, (-\Delta P) = (350 - 101.3) = 248.7 \text{ kN/m}^2,$$

$$V_1 = 0.175 \text{ m}^3 \quad \text{and} \quad A = (2 \times 20 \times 0.3 \times 0.3) = 3.6 \text{ m}^2$$

Thus: $\quad (0.175)^2 + \dfrac{L}{v} \times 3.6 \times 0.175 = \dfrac{(3.6)^2 \times 248.7 \times 10^3 \times 300}{r\mu v}$

or: $\qquad 0.0306 + 0.63(L/v) = 9.68 \times 10^8/r\mu v \qquad \text{(i)}$

For the *constant pressure period*:

$$V = 0.7 \text{ m}^3, \quad V_1 = 0.175 \text{ m}^3,$$

$$(t - t_1) = 1800 \text{ s}, \quad A = 3.6 \text{ m}^2$$

Thus:

$$(0.7^2 - 0.175^2) + 2(L/v) \times 3.6(0.7 - 0.175) = \frac{(2 \times 3.6)^2 \times 248.7 \times 10^3}{r\mu v} \times 1800$$

or: $\qquad 0.459 + 3.78(L/v) = 116.08 \times 10^8/r\mu v \qquad \text{(ii)}$

Solving equations (i) and (ii) simultaneously gives:

$$r\mu v = 210.9 \times 10^8 \quad \text{and} \quad L/v = 0.0243$$

For the *rotary drum filter*:

$$D = 2.2 \text{ m}, L = 1.5 \text{ m}, (-\Delta P) = 70 \text{ kN/m}^2$$

$$A = (2.2\pi \times 1.5) = 10.37 \text{ m}^2$$

$$(-\Delta P) = 70 \times 10^3 \text{ N/m}^2$$

If θ is the time of one revolution, then as the time of filtration is 0.25θ:

$$V^2 + 2A\frac{L}{v}V = \frac{2A^2(-\Delta P)}{r\mu v} \times 0.25\theta$$

$$V^2 + (2 \times 10.37 \times 0.0243V) = \frac{2(10.37)^2 \times 70 \times 10^3 \times 0.25\theta}{210.9 \times 10^8}$$

or: $\qquad V^2 + 0.504V = 1.785 \times 10^{-4}\theta$

The rate of filtration $= V/t = 0.7/(300 + 1800 + 500)$

$$= 2.7 \times 10^{-4} \text{ m}^3/\text{s}$$

Thus:
$$V = 2.7 \times 10^{-4} t$$
and:
$$(2.7 \times 10^{-4} t)^2 + (0.504 \times 2.7 \times 10^{-4})t = (1.785 \times 10^{-4})t$$
from which:
$$t = 580 \text{ s}$$
Hence:
$$\text{speed} = (1/580) = \underline{0.002 \text{ Hz}} \quad (0.12 \text{ rpm})$$

PROBLEM 7.9

It is required to filter a slurry to produce 2.25 m³ of filtrate per working day of 8 hours. The process is carried out in a plate and frame filter press with 0.45 m square frames and a working pressure difference of 348.7 kN/m². The pressure is built up slowly over a period of 300 s and, during this period, the rate of filtration is maintained constant.

When a sample of the slurry is filtered, using a pressure difference of 66.3 kN/m² on a single leaf filter of filtering area 0.05 m², 400 cm³ of filtrate is collected in the first 300 s of filtration and a further 400 cm³ is collected during the following 600 s. Assuming that the dismantling of the filter press, the removal of the cakes and the setting up again of the press takes an overall time of 300 s, plus an additional 180 s for each cake produced, what is the minimum number of frames that need be employed? The resistance of the filter cloth may be taken as the same in the laboratory tests as on the plant.

Solution

For constant pressure filtration on the *leaf filter*:

$$V^2 + 2\frac{L}{v}AV = \frac{2A^2(-\Delta P)t}{r\mu v} \quad \text{(equation 7.18)}$$

When $t = 300$ s, $V = 0.0004$ m³, $A = 0.05$ m², $(-\Delta P) = 66.3$ kN/m², and:

$$(0.0004)^2 + 2(L/v) \times 0.05 \times 0.0004 = \frac{2 \times (0.05)^2 \times 66.3 \times 300}{r\mu v}$$

or:
$$1.6 \times 10^{-7} + 4 \times 10^{-5}(L/v) = 99.4/r\mu v$$

When $t = 900$ s, $V = 800$ cm³ or 0.0008 m³ and substituting these values gives:

$$(6.4 \times 10^{-7}) + (8 \times 10^{-5})(L/v) = 298.4/r\mu v$$

Thus:
$$L/v = 4 \times 10^{-3} \quad \text{and} \quad r\mu v = 3.1 \times 10^8$$

In the filter press

For the constant rate period:

$$V_1^2 + \frac{LA}{v}V_1 = \frac{A^2(-\Delta P)t_1}{r\mu v} \quad \text{(equation 7.17)}$$

$A = 2 \times 0.45n = 0.9n$ where n is the number of frames, $t_1 = 300$ s

Thus: $\quad V_1^2 + (4 \times 10^{-3} \times 0.9)nV_1 = 0.81n^2 \times 348.7 \times (300/3.1) \times 10^8$

or: $\quad V_1^2 + (3.6 \times 10^{-3})nV_1 = (2.73 \times 10^{-4})n^2$

and: $\quad V_1 = 0.0148n$

For the constant pressure period:

$$\left(\frac{V^2 - V_1^2}{2}\right) + \frac{LA}{v}(V - V_1) = \frac{A^2(-\Delta P)}{r\mu v}(t - t_1) \quad \text{(equation 7.18)}$$

Substituting for L/v and $r\mu v$, $t_1 = 300$ and $V_1 = 0.0148n$ gives:

$$\left(\frac{V^2 - 2.2 \times 10^{-4}n^2}{2}\right) + (V - 0.0148n)4 \times 10^{-3} \times 0.9n = \frac{(0.81n^2 \times 348.7)}{(3.1 \times 10^8)}(t_f - 300)$$

or: $\quad 0.5V^2 + 1.1 \times 10^{-4}n^2 + 3.6 \times 10^{-3}nV = 9.11 \times 10^{-7}n^2 t \quad$ (i)

The total cycle time $= (t_f + 300 + 180n)$ s.
Required filtration rate $= 2.25/(8 \times 3600) = 7.81 \times 10^{-5}$ m^3/s.
Volume of filtrate $= V$ m^3.

Thus: $\quad \dfrac{V}{(t_f + 300 + 180n)} = 7.81 \times 10^{-5}$

and: $\quad t_f = 1.28 \times 10^4 V - 300 - 180n \quad$ (ii)

Thus the value of t_f from equation (ii) may be substituted in equation (i) to give:

$$V^2 + V(7.2 \times 10^{-3}n - 2.34 \times 10^{-2}n^2) + (7.66 \times 10^{-4}n^2 + 3.28 \times 10^{-4}n^3) = 0 \quad \text{(iii)}$$

This equation is of the form $V^2 + AV + B = 0$ and may thus be solved to give:

$$V = \frac{-A \pm \sqrt{(A^2 - 4B)}}{2}$$

where A and B are the expressions in parentheses in equation (iii). In order to find the minimum number of frames. dV/dn must be found and equated to zero. From above $(V - a)(V - b) = 0$, where a and b are complex functions of n.
Thus $V = a$ or $V = b$ and dV/dn can be evaluated for each root.
Putting $dV/dn = 0$ gives, for the positive value, $\underline{\underline{n = 13}}$

PROBLEM 7.10

The relation between flow and head for a slurry pump may be represented approximately by a straight line, the maximum flow at zero head being 0.0015 m^3/s and the maximum head at zero flow 760 m of liquid. Using this pump to feed a slurry to a pressure leaf filter,

(a) how long will it take to produce 1 m³ of filtrate, and
(b) what will be the pressure drop across the filter after this time?

A sample of the slurry was filtered at a constant rate of 0.00015 m³/s through a leaf filter covered with a similar filter cloth but of one-tenth the area of the full scale unit and after 625 s the pressure drop across the filter was 360 m of liquid. After a further 480 s the pressure drop was 600 m of liquid.

Solution

For constant rate filtration through the filter leaf:

$$V^2 + \frac{LA}{v}V = \frac{A^2(-\Delta P)t}{r\mu v} \qquad \text{(equation 7.17)}$$

At a constant rate of 0.00015 m³/s when the time = 625 s:

$$V = 0.094 \text{ m}^3, (-\Delta P) = 3530 \text{ kN/m}^2$$

and at $t = 1105$ s: $V = 0.166$ m³   and $(-\Delta P) = 5890$ kN/m²

Substituting these values into equation 7.17 gives:

$$(0.094)^2 + LA/v \times 0.094 = (A^2/r\mu v) \times 3530 \times 625$$

or: $0.0088 + 0.094 LA/v = 2.21 \times 10^6 A^2/r\mu v$

and: $(0.166)^2 + LA/v \times 0.166 = (A^2/r\mu v) \times 5890 \times 1105$

or: $0.0276 + 0.166 LA/v = 6.51 \times 10^6 A^2/r\mu v$

Equations (i) and (ii) may be solved simultaneously to give:

$$LA/v = 0.0154 \quad \text{and} \quad A^2/r\mu v = 4.64 \times 10^{-9}$$

As the filtration area of the full-size plant is 10 times that of the leaf filter then:

$$LA/v = 0.154 \quad \text{and} \quad A^2/r\mu v = 4.64 \times 10^{-7}$$

If the pump develops a head of 760 m of liquid or 7460 kN/m² at zero flow and has zero head at $Q = 0.0015$ m³/s, its performance may be expressed as:

$$(-\Delta P) = 7460 - (7460/0.0015)Q$$

or: $(-\Delta P) = 7460 - 4.97 \times 10^6 Q$ kN/m²

$$\frac{dV}{dt} = \frac{A^2(-\Delta P)}{r\mu v(V + LA/v)} \qquad \text{(equation 7.16)}$$

Substituting for $(-\Delta P)$ and the filtration constants gives:

$$\frac{dV}{dt} = \frac{A^2}{r\mu v} \frac{(7460 - 4.97 \times 10^6 \, dV/dt)}{(V + 0.154)}$$

Since $Q = dV/dt$, then:

$$\frac{dV}{dt} = \frac{4.67 \times 10^{-7}[7460 - 4.97 \times 10^6 (dV/dt)]}{(V + 0.154)}$$

$$(V + 0.154)dV = 3.46 \times 10^{-3} - 2.31 dV/dt$$

The time taken to collect 1 m³ is then given by:

$$\int_0^1 (V + 0.154 + 2.31) dV = \int_0^t (3.46 \times 10^{-3}) dt$$

and: $\underline{\underline{t = 857 \text{ s}}}$

The pressure at this time is found by substituting in equation 7.17 with $V = 1$ m³ and $t = 857$ s². This gives:

$$1^2 + 0.154 \times 1 = 4.64 \times 10^{-7} \times 857(-\Delta P)$$

and: $\underline{\underline{(-\Delta P) = 2902 \text{ kN/m}^2}}$

PROBLEM 7.11

A slurry containing 40 per cent by mass solid is to be filtered on a rotary drum filter 2 m diameter and 2 m long which normally operates with 40 per cent of its surface immersed in the slurry and under a pressure of 17 kN/m². A laboratory test on a sample of the slurry using a leaf filter of area 200 cm² and covered with a similar cloth to that on the drum, produced 300 cm³ of filtrate in the first 60 s and 140 cm³ in the next 60 s, when the leaf was under pressure of 84 kN/m². The bulk density of the dry cake was 1500 kg/m³ and the density of the filtrate was 1000 kg/m³. The minimum thickness of cake which could be readily removed from the cloth was 5 mm.

At what speed should the drum rotate for maximum throughput and what is this throughput in terms of the mass of the slurry fed to the unit per unit time?

Solution

See Volume 2, Example 7.4.

PROBLEM 7.12

A continuous rotary filter is required for an industrial process for the filtration of a suspension to produce 0.002 m³/s of filtrate. A sample was tested on a small laboratory filter of area 0.023 m² to which it was fed by means of a slurry pump to give filtrate at a constant rate of 0.0125 m³/s. The pressure difference across the test filter increased from 14 kN/m² after 300 s filtration to 28 kN/m² after 900 s, at which time the cake thickness had reached 38 mm. What are suitable dimensions and operating conditions for the rotary filter, assuming that the resistance of the cloth used is one-half that on the test filter, and that the vacuum system is capable of maintaining a constant pressure difference of 70 kN/m² across the filter?

Solution

Data from the laboratory filter may be used to find the cloth and cake resistance of the rotary filter. For the laboratory filter operating under constant rate conditions:

$$V_1^2 + \frac{LA}{v}V_1 = \frac{A^2(-\Delta P)t}{r\mu v} \qquad \text{(equation 7.17)}$$

$A = 0.023$ m² and the filtration rate $= 0.0125$ m³/s
At $t = 300$ s, then:

$$(-\Delta P) = 14 \text{ kN/m}^2 \quad \text{and} \quad V_1 = 3.75 \times 10^{-3} \text{ m}^3$$

When $t = 900$ s, then:

$$(-\Delta P) = 28 \text{ kN/m}^2 \quad \text{and} \quad V_1 = 1.125 \times 10^{-2} \text{ m}^3$$

Hence: $(3.75 \times 10^{-3})^2 + (L/v) \times 0.023 \times 3.75 \times 10^{-3} = \dfrac{14}{r\mu v} \times (0.023)^2 \times 300$

or: $\qquad 1.41 \times 10^{-5} + 8.63 \times 10^{-5}(L/v) = 2.22/r\mu v$

and: $(1.25 \times 10^{-2})^2 + (L/v) \times 0.023 \times 1.125 \times 10^{-3} = \dfrac{28 \times (0.023)^2}{r\mu v} \times 900$

$$1.27 \times 10^{-4} + 2.59 \times 10^{-4}(L/v) = 13.33/r\mu v$$

from which: $\qquad L/v = 0.164$ m, and $r\mu v = 7.86 \times 10^4$ kg/m³s

If the cloth resistance is halved by using the rotary filter, $L/v = 0.082$. As the filter operates at constant pressure, then:

$$V^2\left(\frac{2LA}{v}\right) = \frac{2A^2(-\Delta P)t}{r\mu v} \qquad \text{(equation 7.18)}$$

If θ is the time for 1 rev × fraction submerged and V' is volume of filtrate/revolution (given by equation 7.18), the speed $= 0.0167$ Hz (1 rpm) and 20 per cent submergence, then:

$$\theta = (60 \times 0.2) = 12 \text{ s}$$

Thus: $\qquad V'^2 + 2 \times 0.082 AV' = \dfrac{(2A^2 \times 70 \times 12)}{(7.86 \times 10^4)}$

or: $\qquad V'^2 + 0.164 AV' = 0.0214 A^2$

from which: $\qquad (A/V') = 11.7$ m⁻¹

The required rate of filtration $= 0.002$ m³/s.

Thus: $\qquad$ The volume/revolution, $V' = (0.002 \times 60) = 0.12$ m³

$$A = (11.7 \times 0.12) = 1.41 \text{ m}^2$$

If $L = D$, then:

$$\text{area of drum} = \pi DL = \pi D^2 = 1.41 \text{ m}^2$$

and:

$$D = L = \underline{0.67 \text{ m}}$$

The cake thickness on the drum should now be checked.

$v = AL/V$ and from data on the laboratory filter:

$$v = (0.023 \times 0.038)/(1.125 \times 10^{-2}) = 0.078$$

Hence, the cake thickness on the drum, $vV'/A = (0.078/11.7) = 0.0067$ m or $\underline{6.7 \text{ mm}}$ which is acceptable.

PROBLEM 7.13

A rotary drum filter, 1.2 m diameter and 1.2 m long, handles 6.0 kg/s of slurry containing 10 per cent of solids when rotated at 0.005 Hz. By increasing the speed to 0.008 Hz it is found that it can then handle 7.2 kg/s. What will be the percentage change in the amount of wash water which may be applied to each kilogram of cake caused by the increased speed of rotation of the drum, and what is the theoretical maximum quantity of slurry which can be handled?

Solution

For constant pressure filtration:

$$\frac{dV}{dt} = \frac{A^2(-\Delta P)}{r\mu v[V + (LA/v)]} = \frac{a}{V + b} \text{ (say)} \quad \text{(equation 7.16)}$$

or:
$$V^2/2 + bV = at$$

For Case 1:
1 revolution takes $(1/0.005) = 200$ s and the rate $= V_1/200$.

For Case 2:
1 revolution takes $(1/0.008) = 125$ s and the rate $= V_2/125$.

But:
$$\frac{V_1/200}{V_2/125} = \frac{6.0}{7.2}$$

or:
$$V_1/V_2 = 1.33 \quad \text{and} \quad V_2 = 0.75 V_1$$

For case 1, using the filtration equation (7.16):

$$V_1^2 + 2bV_1 = 2a \times 200$$

and for case 2:

$$V_2^2 + 2bV_2 = 2a \times 125$$

Substituting $V_2 = 0.75V_1$ in these two equations allows the filtration constants to found as:

$$a = 0.00375V_1^2 \quad \text{and} \quad b = 0.25V_1$$

The rate of flow of wash water will equal the final rate of filtration so that for case 1:
Wash water rate $= a/(V_1 + b)$.
Wash water per revolution $\propto 200a/(V_1 + b)$.
Wash water/revolution per unit solids $\propto 200a/V_1(V_1 + b)$,

or: $\quad \propto (200 \times 0.00375V_1^2)/V_1(V_1 + 0.25V_1) \propto 0.6$.

Similarly for case 2, the wash water per revolution per unit solids is proportional to:

$$125a/V^2(V_2 + b)$$

which is: $\quad \propto (125 \times 0.00375V_1^2)/0.75V_1(0.75V_1 + 0.25V_1) \propto 0.625$

Hence:
$$\text{per cent increase} = [(0.625 - 0.6)/0.6] \times 100 = \underline{\underline{4.17 \text{ per cent}}}$$

As $0.5V^2 + bV = at$, the rate of filtration V/t is given by:

$$a/(0.5V + b)$$

The highest rate will be achieved as V tends to zero and:

$$(V/t)_{\max} = a/b = 0.00375V_1^2/0.25V_1 = 0.015V_1$$

For case 1, the rate $= (V_1/200) = 0.005V_1$.
Hence the limiting rate is three times the original rate,

that is: $\quad\quad\quad\quad\quad\quad\quad\quad \underline{\underline{18.0 \text{ kg/s}}}$

PROBLEM 7.14

A rotary drum with a filter area of 3 m³ operates with an internal pressure of 71.3 kN/m² below atmospheric and with 30 per cent of its surface submerged in the slurry. Calculate the rate of production of filtrate and the thickness of cake when it rotates at 0.0083 Hz, if the filter cake is incompressible and the filter cloth has a resistance equal to that of 1 mm of cake.

It is desired to increase the rate of filtration by raising the speed of rotation of the drum. If the thinnest cake that can be removed from the drum has a thickness of 5 mm, what is the maximum rate of filtration which can be achieved and what speed of rotation of the drum is required? The voidage of the cake $= 0.4$, the specific resistance of cake $= 2 \times 10^{12}$ m^{-2} the density of solids $= 2000$ kg/m³, the density of filtrate $= 1000$ kg/m³, the viscosity of filtrate $= 10^{-3}$ N s/m² and the slurry concentration $= 20$ per cent by mass solids.

Solution

A 20 per cent slurry contains 20 kg solids/80 kg solution.
Volume of cake $= 20/[2000(1 - 0.4)] = 0.0167$ m³.

Volume of liquid in the cake = (0.167 × 0.4) = 0.0067 m³.
Volume of filtrate = (80/1000) − 0.0067 = 0.0733 m³.

Thus: $v = (0.0167/0.0733) = 0.23$

The rate of filtration is given by:

$$\frac{dV}{dt} = \frac{A^2(-\Delta P)}{r\mu v[V + (LA/v)]} \qquad \text{(equation 7.16)}$$

In this problem:

$A = 3$ m², $(-\Delta P) = 71.3$ kN/m² or (71.3×10^3) N/m²,
$r = 2 \times 10^{12}$ m⁻², $\mu = 1 \times 10^{-3}$ Ns/m², $v = 0.23$ and $L = 1$ mm or 1×10^{-3} m

Thus: $$\frac{dV}{dt} = \frac{(3^2 \times 71.3 \times 10^3)}{0.23 \times 2 \times 10^{12} \times 1 \times 10^{-3}[V + (1 \times 10^{-3} \times 3/0.23)]}$$

$$= \frac{(1.395 \times 10^3)}{(V + 0.013)}$$

From which: $V^2/2 + 0.013V = (1.395 \times 10^{-3})t$

If the rotational speed = 0.0083 Hz, 1 revolution takes (1/0.0083) = 120.5 s and a given element of surface is immersed for (120.5 × 0.3) = 36.2 s. When $t = 36.2$ s, V may be found by substitution to be 0.303 m³.

Hence: rate of filtration = (0.303/120.5) = <u>0.0025 m³/s</u>.

Volume of filtrate for 1 revolution = 0.303 m³.
Volume of cake = (0.23 × 0.303) = 0.07 m³.

Thus: cake thickness = (0.07/3) = 0.023 m or <u>23 mm</u>

As the thinnest cake = 5 mm, volume of cake = (3 × 0.005) = 0.015 m³.
As $v = 0.23$, volume of filtrate = (3 × 0.005)/0.23 = 0.065 m³.

Thus: $(0.065)^2/2 + (0.013 \times 0.065) = 1.395 \times 10^{-3}t$
and: $t = 2.12$ s
Thus: time for 1 revolution = (2.12/0.3) = 7.1 s
and: speed = <u>0.14 Hz</u> (8.5 r.p.m)

Maximum filtrate rate = 0.065 m³ in 7.1 s

or: (0.065/7.1) = <u>0.009 m³/s</u>

PROBLEM 7.15

A slurry containing 50 per cent by mass of solids of density 2600 kg/m³ is to be filtered on a rotary drum filter, 2.25 m in diameter and 2.5 m long, which operates with 35 per cent of its surface immersed in the slurry and under a vacuum of 600 mm Hg. A laboratory test on a sample of the slurry, using a leaf filter with an area of 100 cm² and covered with a cloth similar to that used on the drum, produced 220 cm³ of filtrate in the first minute and 120 cm³ of filtrate in the next minute when the leaf was under a vacuum of 550 mm Hg. The bulk density of the wet cake was 1600 kg/m³ and the density of the filtrate was 1000 kg/m³.

On the assumption that the cake is incompressible and that 5 mm of cake is left behind on the drum, determine the theoretical maximum flowrate of filtrate obtainable. What drum speed will give a filtration rate of 80 per cent of the maximum?

Solution

a) *For the leaf filter*:
$A = 100$ cm² or 0.01 m², $(-\Delta P) = 550$ mm Hg.

when $t = 1$ min, $V = 220$ cm³ $= 0.00022$ m³
when $t = 2$ min, $V = 340$ cm³ $= 0.00034$ m³

These values are substituted into the constant pressure filtration equation:

$$V^2 + \frac{2LAV}{v} = \frac{2(-\Delta P)A^2 t}{r\mu v} \qquad \text{(equation 7.18)}$$

to give the filtration constants as:

$$L/v = 9.4 \times 10^{-3} \text{ m} \quad \text{and} \quad r\mu v = 1.23 \times 10^6 \text{ Ns/m}^4$$

b) *Cake properties:*
The densities are:

$$\text{solids} = 2600 \text{ kg/m}^3, \text{ cake} = 1600 \text{ kg/m}^3 \text{ and filtrate} = 1000 \text{ kg/m}^3.$$

For the cake; with a voidage e:

1 m³ of cake contains $(1 - e)$ m³ of solids and e m³ of liquid

or: $2600(1 - e)$ kg of solids and $1000e$ kg of liquid

Thus the cake density is:

$$1600 = 2600(1 - e) + 1000e$$

and: $e = 0.625$

$$3.8461 \times 10^{-4} = (1/2600) \text{ m}^3 \text{ solids form:}$$

$$(1/2600)(1/0.375) = 1.0256 \times 10^{-3} \text{ m}^3 \text{ cake}$$

and: $(1/2600)(0.625/0.375) = 6.4101 \times 10^{-4}$ m³ liquid.

Thus: $(1/1000) - 6.4101 \times 10^{-4}$ m³ liquid form 1.0256×10^{-3} m³ cake

and: $v = 0.358$

Thus: $L = (9.4 \times 10^{-3} \times 0.358) = 0.365 \times 10^{-3}$ m

and: $1/\mathbf{r}\mu = (8.16 \times 10^{-7} \times 0.358) = 2.92 \times 10^{-7}$

c) *For the rotary filter*

5 mm of cake is left on the drum.

Thus: effective $L = 8.365 \times 10^{-3}$ m

Considering 1 revolution of the filter taking t_r min, then the filtration time is $0.35 t_r$.

$$A = (\pi \times 2.25 \times 2.5) = 17.67 \text{ m}^2 \quad \text{and} \quad -\Delta P = 600 \text{ mm Hg.}$$

Thus, in equation 7.18:

$$V^2 + [(2 \times 17.67 \times 8.365 \times 10^{-3})/0.358]V$$
$$= (2 \times 17.67^2 \times 600 \times 8.16 \times 10^{-7} \times 0.35 t_r)$$

or: $V^2 + 0.826V = 0.107 t_r$

The filtration rate is:

$$V/t_r = 0.107V/(V^2 + 0.826V)$$
$$= 0.107/(V + 0.826)$$

This is a maximum when $V = 0$, that is when the rate is 0.13275 m³/s

The actual rate is:

$$(0.80 \times 0.13275) = 0.1064 \text{ m}^3/\text{s}$$

Thus: $0.107/(V + 0.826) = 0.1064$

and: $V = 0.180$

Hence: $t_r = (1/0.107)(0.180 + 0.826)0.180$

$= 1.69$ min

and: speed of rotation $= (0.35 \times 1.69) = \underline{0.59 \text{ rpm } (0.0099 \text{ Hz})}$

PROBLEM 7.16

A rotary filter which operates at a fixed vacuum gives a desired rate of filtration of a slurry when rotating at 0.033 Hz. By suitable treatment of the filter cloth with a filter aid, its effective resistance is halved and the required filtration rate is now achieved at a rotational speed of 0.0167 Hz (1 rpm). If, by further treatment, it is possible to reduce the effective cloth resistance to a quarter of the original value, what rotational speed is required? If the filter is now operated again at its original speed of 0.033 Hz, by what factor will the filtration rate be increased?

Solution

From equation 7.11:

$$V^2 = 2A^2(-\Delta P)t/(r\mu v)$$
$$V^2 \propto (t/\mathbf{r}) \propto (1/N\mathbf{r})$$

or: $\quad V \propto 1/\sqrt{(N\mathbf{r})}$

Thus: $\quad V/t \propto (1/\sqrt{(N\mathbf{r})})(1/t) \propto (1/\sqrt{(N\mathbf{r})})N \propto \sqrt{(N/\mathbf{r})}$

In this way:

$$V/t = F = k\sqrt{(N/\mathbf{r})} \quad \text{where } k \text{ is a constant.}$$

For a constant value of F:

$$F/k = \sqrt{(0.033/\mathbf{r})} = 0.182/\sqrt{\mathbf{r}}$$

In the second case:

$N = 0.0167$ Hz and the specific resistance is now $0.5\mathbf{r}$

Thus: $\quad F/k = \sqrt{(0.0167/0.5\mathbf{r})} = 0.182/\sqrt{\mathbf{r}}$

which is consistent.

In the third case, the specific resistance is $0.25\mathbf{r}$ and the speed is N Hz

Thus: $\quad 0.182/\sqrt{\mathbf{r}} = \sqrt{(N/0.25\mathbf{r})}$

and: $\quad N = \underline{0.0083 \text{ Hz}}$ (0.5 rpm)

Since: $\quad F_1 = k\sqrt{(0.033/\mathbf{r})}$

For the third case:

$$F_3 = k\sqrt{(0.033/0.25\mathbf{r})}$$

and: $\quad F_3/F_1 = \sqrt{(0.033/0.25\mathbf{r})}/\sqrt{(0.033/\mathbf{r})}$
$$= 2.0$$

in this way, the <u>filtration rate will be doubled</u>.

SECTION 2-8

Membrane Separation Processes

PROBLEM 8.1

Obtain expressions for the optimum concentration for minimum process time in the diafiltration of a solution of protein content S in an initial volume V_0,

(a) If the gel–polarisation model applies.
(b) If the osmotic pressure model applies.

It may be assumed that the extent of diafiltration is given by:

$$V_d = \frac{\text{Volume of liquid permeated}}{\text{Initial feed volume}} = \frac{V_p}{V_0}$$

Solution

See Volume 2, Example 8.1.

PROBLEM 8.2

In the ultrafiltration of a protein solution of concentration 0.01 kg/m³, analysis of data on gel growth rate and wall concentration C_w yields the second order relationship:

$$\frac{dl}{dt} = K_r C_w^2$$

where l is gel thickness, and K_r is a constant, 9.2×10^{-6} m⁷/kg²s.

The water flux through the membrane may be described by:

$$J = \frac{|\Delta P|}{\mu_w R_m}$$

where $|\Delta P|$ is pressure difference, R_m is membrane resistance and μ_w is the viscosity of water.

This equation may be modified for protein solutions to give:

$$J = \frac{|\Delta P|}{\mu_p \left(R_m + \dfrac{l}{P_g} \right)}$$

where P_g is gel permeability, and μ_p is the viscosity of the permeate.

The gel permeability may be estimated from the Carman–Kozeny equation:

$$P_g = \left(\frac{d^2}{180}\right)\left(\frac{e^3}{(1-e)^2}\right)$$

where d is particle diameter and e is the porosity of the gel.
Calculate the gel thickness after 30 minutes of operation.

Data:

	Flux mm/s	$|\Delta P|$ (kN/m²)
	0.02	20
	0.04	40
	0.06	60

Viscosity of water	=	1.3 mNs/m²
Viscosity of permeate	=	1.5 mNs/m²
Diameter of protein molecule	=	20 nm
Operating pressure	=	10 kN/m²
Porosity of gel	=	0.5
Mass transfer coefficient to gel, h_D	=	1.26×10^{-5} m/s

Solution

The gel growth rate as a function of the wall concentration, C_w, is given by:

$$dl/dt = K_r C_w^2$$

where l is the gel thickness, K_r is a constant = 9.2×10^{-6} m⁷/kg s and C_w, the wall concentration given by:

$$C_w = C_f \exp(u/h_D)$$

The permeate flux is given by:

$$J_{\text{soln}} = |\Delta P|/[\mu_p(R_m + l/P_g)]$$

where $|\Delta P|$ is the pressure difference, μ_p the viscosity of the permeate, R_m the membrane resistance and P_g, the gel permeability, which may be estimated from the Carman–Kozeny equation:

$$P_g = \left(\frac{d^2}{180}\right)\left(\frac{e^3}{(1-e)^2}\right)$$

where d is the particle diameter and e the porosity of the gel.

For water:

$$J_w = \Delta P/\mu_w R_m$$

and hence: $R_m = |\Delta P|/J\mu_w = (2.0 \times 10^3)/(1.3 \times 10^{-3} \times 0.02 \times 10^{-3})$

$$= 7.60 \times 10^{11}\, 1/\text{m}$$

Also:
$$P_g = \left(\frac{d^2}{180}\right)\left(\frac{e^3}{(1-e)^2}\right) = \left[\frac{(20 \times 10^{-9})^2}{180}\right]\left[\frac{(0.5)^3}{(1-0.5)^2}\right]$$
$$= 1.11 \times 10^{-18} \text{ m}^2$$

Thus:
$$dl/dt = K_r C_f^2 \exp\{2\Delta P/[h_D \mu_p (R_m + l/P_g)]\}$$

and:
$$\int_0^t dt = \int_0^l dl/[K_r C_f^2 \exp\{2\Delta P/[h_D \mu_p (R_m + l/P_g)]\}]$$

The function of l is plotted against l in Figure 8a and the area under the curve is then plotted in the same figure. It is found that when $t = 30$ min, the area under the curve is 1800 at which $l = 3.25$ μm.

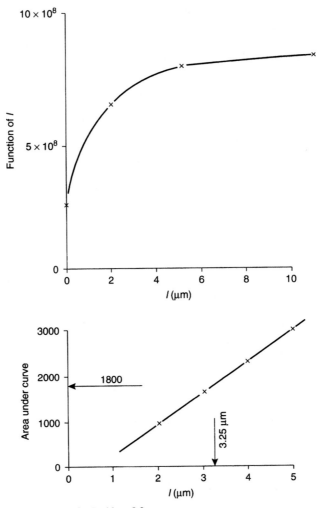

Figure 8a. Graphical integration for Problem 8.2

SECTION 2-9

Centrifugal Separations

PROBLEM 9.1

If a centrifuge is 0.9 m diameter and rotates at 20 Hz, at what speed should a laboratory centrifuge of 150 mm diameter be run if it is to duplicate the performance of the large unit?

Solution

If a particle of mass m is rotating at radius x with an angular velocity ω, it is subjected to a centrifugal force $mx\omega^2$ in a radial direction and a gravitational force mg in a vertical direction. The ratio of the centrifugal to gravitational forces, $x\omega^2/g$, is a measure of the separating power of the machine, and in order to duplicate conditions this must be the same in both machines.

In this case: $x_1 = 0.45$ m, $\omega_1 = (20 \times 2\pi) = 40\pi$ rad/s, $x_2 = 0.075$ m

Thus: $0.45(40\pi)^2/g = 0.075\omega_2^2/g$

$$\omega_2 = \sqrt{[6(40\pi)^2]} = (2.45 \times 40\pi) = 98\pi \text{ rad/s}$$

and the speed of rotation $= (98\pi/2\pi) = \underline{49 \text{ Hz}}$ (2940 rpm)

PROBLEM 9.2

An aqueous suspension consisting of particles of density 2500 kg/m³ in the size range 1–10 μm is introduced into a centrifuge with a basket 450 mm diameter rotating at 80 Hz. If the suspension forms a layer 75 mm thick in the basket, approximately how long will it take for the smallest particle to settle out?

Solution

Where the motion of the fluid with respect to the particle is turbulent, the time taken for a particle to settle from h_1 to distance h_2 from the surface in a radial direction, is given by the application of equation 3.124 as:

$$t = \frac{2}{a'}[(x+h_2)^{0.5} - (x+h_1)^{0.5}]$$

79

where $a' = \sqrt{[3d\omega^2(\rho_s - \rho)/\rho]}$, d is the diameter of the smallest particle $= 1 \times 10^{-6}$ m, ω is the angular velocity of the basket $= (80 \times 2\pi) = 502.7$ rad/s, ρ_s is the density of the solid $= 2500$ kg/m^3, ρ is the density of the fluid $= 1000$ kg/m^3, and x is the radius of the inner surface of the liquid $= 0.150$ m.

Thus: $\quad\quad a' = \sqrt{[3 \times 1 \times 10^{-6} \times 502.7^2(2500 - 1000)/1000]} = 1.066$

and: $\quad\quad t = (2/1.066)[(0.150 + 0.075)^{0.5} - (0.150 + 0)^{0.5}]$

$\quad\quad\quad = 1.876(0.474 - 0.387) = 0.163$ s

This is a very low value, equivalent to a velocity of $(0.075/0.163) = 0.46$ m/s. Because of the very small diameter of the particle, it is more than likely that the conditions are laminar, even at this particle velocity.

For water, taking $\mu = 0.001$ N s/m^2, then:

$$Re = (1 \times 10^{-6} \times 0.46 \times 1000)/0.001 = 0.46$$

and hence by applying equation 3.120:

$t = \{18\mu/[d^2\omega^2(\rho_s - \rho)]\} \ln[(x + h_2)/(x + h_1)]$

$\quad = \{18 \times 0.001/[10^{-12} \times 502.7^2(2500 - 1000)]\} \ln[(0.150 + 0.075)/(0.150 + 0)]$

$\quad = 47.5 \ln(0.225/0.150) = \underline{19.3 \text{ s}}$

PROBLEM 9.3

A centrifuge basket 600 mm long and 100 mm internal diameter has a discharge weir 25 mm diameter. What is the maximum volumetric flow of liquid through the centrifuge such that, when the basket is rotated at 200 Hz, all particles of diameter greater than 1 μm are retained on the centrifuge wall? The retarding force on a particle moving liquid may be taken as $3\pi\mu\, du$, where u is the particle velocity relative to the liquid μ is the liquid viscosity, and d is the particle diameter. The density of the liquid is 1000 kg/m^3, the density of the solid is 2000 kg/m^3 and the viscosity of the liquid is 1.0 mN s/m^2. The inertia of the particle may be neglected.

Solution

With a basket radius of b m, a radius of the inner surface of liquid of x m and h m the distance radially from the surface of the liquid, the equation of motion in the radial direction of a spherical particle of diameter d m under streamline conditions is:

$$(\pi d^3/6)(\rho_s - \rho)(x + h)\omega^2 - 3\pi\mu\, du - (\pi d^3/6)\rho_s\, du/dt = 0 \quad \text{(from equation 3.108)}$$

Replacing u by dh/dt and neglecting the acceleration term gives:

$$(dh/dt) = d^2(\rho_s - \rho)\omega^2(x + h)/18\mu \quad\quad\quad \text{(equation 3.109)}$$

The time any element of material remains in the basket is V'/Q, where Q is the volumetric rate of feed to the centrifuge and V' is the volume of liquid retained in the basket at any time. If the flow rate is adjusted so that a particle of diameter d is just retained when it has to travel through the maximum distance $h = (b - x)$ before reaching the wall, then:

$$h = d^2(\rho_s - \rho)b\omega^2 V'/(18\mu Q)$$

or:
$$Q = d^2(\rho_s - \rho)b\omega^2 V'/(18\mu h)$$

In this case:
$$V' = (\pi/4)(0.1^2 - 0.025^2) \times 0.6 = 0.0044 \text{ m}^3$$
$$h = (0.10 - 0.025)/2$$

Thus:
$$Q = \frac{[(1 \times 10^{-6})^2(2000 - 1000) \times 0.1 \times (200 \times 2\pi)^2 \times 0.0044]}{(18 \times 0.001 \times 0.0375)}$$
$$= \underline{\underline{1.03 \times 10^{-3}}} \text{ m}^3/\text{s} \ (1 \text{ cm}^3/\text{s})$$

PROBLEM 9.4

When an aqueous slurry is filtered in a plate and frame press, fitted with two 50 mm thick frames each 150 mm square at a pressure difference of 350 kN/m², the frames are filled in 3600 s. The liquid in the slurry has the same density as water.

How long will it take to produce the same volume of filtrate as is obtained from a single cycle when using a centrifuge with a perforated basket 300 mm in diameter and 800 mm deep? The radius of the inner surface of the slurry is maintained constant at 75 mm and speed of rotation is 65 Hz (3900 rpm).

It may be assumed that the filter cake is incompressible, that the resistance of the cloth is equivalent to 3 mm of cake in both cases and that the liquid in the slurry has the same density as water.

Solution

See Volume 2, Example 9.3.

PROBLEM 9.5

A centrifuge with a phosphor bronze basket, 380 mm in diameter, is to be run at 67 Hz with a 75 mm layer of liquid of density 1200 kg/m³ in the basket. What thickness of walls are required in the basket? The density of phosphor bronze is 8900 kg/m³ and the maximum safe stress for phosphor bronze is 87.6 MN/m².

Solution

The stress at the walls is given by:

$$S = (r_b/t)(P_c - \rho_m t r_b \omega^2)$$

where:
> r_b is the radius of the basket = (380/2) = 190 mm or 0.19 m
> t is the wall thickness (m)
> ρ_p is the density of the metal = 8900 kg/m³
> ω is the rotational speed of the basket = $(67 \times 2\pi)^2 = 1.77 \times 10^5$ rad/s

and P_c is the pressure at the walls, given by:

$$P_c = 0.5\rho\omega^2(r_b^2 - x^2)$$

where:
> ρ is the density of the fluid = 1200 kg/m³

and x is the radius at the liquid surface = $[380 - (2 \times 75)]/2$
= 115 mm or 0.115 m

Thus: $\quad P_c = (0.5 \times 1200 \times 1.77 \times 10^5)(0.19^2 - 0.115^2)$

$\quad\quad\quad\quad = 2.43 \times 10^6$ N/m²

and, the stress at the walls is:

$$s = (0.19/t)[2.43 \times 10^6 + (8900t \times 0.19 \times 1.77 \times 10t^5)]$$

$$= (0.19/t)(2.43 \times 10^6 + 2.993 \times 10^8 t)$$

$$= (4.62 \times 10^5/t) + 5.69 \times 10^7 \text{ N/m}^2$$

The safe stress = 87.6×10^6 N/m²

Thus: $\quad\quad 87.6 \times 10^6 = (4.62 \times 10^5)/t + 5.63 \times 10^7$

and: $\quad\quad\quad t = 1.51 \times 10^2$ m or <u>15.1 mm</u>

SECTION 2-10

Leaching

PROBLEM 10.1

0.4 kg/s of dry sea-shore sand, containing 1 per cent by mass of salt, is to be washed with 0.4 kg/s of fresh water running countercurrently to the sand through two classifiers in series. It may be assumed that perfect mixing of the sand and water occurs in each classifier and that the sand discharged from each classifier contains one part of water for every two of sand by mass. If the washed sand is dried in a kiln dryer, what percentage of salt will it retain? What wash rate would be required in a single classifier in order to wash the sand to the same extent?

Solution

The problem involves a mass balance around the two stages. If x kg/s salt is in the underflow discharge from stage 1, then:

$$\text{salt in feed to stage 2} = (0.4 \times 1)/100 = 0.004 \text{ kg/s}.$$

The sand passes through each stage and hence the sand in the underflow from stage 1 = 0.4 kg/s, which, assuming constant underflow, is associated with $(0.4/2) = 0.2$ kg/s water. Similarly, 0.2 kg/s water enters stage 1 in the underflow and 0.4 kg/s enters in the overflow. Making a water balance around stage 1, the water in the overflow discharge = 0.4 kg/s.

In the underflow discharge from stage 1, x kg/s salt is associated with 0.2 kg/s water, and hence the salt associated with the 0.4 kg/s water in the overflow discharge = $(x \times 0.4)/0.2 = 2x$ kg/s. This assumes that the overflow and underflow solutions have the same concentration.

In stage 2, 0.4 kg/s water enters in the overflow and 0.2 kg/s leaves in the underflow.

Thus: water in overflow from stage 2 = $(0.4 - 0.2) = 0.2$ kg/s.

The salt entering is 0.004 kg/s in the underflow and $2x$ in the overflow — a total of $(0.004 + 2x)$ kg/s. The exit underflow and overflow concentrations must be the same, and hence the salt associated with 0.2 kg/s water in each stream is:

$$(0.004 + 2x)/2 = (0.002 + x) \text{ kg/s}$$

Making an overall salt balance:

$$0.004 = x + (0.002 + x) \quad \text{and} \quad x = 0.001 \text{ kg/s}$$

This is associated with 0.4 kg/s sand and hence:

$$\text{salt in dried sand} = (0.001 \times 100)/(0.4 + 0.001) = \underline{0.249 \text{ per cent}}$$

The same result may be obtained by applying equation 10.16 over the washing stage:

$$S_{n+1}/S_1 = (R-1)/(R^{n+1} - 1) \quad \text{(equation 10.16)}$$

In this case: $R = (0.4/0.2) = 2, n = 1, S_2 = x, S_1 = (0.002 + x)$ and:

$$x/(0.002 + x) = (2-1)/(2^2 - 1) = 0.33$$

$$x = (0.000667/0.667) = 0.001 \text{ kg/s}$$

and the salt in the sand = $\underline{0.249 \text{ per cent}}$ as before.

Considering a *single stage*:

If y kg/s is the overflow feed of water then, since 0.2 kg/s water leaves in the underflow, the water in the overflow discharge = $(y - 0.2)$ kg/s. With a feed of 0.004 kg/s salt and 0.001 kg/s salt in the underflow discharge, the salt in the overflow discharge = 0.003 kg/s. The ratio (salt/solution) must be the same in both discharge streams or:

$$(0.001)/(0.20 + 0.001) = 0.003/(0.003 + y - 0.2) \quad \text{and} \quad \underline{y = 0.8 \text{ kg/s}}$$

PROBLEM 10.2

Caustic soda is manufactured by the lime-soda process. A solution of sodium carbonate in water containing 0.25 kg/s Na_2CO_3 is treated with the theoretical requirement of lime and, after the reaction is complete, the $CaCO_3$ sludge, containing by mass 1 part of $CaCO_3$ per 9 parts of water is fed continuously to three thickeners in series and is washed countercurrently. Calculate the necessary rate of feed of neutral water to the thickeners, so that the calcium carbonate, on drying, contains only 1 per cent of sodium hydroxide. The solid discharged from each thickener contains one part by mass of calcium carbonate to three of water. The concentrated wash liquid is mixed with the contents of the agitated before being fed to the first thickeners.

Solution

See Volume 2, Example 10.2.

PROBLEM 10.3

How many stages are required for a 98 per cent extraction of a material containing 18 per cent of extractable matter of density 2700 kg/m^3 and which requires 200 volumes of liquid/100 volumes of solid for it to be capable of being pumped to the next stage? The strong solution is to have a concentration of 100 kg/m^3.

Solution

Taking as a basis *100 kg solids fed to the plant*, this contains 18 kg solute and 82 kg inert material. The extraction is 98 per cent and hence $(0.98 \times 18) = 17.64$ kg solute appears in the liquid product, leaving $(18 - 17.64) = 0.36$ kg solute in the washed solid. The concentration of the liquid product is 100 kg/m^3 and hence the volume of the liquid product $= (17.64/100) = 0.1764$ m^3.

Volume of solute in liquid product $= (17.64/2700) = 0.00653$ m^3.

Volume of solvent in liquid product $= (0.1764 - 0.00653) = 0.1699$ m^3.

Mass of solvent in liquid product $= 0.1699\rho$ kg

where ρ kg/m^3 is the density of solvent.

In the washed solids, total solids $= 82$ kg or $(82/2700) = 0.0304$ m^3.

Volume of solution in the washed solids $= (0.0304 \times 200)/100 = 0.0608$ m^3.

Volume of solute in solution $= (0.36/2700) = 0.0001$ m^3.

Volume of solvent in washed solids $= (0.0608 - 0.0001) = 0.0607$ m^3.

and mass of solvent in washed solids $= 0.0607\rho$ kg

Mass of solvent fed to the plant $= (0.0607 + 0.1699)\rho = 0.2306\rho$ kg

The *overall balance* in terms of mass is therefore;

	Inerts	Solute	Solvent
Feed to plant	82	18	—
Wash liquor	—	—	0.2306ρ
Washed solids	82	0.36	0.0607ρ
Liquid product	—	17.64	0.1699ρ

$$\frac{\text{Solvent discharged in the overflow}}{\text{Solvent discharged in the underflow}}, R = (0.2306\rho/0.0607\rho) = 3.80$$

The overflow product contains 100 kg solute/m^3 solution. This concentration is the same as the underflow from the first thickener and hence the material fed to the washing thickeners contains 82 kg inerts and 0.0608 m^3 solution containing $(100 \times 0.0608) = 6.08$ kg solute.

Thus, in equation 10.16:

$$(3.80 - 1)/(3.80^{n+1} - 1) = (0.36/6.08)$$

or: $\qquad 3.80^{n+1} = 48.28$ and $n = 1.89$, say 2 washing thickeners.

Thus a total of <u>3 thickeners</u> is required.

PROBLEM 10.4

Soda ash is mixed with lime and the liquor from the second of three thickeners and passed to the first thickener where separation is effected. The quantity of this caustic solution leaving the first thickener is such as to yield 10 Mg of caustic soda per day of 24 hours. The solution contains 95 kg of caustic soda/1000 kg of water, whilst the sludge leaving each of the thickeners consists of one part of solids to one of liquid.

Determine:

(a) the mass of solids in the sludge,
(b) the mass of water admitted to the third thickener and
(c) the percentages of caustic soda in the sludges leaving the respective thickeners.

Solution

Basis: 100 Mg $CaCO_3$ in the sludge leaving each thickener.

In order to produce 100 Mg $CaCO_3$, 106 Mg Na_2CO_3 must react giving 80 Mg NaOH according to the equation:

$$Na_2CO_3 + Ca(OH)_2 = 2NaOH + CaCO_3.$$

For the purposes of calculation it is assumed that a mixture of 100 Mg $CaCO_3$ and 80 Mg NaOH is fed to the first thickener and w Mg water is the overflow feed to the third thickener. Assuming that x_1, x_2 and x_3 are the ratios of caustic soda to solution by mass in each thickener then the mass balances are made as follows:

	$CaCO_3$	NaOH	Water
Overall			
Underflow feed	100	80	—
Overflow feed	—	—	w
Underflow product	100	$100x_3$	$100(1 - x_3)$
Overflow product	—	$(80 - 100w_3)$	$w - 100(1 - x_3)$
Thickener 1			
Underflow feed	100	80	—
Overflow feed	—	$100(x_1 - x_3)$	$w + 100(x_3 - x_1)$
Underflow product	100	$100x_1$	$100(1 - x_1)$
Overflow product	—	$80 - 100x_3$	$w - 100(1 - x_3)$
Thickener 2			
Underflow feed	100	$100x_1$	$100(1 - x_1)$
Overflow feed	—	$100(x_2 - x_3)$	$w + 100(x_3 - x_2)$
Underflow product	100	$100x_2$	$100(1 - x_2)$
Overflow product	—	$100(x_1 - x_3)$	$w + 100(x_3 - x_1)$
Thickener 3			
Underflow feed	100	$100x_2$	$100(1 - x_2)$
Overflow feed	—	—	w
Underflow product	100	$100x_3$	$100(1 - x_3)$
Overflow product	—	$100(x_2 - x_3)$	$w + 100(x_3 - x_2)$

In the overflow product, 0.095 Mg NaOH is associated with 1 Mg water.

Thus: $\quad x_1 = 0.095/(1 + 0.095) = 0.0868$ Mg/Mg solution $\quad$ (i)

Assuming that equilibrium is attained in each thickener, the concentration of NaOH in the overflow product is equal to the concentration of NaOH in the solution in the underflow product.

Thus:
$$x_3 = [100(x_2 - x_3)]/[100(x_2 - x_3) + w - 100(x_2 - x_3)]$$
$$= 100(x_2 - x_3)/w \quad (ii)$$
$$x_2 = [100(x_1 - x_3)]/[100(x_1 - x_3) + w - 100(x_1 - x_3)]$$
$$= 100(x_1 - x_3)/w \quad (iii)$$
$$x_3 = (80 - 100x_3)/[80 - 100x_3 + w - 100(1 - x_3)]$$
$$= (80 - 100x_3)/(w - 20) \quad (iv)$$

Solving equations (i)–(iv) simultaneously, gives:

$$x_3 = 0.0010 \text{ Mg/Mg}, \quad x_2 = 0.0093 \text{ Mg/Mg}, \quad x_1 = 0.0868 \text{ Mg/Mg}$$

and: $\quad w = 940.5$ Mg/100 Mg $CaCO_3$

The overflow product $= w - 100(1 - x_3) = 840.6$ Mg/100 Mg $CaCO_3$.
The actual flow of caustic solution $= (10/0.0868) = 115$ Mg/day.

Thus: $\quad$ mass of $CaCO_3$ in sludge $= (100 \times 115)/840.6 = \underline{\underline{13.7 \text{ Mg/day}}}$

The mass of water fed to third thickener $= 940.5$ Mg/100 Mg $CaCO_3$

or: $\quad = (940.5 \times 13.7)/100 = \underline{\underline{129 \text{ Mg/day}}}$

The total mass of sludge leaving each thickener $= 200$ Mg/100 Mg $CaCO_3$.

The mass of caustic soda in the sludge $= 100x_1$ Mg/100 Mg $CaCO_3$ and hence the concentration of caustic in sludge leaving,

$$\text{thickener 1} = (100 \times 0.0868 \times 100)/200 = \underline{\underline{4.34 \text{ per cent}}}$$

$$\text{thickener 2} = (100 \times 0.0093 \times 100)/200 = \underline{\underline{0.47 \text{ per cent}}}$$

$$\text{thickener 3} = (100 \times 0.0010 \times 100)/200 = \underline{\underline{0.05 \text{ per cent}}}$$

PROBLEM 10.5

Seeds, containing 20 per cent by mass of oil, are extracted in a countercurrent plant and 90 per cent of the oil is recovered as a solution containing 50 per cent by mass of oil. If the seeds are extracted with fresh solvent and 1 kg of solution is removed in the underflow in association with every 2 kg of insoluble matter, how many ideal stages are required?

Solution

See Volume 2, Example 10.4.

PROBLEM 10.6

It is desired to recover precipitated chalk from the causticising of soda ash. After decanting the liquor from the precipitators the sludge has the composition 5 per cent $CaCO_3$, 0.1 per cent NaOH and the balance water.

1000 Mg/day of this sludge is fed to two thickeners where it is washed with 200 Mg/day of neutral water. The pulp removed from the bottom of the thickeners contains 4 kg of water/kg of chalk. The pulp from the last thickener is taken to a rotary filter and concentrated to 50 per cent solids and the filtrate is returned to the system as wash water. Calculate the net percentage of $CaCO_3$ in the product after drying.

Solution

Basis: 1000 Mg/day sludge fed to the plant

If x_1 and x_2 are the solute/solvent ratios in thickeners 1 and 2 respectively, then the mass balances are:

	$CaCO_3$	NaOH	Water
Overall			
Underflow feed	50	1	949
Overflow feed	—	—	200
Underflow product	50	$200x_2$	200
Overflow product	—	$(1 - 200x_2)$	949
Thickener 1			
Underflow feed	50	1	949
Overflow feed	—	$200(x_1 - x_2)$	200
Underflow product	50	$200x_1$	200
Overflow product	—	$(1 - 200x_2)$	949
Thickener 2			
Underflow feed	50	$200x_1$	200
Overflow feed	—	—	200
Underflow product	50	$200x_2$	200
Overflow product	—	$200(x_1 - x_2)$	200

Assuming that equilibrium is attained, the solute/solvent ratio will be the same in the overflow and underflow products of each thickener and:

$$x_2 = 200(x_1 - x_2)/200 \quad \text{or} \quad x_2 = 0.5x_1$$

and:
$$x_1 = (1 - 200x_2)/949$$

Thus:
$$x_1 = 0.000954 \quad \text{and} \quad x_2 = 0.000477$$

The underflow product contains 50 Mg $CaCO_3$, $(200 \times 0.000477) = 0.0954$ Mg NaOH and 200 Mg water. After concentration to 50 per cent solids, the mass of NaOH in solution

$$= (0.0954 \times 50)/200.0954 = 0.0238 \text{ Mg}$$

and the $CaCO_3$ in dried solids

$$= (100 \times 50)/50.0238 = \underline{\underline{99.95 \text{ per cent}}}$$

This approach ignores the fact that filtrate is returned to the second thickener together with wash water. Taking this into account, the calculation is modified as follows.

The underflow product from the second thickener contains:

$$50 \text{ Mg } CaCO_3, \quad 200x_2 \text{ Mg NaOH} \quad \text{and} \quad 200 \text{ Mg water}$$

After filtration, the 50 Mg $CaCO_3$ is associated with 50 Mg solution of the same concentration and hence this contains:

$$50x_2/(1+x_2) \text{ Mg NaOH} \quad \text{and} \quad 50/(1+x_2) \text{ Mg water}$$

The remainder is returned with the overflow feed to the second thickener. The filtrate returned contains:

$$200x_2 - 50x_2/(1+x_2) \text{ Mg NaOH}$$

and:
$$200 - 50/(1+x_2) \text{ Mg water}$$

The balances are now:

	$CaCO_3$	NaOH	Water
Overall			
Underflow feed	50	1	949
Overflow feed	—	$200x_2 - 50x_2/(1+x_2)$	$400 - 50/(1+x_2)$
Underflow product	50	$200x_2$	200
Overflow product	—	$1 - 50x_2/(1+x_2)$	$1149 - 50/(1+x_2)$
Thickener 1			
Underflow feed	50	1	949
Overflow feed	—	$200x_1 - 50x_2/(1+x_2)$	$400 - 50/(1+x_2)$
Underflow product	50	$200x_1$	200
Overflow product	—	$1 - 50x_2/(1+x_2)$	$1149 - 50/(1+x_{2a})$
Thickener 2			
Underflow feed	50	$200x_1$	200
Overflow feed	—	$200x_2 - 50x_2/(1+x_2)$	$400 - 50/(1+x_2)$
Underflow product	50	$200x_2$	200
Overflow product	—	$200x_1 - 50x_2/(1+x_2)$	$400 - 50/(1+x_2)$

Again, assuming equilibrium is attained, then:

$$x_2 = [200x_1 - 50x_2/(1+x_2)]/[400 - 50/(1+x_2)]$$

and:
$$x_1 = [1 - 50x_2/(1+x_2)]/[1149 - 50/(1+x_2)]$$

Solving simultaneously, then:

$$x_1 = 0.000870 \text{ Mg/Mg} \quad \text{and} \quad x_2 = 0.000435 \text{ Mg/Mg}$$

The solid product leaving the filter contains 50 Mg CaCO$_3$

and: $(50 \times 0.000435)/(1 + 0.000435) = 0.02175$ Mg NaOH in solution.

After drying, the solid product will contain:

$$(100 \times 50)/(50 + 0.02175) = \underline{\underline{99.96 \text{ per cent CaCO}_3}}$$

PROBLEM 10.7

Barium carbonate is to be made by reacting sodium carbonate and barium sulphide. The quantities fed to the reaction agitators per 24 hours are 20 Mg of barium sulphide dissolved in 60 Mg of water, together with the theoretically necessary amount of sodium carbonate.

Three thickeners in series are run on a countercurrent decantation system. Overflow from the second thickener goes to the agitators, and overflow from the first thickener is to contain 10 per cent sodium sulphide. Sludge from all thickeners contains two parts water to one part barium carbonate by mass. How much sodium sulphide will remain in the dried barium carbonate precipitate?

Solution

Basis: 1 day's operation

The reaction is:

$$\text{BaS} + \text{Na}_2\text{CO}_3 = \text{BaCO}_3 + \text{Na}_2\text{S}$$

Molecular masses: 169 106 197 78 kg/kmol.

Thus 20 Mg BaS will react to produce

$$(20 \times 197)/169 = 23.3 \text{ Mg BaCO}_3$$

and: $(20 \times 78)/169 = 9.23$ Mg Na$_2$S

The calculation may be made on the basis of this material entering the washing thickeners together with 60 Mg water. If x_1, x_2, and x_3 are the Na$_2$S/water ratio in the respective thickeners, then the mass balances are:

	BaCO$_3$	Na$_2$S	Water
Overall			
Underflow feed	23.3	9.23	60
Overflow feed	—	—	w (say)
Underflow product	23.3	46.6x_3	46.6
Overflow product	—	9.23 − 46.6x_3	$w + 13.4$

	BaCO$_3$	Na$_2$S	Water
Thickener 1			
Underflow feed	23.3	9.23	60
Overflow feed	—	$46.6(x_1 - x_3)$	w
Underflow product	23.3	$46.6x_1$	46.6
Overflow product	—	$9.23 - 46.6x_3$	$w + 13.4$
Thickener 2			
Underflow feed	23.3	$46.6x_1$	46.6
Overflow feed	—	$46.6(x_2 - x_3)$	w
Underflow product	23.3	$46.6x_2$	46.6
Overflow product	—	$46.6(x_1 - x_3)$	w
Thickener 3			
Underflow feed	23.3	$46.6x_2$	46.6
Overflow feed	—	—	w
Underflow product	23.3	$46.6x_3$	46.6
Overflow product	—	$46.6(x_2 - x_3)$	w

In the overflow product leaving the first thickener:

$$(9.23 - 46.6x_3)/(13.4 + w + 9.23 - 46.6x_3) = 0.10 \tag{i}$$

Assuming equilibrium is attained in each thickener, then:

$$x_1 = (9.23 - 46.6x_3)/(13.4 + w), \tag{ii}$$

$$x_2 = 46.6(x_1 - x_3)/w, \tag{iii}$$

and:
$$x_3 = 46.6(x_2 - x_3)/w \tag{iv}$$

Solving equations (i)–(iv) simultaneously:

$$x_1 = 0.112, \quad x_2 = 0.066, \quad x_3 = 0.030, \quad \text{and} \quad w = 57.1 \text{ Mg/day}$$

In the underflow product from the third thickener, the mass of Na$_2$S is

$$(46.6 \times 0.030) = 1.4 \text{ Mg associated with 23.3 Mg BaCO}_3$$

When this stream is dried, the barium carbonate will contain:

$$(100 \times 1.4)/(1.4 + 23.3) = \underline{\underline{5.7 \text{ per cent sodium sulphide}}}$$

PROBLEM 10.8

In the production of caustic soda by the action of calcium hydroxide on sodium carbonate, 1 kg/s of sodium carbonate is treated with the theoretical quantity of lime. The sodium carbonate is made up as a 20 per cent solution. The material from the extractors is fed to a countercurrent washing system where it is treated with 2 kg/s of clean water. The washing thickeners are so arranged that the ratio of the volume of liquid discharged in the liquid offtake to that discharged with the solid is the same in all the thickeners and is equal to 4.0. How many thickeners must be arranged in series so that not more than 1 per cent of the sodium hydroxide discharged with the solid from the first thickener is wasted?

Solution

The reaction is:

$$Na_2CO_3 + Ca(OH)_2 = 2NaOH + CaCO_3$$
Molecular masses: 106 74 80 100 kg/kmol.

Thus 1 kg/s Na_2CO_3 forms $(1 \times 80)/106 = 0.755$ kg/s NaOH

and: $(1 \times 100)/106 = 0.943$ kg/s $CaCO_3$

In a 20 per cent solution, 1 kg/s Na_2CO_3 is associated with $(1 - 0.20)/0.2 = 4.0$ kg/s water.

If x kg/s NaOH leaves in the underflow product from the first thickener, then $0.01x$ kg/s NaOH should leave in the underflow product from the nth thickener. The amount of NaOH in the overflow from the first thickener is then given from an overall balance as $= (0.755 - 0.01x)$ kg/s:

Since the volume of the overflow product is $4x$, the volume of solution in underflow product, then:

$$(0.755 - 0.01x) = 4x \quad \text{and} \quad x = 0.188 \text{ kg/s}$$

and the NaOH leaving the nth thickener in the underflow $= (0.01 \times 0.188) = 0.00188$ kg/s.

Thus the fraction of solute fed to the washing system which remains associated with the washed solids, $f = (0.00188/0.755) = 0.0025$ kg/kg

In this case $R = 4.0$ and in equation 10.17:

$$n = \{\ln[1 + (4 - 1)/0.0025]\}/(\ln 4) - 1 = 4.11, \quad \text{say 5 washing thickeners}$$

PROBLEM 10.9

A plant produces 100 kg/s of titanium dioxide pigment which must be 99 per cent pure when dried. The pigment is produced by precipitation and the material, as prepared, is contaminated with 1 kg of salt solution containing 0.55 kg of salt/kg of pigment. The material is washed countercurrently with water in a number of thickeners arranged in series. How many thickeners will be required if water is added at the rate of 200 kg/s and the solid discharged from each thickeners removes 0.5 kg of solvent/kg of pigment?

What will be the required number of thickeners if the amount of solution removed in association with the pigment varies with the concentration of the solution in the thickener as follows:

kg solute/kg solution	0	0.1	0.2	0.3	0.4	0.5
kg solution/kg pigment	0.30	0.32	0.34	0.36	0.38	0.40

The concentrated wash liquor is mixed with the material fed to the first thickener.

Solution

Part I

The overall balance in kg/s, is:

	TiO$_2$	Salt	Water
Feed from reactor	100	55	45
Wash liquor added	—	—	200
Washed solid	100	0.1	50
Liquid product	—	54.9	195

The solvent in the underflow from the final washing thickener = 50 kg/s.

The solvent in the overflow will be the same as that supplied for washing = 200 kg/s.

This: $\dfrac{\text{Solvent discharged in overflow}}{\text{Solvent discharged in underflow}} = (200/50) = 4$ for the washing thickeners.

The liquid product from plant contains 54.9 kg of salt in 195 kg of solvent. This ratio will be the same in the underflow from the first thickener.

Thus the material fed to the washing thickeners consists of 100 kg TiO$_2$, 50 kg solvent and $(50 \times 54.9)/195 = 14$ kg salt.

The required number of thickeners for washing is given by equation 10.16 as:

$$(4-1)/(4^{n+1} - 1) = (0.1/14)$$

or: $\qquad 4^{n+1} = 421 \quad \text{giving:} \quad 4 < (n+1) < 5$

Thus 4 washing thickeners or a total of <u>5 thickeners</u> are required.

Part 2

The same nomenclature will be used as in Volume 2, Chapter 10.

By inspection of the data, it is seen that $W_{h+1} = 0.30 + 0.2X_h$.

Thus: $\qquad S_{h+1} = W_{h+1}X_h = 0.30X_h + 0.2X_h^2 = 5W_{h+1}^2 - 1.5W_{h+1}$

Considering the passage of unit quantity of TiO$_2$ through the plant:

$$L_{n+1} = 0, \quad w_{n+1} = 2, \quad X_{n+1} = 0$$

since 200 kg/s pure solvent is used.

$$S_{n+1} = 0.001 \quad \text{and therefore} \quad W_{n+1} = 0.3007.$$
$$S_1 = 0.55 \quad \text{and} \quad W_1 = 1.00$$

Thus the concentration in the first thickener is given by equation 10.23:

$$X_1 = \frac{L_{n+1} + S_1 - S_{n+1}}{W_{n+1} + W_1 - W_{n+1}} = (0 + 0.55 - 0.001)/(2 + 1 - 0.3007) = 0.203$$

From equation 10.26:

$$X_{h+1} = \frac{L_{n+1} - S_{n+1} + S_{h+1}}{W_{n+1} - W_{n+1} + W_{h+1}} = \frac{(0 - 0.001 + S_{h+1})}{(2 - 0.3007 + W_{h+1})} = \frac{(-0.001 + S_{h+1})}{(1.7 + W_{h+1})}$$

Since $X_1 = 0.203$, then $W_2 = (0.30 + 0.2 \times 0.203) = 0.3406$

and: $S_2 = (0.3406 \times 0.203) = 0.0691$

Thus: $X_2 = (0.0691 - 0.001)/(1.7 + 0.3406) = 0.0334$

Since $X_2 = 0.0334$, then $W_3 = (0.30 + 0.2 \times 0.0334) = 0.30668$

and: $S_2 = (0.3067 \times 0.0334) = 0.01025$

Thus: $X_3 = (0.01025 - 0.001)/(1.7 + 0.3067) = 0.00447$

Since $X_3 = 0.00447$, then $W_4 = 0.30089$ and $S_4 = 0.0013$

Hence, by the same method: $X_4 = 0.000150$

Since $X_4 = 0.000150$, then $W_5 = 0.30003$ and $S_5 = 0.000045$.

Thus S_5 is less than S_{n+1} and therefore <u>4 thickeners</u> are required.

PROBLEM 10.10

Prepared cottonseed meats containing 35 per cent of extractable oil are fed to a continuous countercurrent extractor of the intermittent drainage type using hexane as the solvent. The extractor consists of ten sections and the section efficiency is 50 per cent. The entrainment, assumed constant, is 1 kg solution/kg solids. What will be the oil concentration in the outflowing solvent if the extractable oil content in the meats is to be reduced by 0.5 per cent by mass?

Solution

Basis: 100 kg inert cottonseed material

Mass of oil in underflow feed = $(100 \times 0.35)/(1 - 0.35) = 53.8$ kg.

In the underflow product from the plant, mass of inerts = 100 kg and hence mass of oil = $(100 \times 0.005)/(1 - 0.005) = 0.503$ kg.

This is in 100 kg solution and hence the mass of hexane in the underflow product = $(100 - 0.503) = 99.497$ kg.

The overall balance in terms of mass is:

	Inerts	Oil	Hexane
Underflow feed	100	53.8	—
Overflow feed	—	—	h (say)
Underflow product	100	0.503	99.497
Overflow product	—	53.297	$(h - 99.497)$

Since there are ten stages, each 50 per cent efficient, the system may be considered, as a first approximation as consisting of five theoretical stages each of 100 per cent efficiency, in which equilibrium is attained in each stage. On this basis, the underflow from stage 1 contains 100 kg solution in which the oil/hexane ratio $= 53.297/(h - 99.497)$ and hence the amount of oil in this stream is:

$$S_1 = 100[1 - (h - 99.497)/(h - 46.2)] \text{ kg}$$

$$S_{n+1} = 0.503 \text{ kg}$$

With constant underflow, the amount of solution in the overflow from each stage is say, h kg and the solution in the underflow $= 100$ kg.

Thus: $$R = (h/100) = 0.01h$$

and in equation 10.16:

$$0.503/[100 - 100(h - 99.497)/(h - 46.2)] = (0.01h - 1)/[(0.01h)^5 - 1]$$

or: $$(0.503h - 23.24) = (53.30h - 5330)/[(0.01h)^5 - 1]$$

Solving by trial and error: $h = 238$ kg
and in the overflow product:

mass of hexane $= (238 - 99.497) = 138.5$ kg, mass of oil $= 53.3$ kg

and concentration of oil $= (100 \times 53.3)/(53.3 + 138.5) = \underline{\underline{27.8 \text{ per cent}}}$.

PROBLEM 10.11

Seeds containing 25 per cent by mass of oil are extracted in a countercurrent plant and 90 per cent of the oil is to be recovered in a solution containing 50 per cent of oil. It has been found that the amount of solution removed in the underflow in association with every kilogram of insoluble matter, k is given by:

$$k = 0.7 + 0.5y_s + 3y_s^2 \text{ kg/kg}$$

where y_s is the concentration of the overflow solution in terms of mass fraction of solute kg/kg. If the seeds are extracted with fresh solvent, how many ideal stages are required?

Solution

Basis: 100 kg underflow feed to the first stage

The first step is to obtain the underflow line, that is a plot of x_s against x_A. The calculations are made as follows:

y_s	k	Ratio (kg/kg inerts)			Mass fraction	
		oil (ky_s)	solvent $k(1-y_s)$	underflow $(k+1)$	oil (x_A)	solvent (x_s)
0	0.70	0	0.70	1.70	0	0.412
0.1	0.78	0.078	0.702	1.78	0.044	0.394

y_s	k	Ratio (kg/kg inerts)			Mass fraction	
		oil (ky_s)	solvent $k(1-y_s)$	underflow $(k+1)$	oil (x_A)	solvent (x_s)
0.2	0.92	0.184	0.736	1.92	0.096	0.383
0.3	1.12	0.336	0.784	2.12	0.159	0.370
0.4	1.38	0.552	0.828	2.38	0.232	0.348
0.5	1.70	0.850	0.850	2.70	0.315	0.315
0.6	2.08	1.248	0.832	3.08	0.405	0.270
0.7	2.52	1.764	0.756	3.52	0.501	0.215
0.8	3.02	2.416	0.604	4.02	0.601	0.150
0.9	3.58	3.222	0.358	4.58	0.704	0.078
1.0	4.20	4.20	0	5.20	0.808	0

A plot of x_A against x_s is shown in Figure 10a.

Considering the *underflow feed*, the seeds contain 25 per cent oil and 75 per cent inerts, and the point $x_{s1} = 0$, $x_{A1} = 0.25$ is drawn in as x_1.

In the *overflow feed*, pure solvent is used and hence:

$$y_{s \cdot n+1} = 1.0, \quad y_{A \cdot n+1} = 0$$

This point is marked as y_{n+1}.

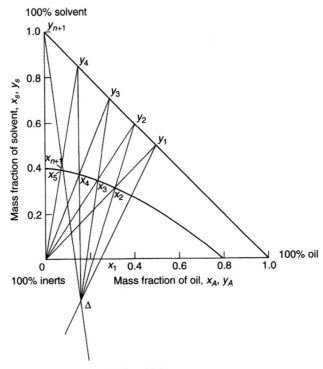

Figure 10a. Graphical construction for Problem 10.11

In the *overflow product*, the oil concentration is 50 per cent and $y_{s1} = 0.50$ and $y_{A1} = 0.50$. This point lies on the hypotenuse and is marked y_1.

90 per cent of the oil is recovered, leaving $25(1 - 0.90) = 2.5$ kg in the underflow product associated with 75 kg inerts; that is:

$$\text{ratio (oil/inerts)} = (2.5/75) = 0.033 = ky_s$$

Thus:
$$0.033 = (0.7y_s + 0.5y_s^2 + 3y_s^3)$$

Solving by substitution gives:

$$y_s = 0.041 \quad \text{and hence} \quad k = (0.033/0.041) = 0.805$$
$$x_A = 0.0173 \quad \text{and} \quad x_s = 0.405$$

This point is drawn as x_{n+1} on the x_s against x_A curve.

The pole point Δ is obtained where $y_{n+1} \cdot x_{n+1}$ and $y_1 \cdot x_1$ extended meet and the construction described in Chapter 10 is then followed.

It is then found that x_{n+1} lies between x_4 and x_5 and hence 4 ideal stages are required.

PROBLEM 10.12

Halibut oil is extracted from granulated halibut livers in a countercurrent multi-batch arrangement using ether as the solvent. The solids charge contains 0.35 kg oil/kg of exhausted livers and it is desired to obtain a 90 per cent oil recovery. How many theoretical stages are required if 50 kg of ether are used/100 kg of untreated solids. The entrainment data are:

Concentration of overflow (kg oil/kg solution)	0	0.1	0.2	0.3	0.4	0.5	0.6	0.67
Entrainment (kg solution/ kg extracted livers)	0.28	0.34	0.40	0.47	0.55	0.66	0.80	0.96

Solution

See Volume 2, Example 10.5.

SECTION 2-11

Distillation

PROBLEM 11.1

A liquid containing four components, **A, B, C** and **D**, with 0.3 mole fraction each of **A, B** and **C**, is to be continuously fractionated to give a top product of 0.9 mole fraction **A** and 0.1 mole fraction **B**. The bottoms are to contain not more than 0.5 mole fraction **A**. Estimate the minimum reflux ratio required for this separation, if the relative volatility of **A** to **B** is 2.0.

Solution

The given data may be tabulated as follows:

	Feed	Top	Bottoms
A	0.3	0.9	0.05
B	0.3	0.1	—
C	0.3	—	—
D	0.1	—	—

The Underwood and Fenske equations may be used to find the minimum number of plates and the minimum reflux ratio for a binary system. For a multicomponent system n_m may be found by using the two key components in place of the binary system and the relative volatility between those components in equation 11.56 enables the minimum reflux ratio R_m to be found. Using the feed and top compositions of component **A**:

$$R_m = \frac{1}{\alpha - 1}\left[\left(\frac{x_d}{x_f}\right) - \alpha\frac{(1 - x_d)}{(1 - x_f)}\right] \quad \text{(equation 11.50)}$$

Thus: $\quad R_m = \dfrac{1}{2 - 1}\left[\left(\dfrac{0.9}{0.3}\right) - 2\dfrac{(1 - 0.9)}{(1 - 0.3)}\right] = \underline{\underline{2.71}}$

PROBLEM 11.2

During the batch distillation of a binary mixture in a packed column the product contained 0.60 mole fraction of the more volatile component when the concentration in the still was 0.40 mole fraction. If the reflux ratio used was 20:1, and the vapour composition y is related to the liquor composition x by the equation $y = 1.035x$ over the range of concentration concerned, determine the number of ideal plates represented by the column. x and y are in mole fractions.

Solution

It is seen in equation 11.48, the equation of the operating line, that the slope is given by $R/(R+1)(=L/V)$ and the intercept on the y-axis by:

$$x_d/(R+1) = (D/V_n)$$

$$y_n = \frac{R}{R+1}x_{n+1} + \frac{x_d}{R+1} \qquad \text{(equation 11.41)}$$

In this problem, the equilibrium curve over the range $x = 0.40$ to $x = 0.60$ is given by $y = 1.035x$ and it may be drawn as shown in Figure 11a. The intercept of the operating line on the y-axis is equal to $x_d/(R+1) = 0.60/(20+1) = 0.029$ and the operating line is drawn through the points (0.60, 0.60) and (0, 0.029) as shown.

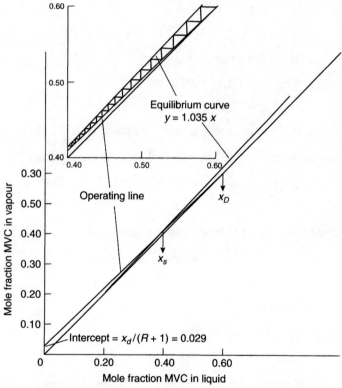

Figure 11a. Construction for Problem 11.2

In this particular example, all these lines are closely spaced and the relevant section is enlarged in the inset of Figure 11a. By stepping off the theoretical plates as in the McCabe–Thiele method, it is seen that 18 theoretical plates are represented by the column.

PROBLEM 11.3

A mixture of water and ethyl alcohol containing 0.16 mole fraction alcohol is continuously distilled in a plate fractionating column to give a product containing 0.77 mole fraction alcohol and a waste of 0.02 mole fraction alcohol. It is proposed to withdraw 25 per cent of the alcohol in the entering stream as a side stream containing 0.50 mole fraction of alcohol.

Determine the number of theoretical plates required and the plate from which the side stream should be withdrawn if the feed is liquor at its boiling point and a reflux ratio of 2 is used.

Solution

Taking 100 kmol of feed to the column as a basis, 16 kmol of alcohol enter, and 25 per cent, that is 4 kmol, are to be removed in the side stream. As the side-stream composition is to be 0.5, that stream contains 8 kmol.

An overall mass balance gives:

$$F = D + W + S$$

That is: $\quad 100 = D + W + 8 \quad$ or $\quad 92 = D + W$

A mass balance on the alcohol gives:

$$(100 \times 0.16) = 0.77D + 0.02W + 4$$

or: $\quad 12 = 0.77D + 0.02W.$

from which: distillate, $D = 13.55$ kmol and bottoms, $W = 78.45$ kmol.

In the top section between the side-stream and the top of the column:

$$R = L_n/D = 2, \text{ and hence } L_n = (2 \times 13.55) = 27.10 \text{ kmol}$$

$$V_n = L_n + D \quad \text{and} \quad V_n = (27.10 + 13.55) = 40.65 \text{ kmol}$$

For the section between the feed and the side stream:

$$V_s = V_n = 40.65, \quad L_n = S + L_s$$

and: $\quad L_s = (27.10 - 8) = 19.10$ kmol

At the bottom of the column:

$$L_m = L_s + F = (19.10 + 100) = 119.10, \text{ if the feed is at its boiling-point.}$$

$$V_m = L_m - W = (119.10 - 78.45) = 40.65 \text{ kmol.}$$

The slope of the operating line is always L/V and thus the slope in each part of the column can now be calculated. The top operating line passes through the point (x_d, x_d) and has a slope of $(27.10/40.65) = 0.67$. This is shown in Figure 11b and it applies until $x_s = 0.50$ where the slope becomes $(19.10/40.65) = 0.47$. The operating line in the bottom of the column applies from $x_f = 0.16$ and passes through the point (x_w, x_w) with a slope of $(119.10/40.65) = 2.92$.

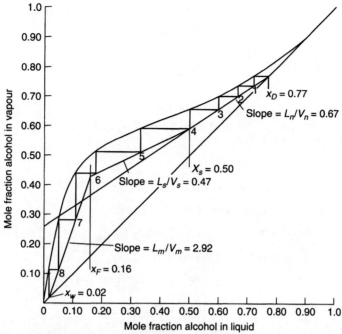

Figure 11b. Graphical construction for Problem 11.3

The steps corresponding to the theoretical plates may be drawn in as shown, and 8 plates are required with the side stream being withdrawn from the fourth plate from the top.

PROBLEM 11.4

In a mixture to be fed to a continuous distillation column, the mole fraction of phenol is 0.35, o-cresol is 0.15, m-cresol is 0.30 and xylenols is 0.20. A product is required with a mole fraction of phenol of 0.952, o-cresol 0.0474 and m-cresol 0.0006. If the volatility to o-cresol of phenol is 1.26 and of m-cresol is 0.70, estimate how many theoretical plates would be required at total reflux.

Solution

The data may be tabulated in terms of mole fractions as follows.

Component	Feed	Top	Bottoms	α
P	0.35	0.952		1.26
O	0.15	0.0474		1.0
M	0.30	0.0006		0.7
X	0.20	—		
	1.0000			

101

Fenske's equation may be used to find the minimum number of plates.
Thus the number of plates at total reflux is given by:

$$n + 1 = \frac{\log[(x_A/x_B)_d(x_B/x_A)_s]}{\log \alpha_{AB}} \quad \text{(equation 11.58)}$$

For multicomponent systems, components **A** and **B** refer to the light and heavy keys respectively. In this problem, o-cresol is the light key and m-cresol is the heavy key. A mass balance may be carried out in order to determine the bottom composition. Taking as a basis, 100 kmol of feed, then:

$$100 = D + W$$

For phenol: $\quad (100 \times 0.35) = 0.952D + x_{wp}W$

If x_{wp} is zero then: $\quad D = 36.8 \quad W = 63.2$

For o-cresol: $\quad (100 \times 0.15) = (0.0474 \times 36.8) + (x_{wo} \times 63.2)$ and $x_{wo} = 0.21$

For m-cresol: $\quad (100 \times 0.30) = (0.0006 \times 36.8) + (x_{wm} \times 63.2)$ and $x_{wm} = 0.474$

By difference: $\quad x_{wx} = 0.316 \quad \alpha_{om} = (1/0.7) = 1.43$

Hence, substituting into Fenske's equation gives:

$$n + 1 = \frac{\log[(0.0474/0.0006)(0.474/0.21)]}{\log 1.43}$$

and: $\quad \underline{\underline{n = 13.5}}$

PROBLEM 11.5

A continuous fractionating column, operating at atmospheric pressure, is to be designed to separate a mixture containing 15.67 per cent CS_2 and 84.33 per cent CCl_4 into an overhead product containing 91 per cent CS_2 and a waste of 97.3 per cent CCl_4 all by mass. A plate efficiency of 70 per cent and a reflux of 3.16 kmol/kmol of product may be assumed. Using the following data, determine the number of plates required.

The feed enters at 290 K with a specific heat capacity of 1.7 kJ/kg K and a boiling point of 336 K. The latent heats of CS_2 and CCl_4 are 25.9 kJ/kmol.

CS_2 in the vapour (Mole per cent)	0	8.23	15.55	26.6	33.2	49.5	63.4	74.7	82.9	87.8	93.2
CS_2 in the liquid (Mole per cent)	0	2.36	6.15	11.06	14.35	25.85	33.0	53.18	66.30	75.75	86.04

Solution

The equilibrium data are shown in Figure 11c and the problem may be solved using the method of McCabe and Thiele. All compositions are in terms of mole fractions so that:

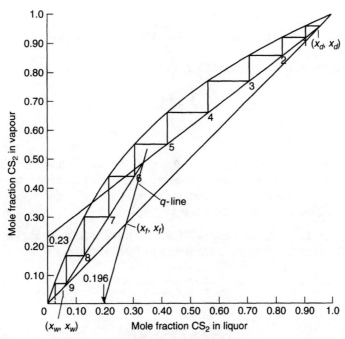

Figure 11c. Equilibrium data for Problem 11.5

Top product: $\quad x_d = \dfrac{(91/76)}{(91/76) + (9/154)} = 0.953$

Feed: $\quad x_f = \dfrac{(15.67/76)}{(15.67/76) + (84.33/154)} = 0.274$

Bottom product: $\quad x_w = \dfrac{(2.7/76)}{(2.7/76) + (97.3/154)} = 0.053$

In this problem, the feed is not at its boiling-point so the slope of the q-line must be determined in order to locate the intersection of the operating lines.

q is defined as the heat required to vaporise 1 kmol of feed/molar latent heat of feed, or

$$q = (\lambda + H_{fs} - H_f)/\lambda$$

where λ is the molar latent heat. H_{fs} is the enthalpy of 1 kmol of feed at its boiling-point, and H_f is the enthalpy of 1 kmol of feed.

The feed composition is 27.4 per cent CS_2 and 72.6 per cent CCl_4 so that the mean molecular mass of the feed is given by:

$$(0.274 \times 76) + (0.726 \times 154) = 132.6 \text{ kg/kmol}$$

Taking a datum of 273 K:

$$H_f = 1.7 \times 132.6(290 - 273) = 3832 \text{ kJ/kmol}$$

$$H_{fs} = 1.7 \times 132.6(336 - 273) = 14{,}200 \text{ kJ/kmol}$$

$$\lambda = 25{,}900 \text{ kJ/kmol}$$

Thus: $\quad q = (25{,}900 + 14{,}200 - 3832)/25{,}900 = 1.4$

The intercept of the q-line on the x-axis is shown from equation 11.46 to be x_f/q or:

$$y_q = \left(\frac{q}{q-1}\right)x_q - \left(\frac{x_f}{q-1}\right) \qquad \text{(equation 11.46)}$$

$$x_f/q = (0.274/1.4) = 0.196$$

Thus the q-line is drawn through (x_f, x_f) and $(0.196, 0)$ as shown in Figure 11c. As the reflux ratio is given as 3.16, the top operating line may be drawn through (x_d, x_d) and $(0, x_d/4.16)$. The lower operating line is drawn by joining the intersection of the top operating line and the q-line with the point (x_w, x_w).

The theoretical plates may be stepped off as shown and 9 theoretical plates are shown. If the plate efficiency is 70 per cent, the number of actual plates $= (9/0.7) = 12.85$,

Thus: $\qquad\qquad\qquad$ <u>13 plates are required</u>

PROBLEM 11.6

A batch fractionation is carried out in a small column which has the separating power of 6 theoretical plates. The mixture consists of benzene and toluene containing 0.60 mole fraction of benzene. A distillate is required, of constant composition, of 0.98 mole fraction benzene, and the operation is discontinued when 83 per cent of the benzene charged has been removed as distillate. Estimate the reflux ratio needed at the start and finish of the distillation, if the relative volatility of benzene to toluene is 2.46.

Solution

The equilibrium data are calculated from the relative volatility by the equation:

$$y_A = \frac{\alpha x_A}{1 + (\alpha - 1)x_A} \qquad \text{(equation 11.16)}$$

to give:

x_A	0	0.1	0.2	0.3	0.4	0.5	0.6	0.7	0.8	0.9	1.0
y_A	0	0.215	0.380	0.513	0.621	0.711	0.787	0.852	0.908	0.956	1.0

If a constant product is to be obtained from a batch still, the reflux ratio must be constantly increased. Initially S_1 kmol of liquor are in the still with a composition x_{s1} of the MVC and a reflux ratio of R_1 is required to give the desired product composition x_d. When S_2 kmol remain in the still of composition x_{s2}, the reflux ratio has increased to R_2 when the amount of product is D kmol.

From an overall mass balance: $\quad S_1 - S_2 = D$

For the MVC: $\quad S_1 x_{s1} - S_2 x_{s2} = D x_d$

from which:
$$D = S_1 \frac{(x_{s1} - x_{s2})}{(x_d - x_{s2})} \quad \text{(equation 11.98)}$$

In this problem, $x_{s1} = 0.6$ and $x_d = 0.98$ and there are 6 theoretical plates in the column. It remains, by using the equilibrium data, to determine values of x_{s2} for selected reflux ratios. This is done graphically by choosing an intercept on the y-axis, calculating R, drawing in the resulting operating line, and stepping off in the normal way 6 theoretical plates and finding the still composition x_{s2}.

This is shown in Figure 11d for two very different reflux ratios and the procedure is repeated to give the following table.

Intercept on y-axis (ϕ)	Reflux ratio ($\phi = x_d/R + 1$)	x_{s2}
0.45	1.18	0.725
0.40	1.20	0.665
0.30	2.27	0.545
0.20	3.90	0.46
0.10	8.8	0.31
0.05	18.6	0.26

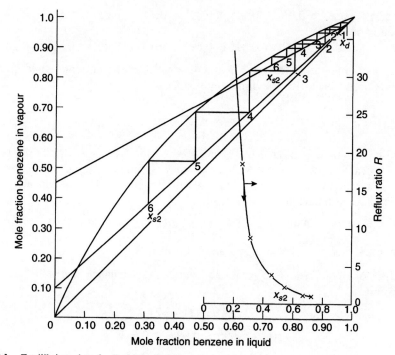

Figure 11d. Equilibrium data for Problem 11.6

From the inset plot of x_{s2} against R in Figure 11d:

At the start: $x_{s2} = 0.6$ and $\underline{R = 1.7}$.

At the end: x_{s2} is calculated using equation 11.98 as follows.

If $S_1 = 100$ kmol,

$$\text{kmol of benzene initially} = (100 \times 0.60) = 60 \text{ kmol.}$$

$$\text{kmol of benzene removed} = (0.83 \times 60) = 49.8 \text{ kmol.}$$

Thus: $\qquad D = (49.8/0.98) = 50.8$

and: $\qquad 50.8 = 100 \dfrac{(0.6 - x_{s2})}{(0.98 - x_{s2})}$

from which: $\qquad x_{s2} = 0.207 \quad \text{and} \quad \underline{R = 32}$

PROBLEM 11.7

A continuous fractionating column is required to separate a mixture containing 0.695 mole fraction n-heptane (C_7H_{16}) and 0.305 mole fraction n-octane (C_8H_{18}) into products of 99 mole per cent purity. The column is to operate at 101.3 kN/m² with a vapour velocity of 0.6 m/s. The feed is all liquid at its boiling-point, and this is supplied to the column at 1.25 kg/s. The boiling-point at the top of the column may be taken as 372 K, and the equilibrium data are:

mole fraction of heptane in vapour	0.96	0.91	0.83	0.74	0.65	0.50	0.37	0.24
mole fraction of heptane in liquid	0.92	0.82	0.69	0.57	0.46	0.32	0.22	0.13

Determine the minimum reflux ratio required. What diameter column would be required if the reflux used were twice the minimum possible?

Solution

The equilibrium curve is plotted in Figure 11e. As the feed is at its boiling-point, the q-line is vertical and the minimum reflux ratio may be found by joining the point (x_d, x_d) with the intersection of the q-line and the equilibrium curve. This line when produced to the y-axis gives an intercept of 0.475.

Thus: $\qquad 0.475 = x_D/(R_m + 1) \quad \text{and} \quad R_m = \underline{1.08}$

If $2R_m$ is used, then: $\qquad R = 2.16 \quad \text{and} \quad L_n/D = 2.16$

Taking 100 kmol of feed, as a basis, an overall mass balance and a balance for the n-heptane give:

$$100 = (D + W)$$

and: $\qquad 100 \times 0.695 = 0.99 D + 0.01 W$

since 99 per cent n-octane is required.

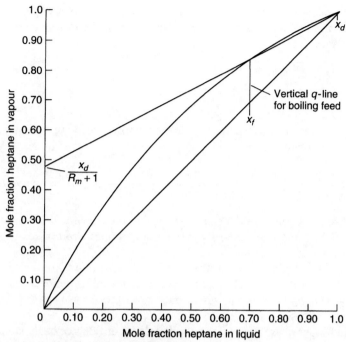

Figure 11e. Geometrical construction for Problem 11.7

Hence: $\quad D = 69.9 \quad$ and $\quad W = 30.1$

and: $\quad L_n = 2.16D = 151 \quad$ and $\quad V_n = L_n + D = 221$

The mean molecular mass of the feed $= (0.695 \times 100) + (0.305 \times 114) = 104.3$ kg/kmol.

Thus: $\quad$ feed rate $= (1.25/104.3) = 0.0120$ kmol/s

The vapour flow at the top of the column $= (221/100) \times 0.0120 = 0.0265$ kmol/s.
The vapour density at the top of the column $= (1/22.4)(273/372) = 0.0328$ kmol/m^3.
Hence the volumetric vapour flow $= (0.0265/0.0328) = 0.808$ m^3/s.
If the vapour velocity $= 0.6$ m/s, the area required $= (0.808/0.6) = 1.35$ m^2 equivalent to a column diameter of $[(4 \times 1.35)/\pi]^{0.5} = \underline{1.31 \text{ m}}$.

PROBLEM 11.8

The vapour pressures of chlorobenzene and water are:

Vapour pressure (kN/m^2)	13.3	6.7	4.0	2.7
(mm Hg)	100	50	30	20
Temperatures, (K)				
Chlorobenzene	343.6	326.9	315.9	307.7
Water	324.9	311.7	303.1	295.7

A still is operated at 18 kN/m² and steam is blown continuously into it. Estimate the temperature of the boiling liquid and the composition of the distillate if liquid water is present in the still.

Solution

For steam distillation, assuming the gas laws to apply, the composition of the vapour produced may be obtained from:

$$\left(\frac{m_A}{M_A}\right) \Big/ \left(\frac{m_B}{M_B}\right) = \frac{P_A}{P_B} = \frac{y_A}{y_B} = \frac{P_A}{(P - P_A)} \quad \text{(equation 11.120)}$$

where the subscript **A** refers to the component being recovered and **B** to steam, and m is the mass, M is the molecular mass, P_A and P_B are the partial pressures of **A** and **B** and P is the total pressure.

If there is no liquid phase present, then from the phase rule there will be two degrees of freedom. Thus both the total pressure and the operating temperature can be fixed independently, and $P_B = P - P_A$ (which must not exceed the vapour pressure of pure water if no liquid phase is to appear).

With a liquid water phase present, there will only be one degree of freedom, and setting the temperature or pressure fixes the system and the water and the other component each exert a partial pressure equal to its vapour pressure at the boiling-point of the mixture. In this case, the distillation temperature will always be less than that of boiling water at the total pressure in question. Consequently, a high-boiling organic material may be steam-distilled at temperatures below 373 K at atmospheric pressure. By using reduced operating pressures, the distillation temperature may be reduced still further, with a consequent economy of steam.

A convenient method of calculating the temperature and composition of the vapour, for the case where the liquid water phase is present, is by using Figure 11.47 in Volume 2 where the parameter $(P - P_B)$ is plotted for total pressures of 101, 40 and 9.3 kN/m² and the vapour pressures of a number of other materials are plotted directly against temperature. The intersection of the two appropriate curves gives the temperature of distillation and the molar ratio of water to organic material is given by $(P - P_A)/P_A$.

The relevance of the method to this problem is illustrated in Figure 11f where the vapour pressure of chlorobenzene is plotted as a function of temperature. On the same graph $(P - P_B)$ is plotted where $P = 18$ kN/m² (130 mm Hg) and P_B is the vapour pressure of water at the particular temperature. These curves are seen to intersect at the distillation temperature of <u>323 K</u>.

The composition of the distillate is found by substitution in equation 11.120 since $P_A = 5.5$ kN/m² (41 mmHg) at 323 K.

Hence: $\quad \dfrac{y_A}{y_B} = \dfrac{P_A}{(P - P_A)} = \dfrac{5.5}{(18 - 5.5)} = \underline{\underline{0.44}}$

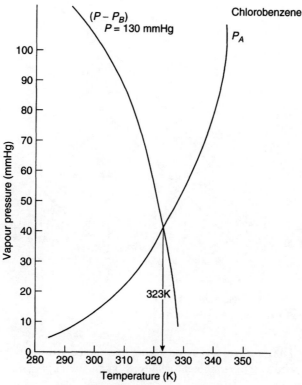

Figure 11f. Vapour pressure as a function of temperature, Problem 11.8

PROBLEM 11.9

The following values represent the equilibrium conditions in terms of mole fraction of benzene in benzene–toluene mixtures at their boiling-point:

Liquid	0.521	0.38	0.26	0.15
Vapour	0.72	0.60	0.45	0.30

If the liquid compositions on four adjacent plates in a column were 0.18, 0.28, 0.41 and 0.57 under conditions of total reflux, determine the plate efficiencies.

Solution

The equilibrium data are plotted in Figure 11g over the range given and a graphical representation of the plate efficiency is shown in the inset. The efficiency E_{Ml} in terms of the liquid compositions is defined by:

$$E_{Ml} = \frac{(x_{n+1} - x_n)}{(x_{n+1} - x_e)} \quad \text{(equation 11.125)}$$

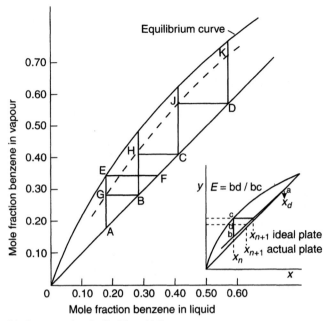

Figure 11g. Graphical construction for Problem 11.9

In the inset, the line ab represents an operating line and bc is the enrichment achieved on a theoretical plate. bd is the enrichment achieved on an actual plate so that the efficiency is then the ratio ba/bc.

Referring to the data given, at total reflux, the conditions on actual plates in the column are shown as points A, B, C, and D. Considering point A, if equilibrium were achieved on that plate, point E would represent the vapour composition and point F the liquid composition on the next plate. The liquid on the next plate is determined by B however so that the line AGE may be located and the efficiency is given by AG/AE = 0.59 or 59 per cent

In an exactly similar way, points H, J, and K are located to give efficiencies of 66 per cent, 74 per cent, and 77 per cent.

PROBLEM 11.10

A continuous rectifying column handles a mixture consisting of 40 per cent of benzene by mass and 60 per cent of toluene at the rate of 4 kg/s, and separates it into a product containing 97 per cent of benzene and a liquid containing 98 per cent toluene. The feed is liquid at its boiling-point.

(a) Calculate the mass flows of distillate and waste liquor.
(b) If a reflux ratio of 3.5 is employed, how many plates are required in the rectifying part of the column?

(c) What is the actual number of plates if the plate-efficiency is 60 per cent?

Mole fraction of benzene in liquid	0.1	0.2	0.3	0.4	0.5	0.6	0.7	0.8	0.9
Mole fraction of benzene in vapour	0.22	0.38	0.51	0.63	0.7	0.78	0.85	0.91	0.96

Solution

The equilibrium data are plotted in Figure 11h. As the compositions are given as mass per cent, these must first be converted to mole fractions before the McCabe–Thiele method may be used.

Mole fraction of benzene in feed, $x_f = \dfrac{(40/78)}{(40/78) + (60/92)} = 0.440$

Similarly: $x_d = 0.974$ and $x_w = 0.024$

As the feed is a liquid at its boiling-point, the q-line is vertical and may be drawn at $x_f = 0.44$.

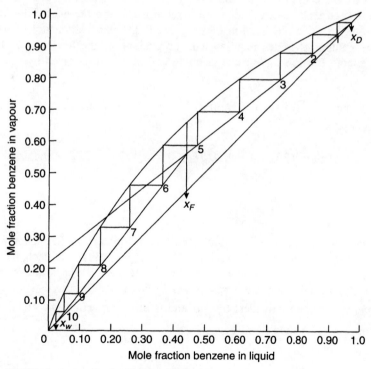

Figure 11h. Graphical construction for Problem 11.10

(a) A mass balance over the column and on the more volatile component in terms of the mass flow rates gives:

$$4.0 = W' + D'$$

$$(4 \times 0.4) = 0.02W' + 0.97D'$$

from which: bottoms flowrate, $W' = \underline{\underline{2.4 \text{ kg/s}}}$

and: top product rate, $D' = \underline{\underline{1.6 \text{ kg/s}}}$

(b) If $R = 3.5$, the intercept of the top operating line on the y-axis is given by $x_d/(R+1) = (0.974/4.5) = 0.216$, and thus the operating lines may be drawn as shown in Figure 11h. The plates are stepped off as shown and $\underline{\underline{10 \text{ theoretical plates}}}$ are required.

(c) If the efficiency is 60 per cent, the number of actual plates $= (10/0.6)$

$$= 16.7 \text{ or } \underline{\underline{17 \text{ actual plates}}}$$

PROBLEM 11.11

A distillation column is fed with a mixture of benzene and toluene, in which the mole fraction of benzene is 0.35. The column is to yield a product in which the mole fraction of benzene is 0.95, when working with a reflux ratio of 3.2, and the waste from the column is not to exceed 0.05 mole fraction of benzene. If the plate efficiency is 60 per cent, estimate the number of plates required and the position of the feed point. The relation between the mole fraction of benzene in liquid and in vapour is given by:

Mole fraction of benzene in liquid	0.1	0.2	0.3	0.4	0.5	0.6	0.7	0.8	0.9
Mole fraction of benzene in vapour	0.20	0.38	0.51	0.63	0.71	0.78	0.85	0.91	0.96

Solution

The solution to this problem is very similar to that of Problem 11.10 except that the data are presented here in terms of mole fractions. Following a similar approach, the theoretical plates are stepped off and it is seen from Figure 11i that 10 plates are required. Thus $(10/0.6) = 16.7$ actual plates are required and $\underline{\underline{17}}$ would be employed.

The feed tray lies between ideal trays 5 and 6, and in practice, the $\underline{\underline{\text{eighth actual tray}}}$ from the top would be used.

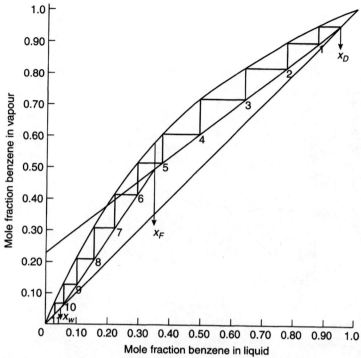

Figure 11i. Graphical construction for Problem 11.11

PROBLEM 11.12

The relationship between the mole fraction of carbon disulphide in the liquid and in the vapour during the distillation of a carbon disulphide–carbon tetrachloride mixture is:

x	0	0.20	0.40	0.60	0.80	1.00
y	0	0.445	0.65	0.795	0.91	1.00

Determine graphically the theoretical number of plates required for the rectifying and stripping portions of the column.

The reflux ratio = 3, the slope of the fractionating line = 1.4, the purity of product = 99 per cent, and the concentration of carbon disulphide in the waste liquors = 1 per cent.

What is the minimum slope of the rectifying line in this case?

Solution

The equilibrium data are plotted in Figure 11j. In this problem, no data are provided on the composition or the nature of the feed so that conventional location of the q-line

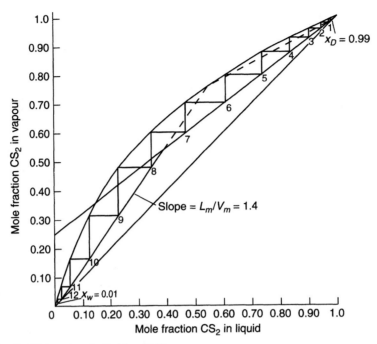

Figure 11j. Equilibrium data for Problem 11.12

is impossible. The rectifying line may be drawn, however, as both the top composition and the reflux ratio are known. The intercept on the y-axis is given by $x_d/(R+1) = (0.99/4) = 0.248$.

The slope of the lower operating line is given as 1.4. Thus the line may be drawn through the point (x_w, x_w) and the number of theoretical plates may be determined, as shown, as <u>12</u>.

The minimum slope of the rectifying line corresponds to an infinite number of theoretical stages. If the slope of the stripping line remains constant, then production of that line to the equilibrium curve enables the rectifying line to be drawn as shown dotted in Figure 11j.

The slope of this line may be measured to give $\underline{L_n/V_n = 0.51}$

PROBLEM 11.13

A fractionating column is required to distill a liquid containing 25 per cent benzene and 75 per cent toluene by mass, to give a product of 90 per cent benzene. A reflux ratio of 3.5 is to be used, and the feed will enter at its boiling point. If the plates used are 100 per cent efficient, calculate by the Lewis–Sorel method the composition of liquid on the third plate, and estimate the number of plates required using the McCabe–Thiele method.

Solution

The equilibrium data for this problem are plotted as Figure 11k. Converting mass per cent to mole fraction gives $x_f = 0.282$ and $x_d = 0.913$. There are no data given on the bottom product so that usual mass balances cannot be applied. The equation of the top operating line is:

$$y_n = \frac{L_n}{V_n} x_{n+1} + \frac{D}{V_n} x_d \qquad \text{(equation 11.35)}$$

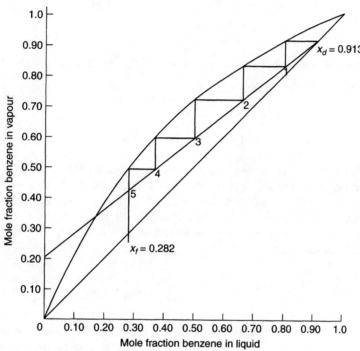

Figure 11k. Equilibrium data for Problem 11.13

L_n/V_n is the slope of the top operating line which passes through the points (x_d, x_d) and $0, x_d/(R+1)$. This line is drawn in Figure 11k and its slope measured or calculated as 0.78. The reflux ratio which is equal to L_n/D is given as 3.5, so that D/V_n may then be found since:

$$\frac{D}{V_n} = \frac{D}{L_n} = \left(\frac{1}{3.5} \times 0.78\right) = 0.22$$

Thus: $\qquad y_n = 0.78_{n+1} + 0.22 x_d = 0.78_{n+1} + 0.20$

The composition of the vapour y_t leaving the top plate must be the same as the top product x_d since all the vapour is condensed. The composition of the liquid on the top plate x_t is found from the equilibrium curve since it is in equilibrium with vapour of composition $y_t = x_d = 0.913$.

Thus: $\qquad x_t = 0.805$

The composition of the vapour rising to the top plate y_{t-1} is found from the equation of the operating line. That is:

$$y_{t-1} = (0.78 \times 0.805) + 0.20 = 0.828$$

x_{t-1} is in equilibrium with y_{t-1} and is found to be 0.66 from the equilibrium curve.

Then: $\qquad\qquad\qquad y_{t-2} = (0.78 \times 0.66) + 0.20 = 0.715$

Similarly: $\qquad\qquad\quad x_{t-2} = 0.50$

and: $\qquad\qquad\qquad\; y_{t-3} = (0.78 \times 0.50) + 0.20 = 0.557$

and: $\qquad\qquad\qquad\; \underline{\underline{x_{t-3} = 0.335}}$

The McCabe–Thiele construction in Figure 11k shows that $\underline{\underline{5 \text{ theoretical plates}}}$ are required in the rectifying section.

PROBLEM 11.14

A 50 mole per cent mixture of benzene and toluene is fractionated in a batch still which has the separating power of 8 theoretical plates. It is proposed to obtain a constant quality product containing 95 mole per cent benzene, and to continue the distillation until the still has a content of 10 mole per cent benzene. What will be the range of reflux ratios used in the process? Show graphically the relation between the required reflux ratio and the amount of distillate removed.

Solution

If a constant product is to be obtained from a batch still, the reflux ratio must be constantly increased. Initially S_1 kmol of liquor is in the still with a composition x_{s_1} of the MVC and a reflux ratio of R_1 is required to give the desired product composition x_d. When S_2 kmol remain in the still of composition x_{s_2}, the amount of product is D kmol and the reflux ratio has increased to R_2.

From an overall mass balance: $\quad (S_1 - S_2) = D$

For the MVC: $\qquad\qquad\qquad S_1 x_{s_1} - S_2 x_{s_2} = D x_d$

from which: $\qquad\qquad\qquad D = S_1 \dfrac{(x_{s_1} - x_{s_2})}{(x_d - x_{s_2})} \qquad\qquad$ (equation 11.98)

In this problem, $x_{s_1} = 0.5$ and $x_d = 0.95$ and there are 8 theoretical plates in the column. It remains, by using the equilibrium data, to determine values of x_{s_2} for selected reflux ratios. This is done graphically by choosing an intercept on the y-axis, calculating R, drawing in the resulting operating line, and stepping off in the usual way 8 theoretical plates and finding the still composition x_{s_2} and hence D. The results of this process are as follows for $S_1 = 100$ kmol.

$\phi = x_d/(R+1)$	R	x_{s_2}	D
0.4	1.375	0.48	4.2
0.35	1.71	0.405	17.3
0.30	2.17	0.335	26.8
0.25	2.80	0.265	34.3
0.20	3.75	0.195	40.3
0.15	5.33	0.130	45.1
0.10	8.50	0.090	47.7

The initial and final values of R are most easily determined by plotting R against x_{s_2} as shown in Figure 11l. The initial value of R corresponds to the initial still composition of 0.50 and is seen to be 1.3 and, at the end of the process when $x_{s_2} = 0.1$, $R = 7.0$.

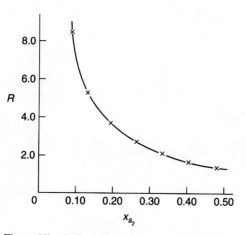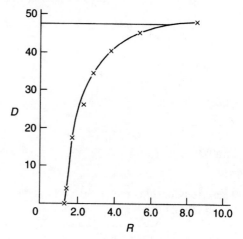

Figure 11l. Reflux ratio data for Problem 11.14

Figure 11l includes a plot of reflux ratio against the quantity of distillate. When $R = 7.0$, $D = 47.1$ kmol/100 kmol charged initially.

PROBLEM 11.15

The vapour composition on a plate of a distillation column is:

	C_1	C_2	$i - C_3$	$n - C_3$	$i - C_4$	$n - C_4$
mole fraction	0.025	0.205	0.210	0.465	0.045	0.050
relative volatility	36.5	7.4	3.0	2.7	1.3	1.0

What will be the composition of the liquid on the plate of it is in equilibrium with the vapour?

Solution

In a mixture of **A, B, C, D**, and so on, if the mole fractions in the liquid are x_A, x_B, x_C, x_D, and so on, and in the vapour $y_A, y_B, y_C,$ and y_D, then:

$$x_A + x_B + x_C + \cdots = 1$$

Thus:
$$\frac{x_A}{x_B} + \frac{x_B}{x_B} + \frac{x_C}{x_B} + \cdots = \frac{1}{x_B}$$

But:
$$\frac{x_A}{x_B} = \frac{y_A}{y_B \alpha_{AB}}$$

Thus:
$$\frac{y_A}{y_B \alpha_{AB}} + \frac{y_B}{y_B \alpha_{BB}} + \frac{y_C}{y_B \alpha_{CB}} + \cdots = \frac{1}{x_B}$$

or:
$$\sum \left(\frac{y_A}{\alpha_{AB}}\right) = \frac{y_B}{x_B}$$

But:
$$y_B = \frac{y_A x_B}{\alpha_{AB} x_A}$$

and substituting:
$$\sum \left(\frac{y_A}{\alpha_{AB}}\right) = \frac{y_A x_B}{\alpha_{AB} x_A x_B}$$

Thus
$$x_A = \frac{(y_A/\alpha_{AB})}{\sum(y_A/\alpha_{AB})}$$

Similarly:
$$x_B = \frac{(y_B/\alpha_{BB})}{\sum(y_A/\alpha_{AB})} \quad \text{and} \quad x_C = \frac{(y_C/\alpha_{BC})}{\sum(y_A/\alpha_{AB})}$$

These relationships may be used to solve this problem and the calculation is best carried out in tabular form as follows.

Component	y_i	α	y_i/α	$x_i = (y_i/\alpha)/\sum(y_i/\alpha)$
C_1	0.025	36.5	0.00068	0.002
C_2	0.205	7.4	0.0277	0.078
$i - C_3$	0.210	3.0	0.070	0.197
$n - C_3$	0.465	2.7	0.1722	0.485
$i - C_4$	0.045	1.3	0.0346	0.097
$n - C_4$	0.050	1.0	0.050	0.141
			$\sum(y_i/\alpha) = 0.355$	1.000

PROBLEM 11.16

A liquor of 0.30 mole fraction of benzene and the rest toluene is fed to a continuous still to give a top product of 0.90 mole fraction benzene and a bottom product of 0.95 mole fraction toluene.

If the reflux ratio is 5.0, how many plates are required:

(a) if the feed is saturated vapour?
(b) if the feed is liquid at 283 K?

Solution

In this problem, the q-lines have two widely differing slopes and the effect of the feed condition is to alter the number of theoretical stages as shown in Figure 11m.

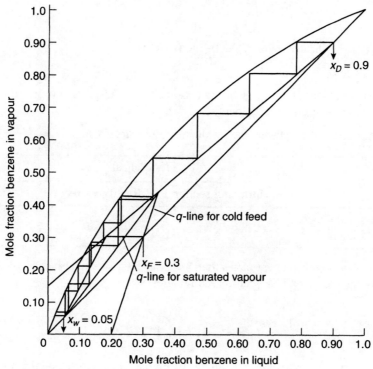

Figure 11m. Equilibrium data for Problem 11.16

$$q = \frac{\lambda + H_{fs} - H_f}{\lambda}$$

where λ is the molar latent heat of vaporisation, H_{fs} is the molar enthalpy of the feed at its boiling-point, and H_f is the molar enthalpy of the feed.

For benzene and toluene: $\quad\quad\quad\quad \lambda = 30$ MJ/kmol

and: $\quad\quad\quad\quad$ specific heat capacity $= 1.84$ kJ/kg K.

The boiling-points of benzene and toluene are 353.3 and 383.8 K respectively.

(a) If the feed is a saturated vapour, $q = 0$.
(b) If the feed is a cold liquid at 283 K, the mean molecular mass is:

$$(0.3 \times 78) + (0.7 \times 92) = 87.8 \text{ kg/kmol}$$

and the mean boiling-point $= (0.3 \times 353.3) \times (0.7 \times 383.8) = 374.7$ K.

Using a datum of 273 K:

$$H_{fs} = 1.84 \times 87.8(374.7 - 273) = 16{,}425 \text{ kJ/kmol} \quad \text{or} \quad 16.43 \text{ MJ/kmol}$$

$$H_f = 1.84 \times 87.8(283 - 273) = 1615 \text{ kJ/kmol} \quad \text{or} \quad 1.615 \text{ MJ/kmol}$$

Thus: $\quad q = (30 + 16.43 - 1.615)/30 = 1.49$.

From equation 11.46, the slope of the q-line is $q/(q-1)$.

Hence the slope $= (1.49/0.49) = 3.05$.

Thus for (a) and (b) the slope of the q-line is zero and 3.05 respectively, and in Figure 11m these lines are drawn through the point (x_f, x_f).

By stepping off the ideal stages, the following results are obtained:

Feed	Theoretical plates		
	Stripping section	Rectifying section	Total
Saturated vapour	4	5	9
Cold liquid	4	3	7

Thus a cold feed requires fewer plates than a vapour feed although the capital cost saving is offset by the increased heat load on the reboiler.

PROBLEM 11.17

A mixture of alcohol and water containing 0.45 mole fraction of alcohol is to be continuously distilled in a column to give a top product of 0.825 mole fraction alcohol and a liquor at the bottom containing 0.05 mole fraction alcohol. How many theoretical plates are required if the reflux ratio used is 3? Indicate on a diagram what is meant by the Murphree plate efficiency.

Solution

This example is solved by a simple application of the McCabe–Thiele method and is illustrated in Figure 11n, where it is seen that 10 theoretical plates are required. The Murphree plate efficiency is discussed in the solution to Problem 11.9.

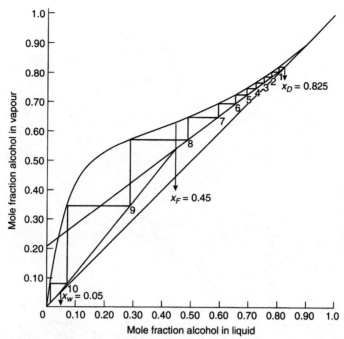

Figure 11n. Graphical construction for Problem 11.17

PROBLEM 11.18

It is desired to separate 1 kg/s of an ammonia solution containing 30 per cent NH_3 by mass into 99.5 per cent liquid NH_3 and a residual weak solution containing 10 per cent NH_3. Assuming the feed to be at its boiling point, a column pressure of 1013 kN/m^2, a plate efficiency of 60 per cent and that an 8 per cent excess over the minimum reflux requirements is used, how many plates must be used in the column and how much heat is removed in the condenser and added in the boiler?

Solution

Taking a material balance for the whole throughput and for the ammonia gives:

$$D' + W' = 1.0$$

and: $(0.995 D' + 0.1 W') = (1.0 \times 0.3)$

Thus: $D' = 0.22$ kg/s and $W' = 0.78$ kg/s

The enthalpy-composition chart for this system is shown in Figure 11o. It is assumed that the feed F and the bottom product W are liquids at their boiling-points.

Location of the poles N *and* M
N_m for minimum reflux is found by drawing a tie line through F, representing the feed, to cut the line $x = 0.995$ at N_m.

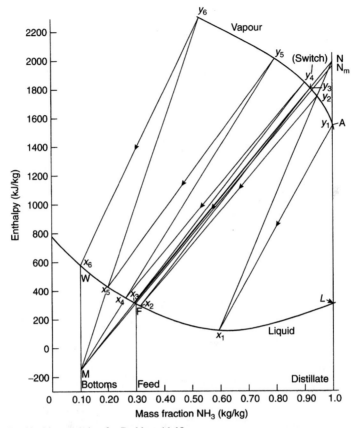

Figure 11o. Graphical construction for Problem 11.18

The minimum reflux ratio is given by:

$$R_m = \frac{\text{length } N_m A}{\text{length } AL} = \frac{(1952 - 1547)}{(1547 - 295)} = 0.323$$

Since the actual reflux is 8 per cent above the minimum,

$$NA = 1.08 N_m A = (1.08 \times 405) = 437$$

Point N therefore has an ordinate $(437 + 1547) = 1984$ and abscissa 0.995.
Point M is found by drawing NF to cut the line $x = 0.10$, through W, at M.
The number of theoretical plates is found, as on the diagram, to be 5+.
The number of plates to be provided $= (5/0.6) = 8.33$, say 9.
The feed is introduced just below the third ideal plate from the top, or just below the fifth actual plate.
The heat input at the boiler per unit mass of bottom product is given by:

$$Q_B/W = 582 - (-209) = 791 \text{ kJ/kg (from equation 11.88)}.$$

The heat input to boiler $= (791 \times 0.78) = \underline{\underline{617 \text{ kW}}}$.

The condenser duty = length NL × D

$$= (1984 - 296) \times 0.22 = \underline{372 \text{ kW}}.$$

PROBLEM 11.19

A mixture of 60 mole per cent benzene, 30 per cent of toluene and 10 per cent xylene is handled in a batch still. If the top product is to be 99 per cent benzene, determine:

(a) the liquid composition on each plate at total reflux,
(b) the composition on the 2nd and 4th plates for $R = 1.5$,
(c) as for (b) but $R = 5$,
(d) as for (c) but $R = 8$ and for the condition when the mol per cent benzene in the still is 10,
(e) as for (d) but with $R = 5$.

The relative volatility of benzene to toluene may be taken as 2.4, and the relative volatility of xylene to toluene as 0.43.

Solution

Although this problem is one of multicomponent batch distillation, the product remains of constant composition so that normal methods can be used for plate-to-plate calculations at a point value of the varying reflux ratio.

(a) At total reflux, for each component the operating line is:

$$y_n = x_{n+1}$$

Also:
$$y = \alpha x \Big/ \sum (\alpha x)$$

The solution is given in tabular form as:

	α	x_s	αx_s	$y_s = x_1$	αx_1	$y_1 = x_2$	αx_2	$y_2 = x_3$	Similarly x_4	x_5
B	2.4	0.60	1.44	0.808	1.164	0.867	2.081	0.942	0.975	0.989
T	1.0	0.30	0.30	0.168	0.168	0.125	0.125	0.057	0.025	0.011
X	0.43	0.10	0.043	0.024	0.010	0.008	0.003	0.001	—	—
			1.783	1.000	1.342	1.000	2.209	1.000	1.000	1.000

(b) The operating line for the rectifying section is:

$$y_n = \frac{L_n}{V_n} x_{n+1} + \frac{D}{V_n} x_d$$

$$R = L_n/D \quad \text{and} \quad V = L_n + D$$

Thus:
$$y_n = \frac{R}{R+1} x_{n+1} + \frac{x_d}{R+1}$$

If $R = 1.5$: for benzene, $y_{nb} = 0.6x_{n+1} + 0.396$,

toluene, $y_{nt} = 0.6x_{n+1} + 0.004$, and xylene, $y_{nx} = 0.6x_{n+1}$

The liquid composition on each plate is then found from these operating lines.

	y_s	x_1	αx_1	y_1	x_2	αx_2	y_2	x_3	αx_3	y_3	x_4
B	0.808	0.687	1.649	0.850	0.757	1.817	0.848	0.754	1.810	0.899	0.838
T	0.168	0.273	0.273	0.141	0.228	0.228	0.107	0.171	0.171	0.085	0.135
X	0.024	0.040	0.017	0.009	0.015	0.098	0.045	0.075	0.032	0.016	0.027
	1.000	1.000	1.939	1.000	1.000	2.143	1.000	1.000	2.013	1.000	1.000

(c) If $R = 5$, the operating line equations become:

$$y_{nb} = 0.833x_{n+1} + 0.165$$

$$y_{nt} = 0.833x_{n+1} + 0.0017$$

$$y_{nx} = 0.833x_{n+1}$$

	y_s	x_1	αx_1	y_1	x_2	αx_2	y_2	x_3	αx_3	y_3	x_4
B	0.808	0.772	1.853	0.897	0.879	2.110	0.947	0.939	2.254	0.974	0.971
T	0.168	0.200	0.200	0.097	0.114	0.114	0.051	0.059	0.059	0.025	0.028
X	0.024	0.028	0.012	0.006	0.007	0.003	0.002	0.002	0.001	0.001	0.001
	1.000	1.000	2.065	1.000	1.000	2.227	1.000	1.000	2.314	1.000	1.000

(d) When the benzene content is 10 per cent in the still, a mass balance gives the kmol of distillate removed, assuming 100 kmol initially, as:

$$D = 100(0.6 - 0.1)/(0.99 - 0.1) = 56.2 \text{ kmol}$$

Thus 43.8 kmol remain of which 4.38 are benzene, $x_b = 0.10$

29.42 are toluene, $x_t = 0.67$

and: 10.00 are xylene, $x_x = 0.23$

If $R = 8$, the operating lines become:

$$y_{nb} = 0.889x_{n+1} + 0.11, \quad y_{nt} = 0.889x_{n+1} + 0.001 \quad \text{and} \quad y_{nx} = 0.889x_{n+1}$$

	x_s	αx_s	y_s	x_1	αx_1	y_1	x_2	αx_2	y_2	x_3	αx_3	y_3	x_4
B	0.10	0.24	0.24	0.146	0.350	0.307	0.222	0.533	0.415	0.343	0.823	0.560	0.506
T	0.67	0.67	0.66	0.741	0.741	0.650	0.730	0.730	0.569	0.639	0.639	0.435	0.488
X	0.23	0.10	0.10	0.113	0.049	0.043	0.048	0.021	0.016	0.018	0.008	0.005	0.006
		1.01	1.00	1.000	1.140	1.000	1.000	1.284	1.000	1.000	1.470	1.000	1.000

(e) Exactly the same procedure is repeated for this part of the question, when the operating lines become:

$$y_{nb} = 0.833x_{n+1} + 0.165, \quad y_{nt} = 0.833x_{n+1} + 0.0017 \quad \text{and} \quad y_{nx} = 0.833x_{n+1}$$

PROBLEM 11.20

A continuous still is fed with a mixture of 0.5 mole fraction of the more volatile component, and gives a top product of 0.9 mole fraction of the more volatile component and a bottom product containing 0.10 mole fraction.

If the still operates with an L_n/D ratio of 3.5:1, calculate by Sorel's method the composition of the liquid on the third theoretical plate from the top:

(a) for benzene–toluene, and
(b) for n-heptane–toluene.

Solution

A series of mass balances as described in other problems enables the flows within the column to be calculated as follows.

For a basis of 100 kmol of feed and a reflux ratio of 3.5:

$$D = 50, \quad L_n = 175, \quad V_n = 225 \text{ kmol}$$

The top operating line equation is then:

$$y_n = 0.778x_{n+1} + 0.20$$

(a) Use is made of the equilibrium data from other examples involving benzene and toluene.

The vapour leaving the top plate has the same composition as the top product, or $y_t = 0.9$. From the equilibrium data, $x_t = 0.78$.

Thus: $\quad y_{t-1} = (0.778 \times 0.78) + 0.20 = 0.807$

and x_{t-1}, from equilibrium data $= 0.640$

Similarly: $\quad y_{t-2} = (0.778 \times 0.640) + 0.20 = 0.698 \quad$ and $\quad x_{t-2} = 0.49$

$y_{t-3} = (0.778 \times 0.490) + 0.20 = 0.581 \quad$ and $\quad \underline{\underline{x_{t-3} = 0.36}}$

(b) Vapour pressure data from Perry[1] for n-heptane and toluene are plotted in Figure 11p. These data may be used to calculate an mean value of the relative volatility α from:

$$\alpha = P_H^0 / P_T^0$$

---

[1] PERRY, R. H., GREEN, D. W. and MALONEY, J. O.: *Perry's Chemical Engineers' Handbook*, 6th edn. (McGraw-Hill, New York, 1987).

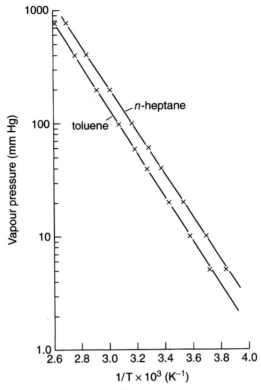

Figure 11p. Vapour pressure data for Problem 11.20

T	$10^3/T$	P_H^0	P_T^0	α	
370	2.7	730	510	1.43	
333	3.0	200	135	1.48	
303	3.3	55	35	1.57	Mean $\alpha = 1.52$
278	3.6	15	9.4	1.60	

As an alternative to drawing the equilibrium curve for the system, point values may be calculated from:

$$x = \frac{y}{\alpha - (\alpha - 1)y} \quad \text{(equation 11.16)}$$

Thus: $y_t = 0.9$ and $x_t = 0.9/(1.52 - 0.52 \times 0.9) = 0.856$

From the same operating line equation:

$$y_{t-1} = (0.778 \times 0.856) + 0.20 = 0.865$$

Similarly: $x_{t-1} = 0.808$, $y_{t-2} = 0.829$, $x_{t-2} = 0.761$,

$y_{t-3} = 0.792$ and $\underline{\underline{x_{t-3} = 0.715}}$

PROBLEM 11.21

A mixture of 40 mole per cent benzene with toluene is distilled in a column to give a product of 95 mole per cent benzene and a waste of 5 mole per cent benzene, using a reflux ratio of 4.

(a) Calculate by Sorel's method the composition on the second plate from the top.
(b) Using the McCabe and Thiele method, determine the number of plates required and the position of the feed if supplied to the column as liquid at the boiling-point.
(c) Find the minimum reflux ratio possible.
(d) Find the minimum number of plates.
(e) If the feed is passed in at 288 K find the number of plates required using the same reflux ratio.

Solution

The equilibrium data for benzene and toluene are plotted in Figure 11q.

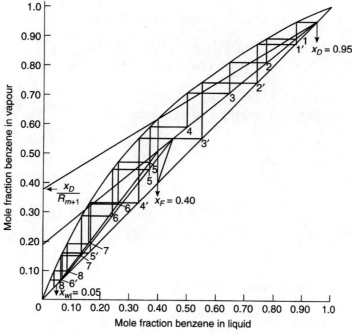

Figure 11q. Equilibrium data for Problem 11.21

$$x_f = 0.40, \quad x_d = 0.95, \quad x_w = 0.05, \quad \text{and} \quad R = 4.0$$

and mass balances may be carried out to calculate the operating line equations. Taking a basis of 100 kmol, then:

$$100 = D + W$$

and: $(100 \times 0.4) = 0.95D + 0.05W$

from which: $D = 38.9$ and $W = 61.1$ kmol

$L_n/D = 4$ so that $L_n = 155.6$ kmol

$V_n = L_n + D = 194.5$ kmol

(a) The top operating line equation is:

$$y_n = \frac{L_n}{V_n}x_{n+1} + \frac{D}{V_n}x_d \qquad \text{(equation 11.35)}$$

or: $y_n = (155.6/194.5)x_{n+1} + (38.9/194.5) \times 0.95$

$y_n = 0.8x_{n+1} + 0.19$

Vapour y_t leaving the top tray has a composition, $x_d = 0.95$ so that x_t, the liquid composition on the top tray, is found from the equilibrium curve to be equal to 0.88. y_{t-1} is found from the operating line equation as:

$y_{t-1} = (0.8 \times 0.88) + 0.19 = 0.894$

$x_{t-1} = 0.775$, from the equilibrium curve.

Thus: $y_{t-2} = (0.8 \times 0.775) + 0.19 = 0.810$

$x_{t-2} = \underline{0.645}$, from the equilibrium curve.

(b) The steps in the McCabe–Thiele determination are shown in Figure 11q where 8 theoretical plates are required with a boiling liquid feed.

(c) The minimum reflux ratio corresponds to an infinite number of plates. This condition occurs when the top operating line passes through the intersection of the q-line and the equilibrium curve. This line is seen to give an intersection on the y-axis equal to 0.375.

Thus: $0.375 = x_d/(R_M + 1)$ and $\underline{R_M = 1.53}$

(d) The minimum number of plates occurs at total reflux and may be determined by stepping between the equilibrium curve and the diagonal $y = x$ to give 6 theoretical plates as shown.

Alternatively, Fenske's equation may be used.

Thus: $n + 1 = \log \dfrac{[(x_A/x_B)_d(x_B/x_A)_s]}{\log \alpha_{AB}}$ (equation 11.58)

$= \dfrac{\log(0.95/0.05)(0.95/0.05)}{\log 2.4}$ and $n = 5.7$ or $\underline{6 \text{ plates}}$

(e) If a cold feed is introduced, the q-line is no longer vertical. The slope of the line may be calculated as shown in Problem 11.16. In this problem, q is found to be 1.45 and the q-line has a slope of 3.22. This line is shown in Figure 11q and the number of theoretical plates is found to be unchanged at $\underline{8}$.

PROBLEM 11.22

Determine the minimum reflux ratio using Fenske's equation and Colburn's rigorous method for the following three systems:

(a) 0.60 mole fraction C_6, 0.30 mole fraction C_7, and 0.10 mole fraction C_8 to give a product of 0.99 mole fraction C_6.

			Mole fraction	Relative volatility, α	x_d
(b) Components	A		0.3	2	1.0
	B		0.3	1	—
	C		0.4	0.5	—
(c) Components	A		0.25	2	1.0
	B		0.25	1	—
	C		0.25	0.5	—
	D		0.25	0.25	—

Solution

(a) Under conditions where the relative volatility remains constant, Underwood developed the following equations from which R_m may be calculated:

$$\frac{\alpha_A x_{fA}}{\alpha_A - \theta} + \frac{\alpha_B x_{fB}}{\alpha_B - \theta} + \frac{\alpha_C x_{fC}}{\alpha_C - \theta} + \cdots = 1 - q \quad \text{(equation 11.114)}$$

and:

$$\frac{\alpha_A x_{dA}}{\alpha_A - \theta} + \frac{\alpha_B x_{dB}}{\alpha_B - \theta} + \frac{\alpha_C x_{dC}}{\alpha_C - \theta} + \cdots = R_m + 1 \quad \text{(equation 11.115)}$$

where $x_{fA}, x_{fB}, x_{dA}, x_{dB}$, and so on, are the mole fractions of components **A** and **B**, and so on, in the feed and distillate with **A** the light key and **B** the heavy key.

$\alpha_A, \alpha_B, \alpha_C$, etc., are the volatilities with respect to the least volatile component.

θ is the root of equation 11.114 and lies between the values of α_A and α_B. Thus θ may be calculated from equation 11.114 and substituted into 11.115 to give R_m.

(b) Colburn's method allows the value of R_m to be calculated from approximate values of the pinch compositions of the key components. This value may then be checked again empirical relationships as shown in Example 11.15 in Volume 2.

The method is long and tedious and only the first approximation will be worked here.

$$R_m = \frac{1}{\alpha_{AB} - 1} \left[\left(\frac{x_{dA}}{x_{nA}}\right) - \alpha_{AB} \left(\frac{x_{dB}}{x_{nB}}\right) \right] \quad \text{(equation 11.108)}$$

where x_{dA} and x_{nA} are the top and pinch compositions of the light key component, x_{dB} and x_{nB} are the top and pinch compositions of the heavy key component, and α_{AB} is the volatility of the light key relative to the heavy key component.

The difficulty in using this equation is that the values of x_{nA} and x_{nB} are known only for special cases where the pinch coincides with the feed composition. Colburn has suggested that an approximate value for x_{nA} is given by:

$$x_{nA} \text{ (approx.)} = \frac{r_f}{(1+r_f)(1+\Sigma \alpha x_{fh})} \qquad \text{(equation 11.109)}$$

and:
$$x_{nB} \text{ (approx.)} = \frac{x_{nA}}{r_f} \qquad \text{(equation 11.110)}$$

where r_f is the estimated ratio of the key components on the feed plate.

For an all liquid feed at its boiling-point, r_f is equal to the ratio of the key components in the feed. Otherwise r_f is the ratio of the key components in the liquid part of the feed, x_{fh} is the mole fraction of each component in the liquid portion of feed heavier than the heavy key, and α is the volatility of the component relative to the heavy key.

(a) Relative volatility data are required and it will be assumed that $\alpha_{6,8} = 5$, $\alpha_{6,7} = 2.5$, and also that $x_{d7} = 0.01$ and that $q = 1$.

Substituting in Underwood's equation gives:

$$\left(\frac{5 \times 0.60}{5-\theta}\right) + \left(\frac{2.5 \times 0.30}{2.5-\theta}\right) + \left(\frac{0.1 \times 1}{1-\theta}\right) = 1 - q = 0$$

from which by trial and error, $\theta = 3.1$.

Then:
$$\left(\frac{5 \times 0.99}{5-3.1}\right) + \left(\frac{2.5 \times 0.01}{2.5-3.1}\right) = R_m + 1 \quad \text{and} \quad \underline{\underline{R_m = 1.57}}$$

In Colburn's equation, using C_6 and C_7 as the light and heavy keys respectively:

$$r_f = (0.6/0.3) = 2.0$$

$$\Sigma \alpha x_{fh} = (1/2.5) \times 0.1 = 0.04, \quad x_{nA} = 2/(3 \times 1.04) = 0.641$$

$$\alpha_{AB} = (5.0/2.5) = 2.0, \quad x_{nB} = (0.641/2) = 0.32$$

$$R_m = \frac{1}{1}\left[\left(\frac{0.99}{0.641}\right) - \left(\frac{2 \times 0.01}{0.32}\right)\right] \quad \text{and} \quad \underline{\underline{R_m = 1.48}}$$

(b) The light key = **A**, the heavy key = **B**, and $\alpha_A = 4$, $\alpha_B = 2$, $\alpha_C = 1$, and if q is assumed to equal 1, then substitution into Underwood's equations gives $\theta = 2.8$ and $\underline{\underline{R_m = 2.33}}$.

In Colburn's method, $x_{dB} = 0$ and $r_f = (0.3/0.3) = 1.0$.

$$\Sigma \alpha n_{fh} = (1/2) \times 0.4 = 0.2$$

$$x_{nA} = 1/(2 \times 1.2) = 0.417$$

$$\alpha_{AB} = (4/2) = 2.0 \quad \text{and} \quad \underline{\underline{R_m = 2.40}}$$

(c) In this case, $\alpha_A = 8$, $\alpha_B = 4$, $\alpha_C = 2$, $\alpha_D = 1$. If $q = 1$, then θ is found from Underwood's equation to be equal to 5.6 and $\underline{\underline{R_m = 2.33}}$.

In Colburn's method, $x_{dB} = 0$ and $r_f = 1.0$, the light key = **A**, the heavy key = **B**.

$$\Sigma \alpha x_{fh} = (0.5 \times 0.25) + (0.25 \times 0.25) = 0.188$$

$$x_{nA} = 1/(2 \times 1.188) = 0.421$$

$$\alpha_{AB} = 2 \quad \text{and} \quad \underline{R_m = 2.38}$$

In all cases, good agreement is shown between the two methods.

PROBLEM 11.23

A liquor consisting of phenol and cresols with some xylenols is fractionated to give a top product of 95.3 mole per cent phenol. The compositions of the top product and of the phenol in the bottoms are:

	Compositions (mole per cent)		
	Feed	Top	Bottom
phenol	35	95.3	5.24
o-cresol	15	4.55	—
m-cresol	30	0.15	—
xylenols	20	—	—
	100	100	

If a reflux ratio of 10 is used,

(a) Complete the material balance over the still for a feed of 100 kmol.
(b) Calculate the composition on the second plate from the top.
(c) Calculate the composition on the second plate from the bottom.
(d) Calculate the minimum reflux ratio by Underwood's equation and by Colburn's approximation.

The heavy key is m-cresol and the light key is phenol.

Solution

(a) An overall mass balance and a phenol balance gives, on a basis of 100 kmol:

$$100 = D + W$$

and: $\quad (100 \times 0.35) = 0.953D + 0.0524W$

from which: $\quad D = 33.0 \text{ kmol} \quad \text{and} \quad W = 67.0 \text{ kmol}.$

Balances on the remaining components give the required bottom product composition as:

$\quad$ o-cresol: $\quad (100 \times 0.15) = (0.0455 \times 33) + 67x_{wo} \quad \text{and} \quad x_{wo} = \underline{0.2017}$

$\quad$ m-cresol: $\quad (100 \times 0.30) = (0.0015 \times 33) + 67x_{wm} \quad \text{and} \quad x_{wm} = \underline{0.4472}$

$\quad$ xylenols: $\quad (100 \times 0.20) = 0 + 67x_{wx}, \quad \text{and} \quad x_{wx} = \underline{0.2987}$

(b)
$$L_n/D = 10 \quad \text{and} \quad L_n = 330 \text{ kmol}$$
$$V_n = L_n + D \quad \text{and} \quad V_n = 363 \text{ kmol}$$

The equation of the top operating line is:

$$y_n = \frac{L_n}{V_n}x_{n+1} + \frac{D}{V_n}x_d = (330/363)x_{n+1} + (33/330)x_d = 0.91x_{n+1} + 0.091x_d$$

The operating lines for each component then become:

phenol: $\quad y_{np} = 0.91x_{n+1} + 0.0867$

o-cresol: $\quad y_{no} = 0.91x_{n+1} + 0.0414$

m-cresol: $\quad y_{nm} = 0.91x_{n+1} + 0.00014$

xylenols: $\quad y_{nx} = 0.91x_{n+1}$

Mean α-values are taken from the data given in Volume 2, Table 11.2 as:

$$\alpha_{PO} = 1.25, \quad \alpha_{OO} = 1.0, \quad \alpha_{MO} = 0.63, \quad \alpha_{XO} = 0.37$$

The solution may be set out as a table as follows, using the operating line equations and the equation:

$$x = \frac{y/\alpha}{\sum(y/\alpha)}$$

	$y = x_d$	y_t/α	x_t	y_{t-1}	y_{t-1}/α	x_{t-1}
phenol	0.953	0.762	0.941	0.943	0.754	0.928
o-cresol	0.0455	0.0455	0.056	0.054	0.054	0.066
m-cresol	0.0015	0.0024	0.003	0.003	0.005	0.006
xylenols	—	—	—	—	—	—
	$\Sigma(y_t/\alpha) = 0.8099$	1.000	1.000	0.813	1.000

(c) In the bottom of the column:

$$L_m = L_n + F = 430 \text{ kmol}$$
$$V_m = L_m - W = 363 \text{ kmol}$$

and: $\quad y_m = \frac{L_m}{V_m}x_{n+1} - \frac{W}{V_m}x_w = 1.185x_{m+1} - 0.185x_w$

Hence for each component:

phenol: $\quad y_{mp} = 1.185x_{m+1} - 0.0097$

o-cresol: $\quad y_{mo} = 1.185x_{m+1} - 0.0373$

m-cresol: $\quad y_{mm} = 1.185x_{m+1} - 0.0827$

xylenols: $\quad y_{mx} = 1.185x_{m+1} - 0.0553$

Using these operating lines to calculate x and also $y = \alpha x / \Sigma \alpha x$ gives the following data:

	x_s	αx_s	y_s	x_1	αx_1	y_1	x_2
phenol	0.0524	0.066	0.100	0.093	0.116	0.156	0.140
o-cresol	0.2017	0.202	0.305	0.289	0.289	0.387	0.358
m-cresol	0.4472	0.282	0.427	0.430	0.271	0.363	0.376
xylenols	0.2987	0.111	0.168	0.188	0.070	0.094	0.126
	1.000	$\Sigma = 0.661$	1.000	1.000	$\Sigma = 0.746$	1.000	1.000

(d) Underwood's equations defined in, Problem 11.22, are used with $\alpha_p = 3.4$, $\alpha_o = 2.7, \alpha_m = 1.7, \alpha_x = 1.0$ to give:

$$\left(\frac{3.4 \times 0.35}{3.4 - \theta}\right) + \left(\frac{2.7 \times 0.15}{2.7 - \theta}\right) + \left(\frac{1.7 \times 0.30}{1.7 - \theta}\right) + \left(\frac{1.0 \times 0.20}{1 - \theta}\right) = (1 - q) = 0$$

$3.4 > \theta > 1.7$ and θ is found by trial and error to be 2.06.

Then: $\left(\dfrac{3.4 \times 0.953}{3.4 - 2.06}\right) + \left(\dfrac{2.7 \times 0.0455}{2.7 - 2.06}\right) + \left(\dfrac{1.7 \times 0.0015}{1.7 - 2.06}\right) = R_{m+1}$

and: $\underline{R_{m+1} = 1.60}$

Colburn's equation states that:

$$R_m = \frac{1}{\alpha_{AB} - 1}\left[\left(\frac{x_{dA}}{x_{nA}}\right) - \alpha_{AB}\left(\frac{x_{dB}}{x_{nB}}\right)\right] \qquad \text{(equation 11.108)}$$

$$x_{nA} = \frac{r_f}{(1 + r_f)(1 + \Sigma x_{fh})} \qquad \text{(equation 11.109)}$$

$$x_{nB} = x_{nA}/r_f \qquad \text{(equation 11.110)}$$

where **A** and **B** are the light and heavy keys, that is phenol and m-cresol.

$$r_f = (0.35/0.30) = 1.17$$
$$\Sigma \alpha x_{fh} = (0.37/0.63) \times 0.20 = 0.117$$
$$x_{nA} = 1.17/(2.17 \times 1.117) = 0.482$$
$$x_{nB} = (0.482/1.17) = 0.413$$
$$\alpha_{AB} = (1.25/0.63) = 1.98$$

Thus: $R_m = \dfrac{1}{0.98}\left[\left(\dfrac{0.953}{0.482}\right) - \left(\dfrac{1.98 \times 0.0015}{0.413}\right)\right] = \underline{1.95}$

This is the first approximation by Colburn's method and provides a good estimate of R_m.

PROBLEM 11.24

A continuous fractionating column is to be designed to separate 2.5 kg/s of a mixture of 60 per cent toluene and 40 per cent benzene, so as to give an overhead of 97 per cent benzene and a bottom product containing 98 per cent toluene by mass. A reflux ratio of 3.5 kmol of reflux/kmol of product is to be used and the molar latent heat of benzene and toluene may be taken as 30 MJ/kmol. Calculate:

(a) The mass flow of top and bottom products.
(b) The number of theoretical plates and position of the feed if the feed is liquid at 295 K, of specific heat capacity 1.84 kJ/kg K.
(c) How much steam at 240 kN/m² is required in the still.
(d) What will be the required diameter of the column if it operates at atmospheric pressure and a vapour velocity of 1 m/s.
(e) If the vapour velocity is to be 0.75 m/s, based on free area of column, determine the necessary diameter of the column.
(f) The minimum possible reflux ratio, and the minimum number of plates for a feed entering at its boiling-point.

Solution

(a) An overall mass balance and a benzene balance permit the mass of product and waste to be calculated directly:

$$2.5 = D' + W'$$

and:
$$2.5 \times 0.4 = 0.97 D' + 0.02 W'$$

from which: $\underline{\underline{W' = 1.5 \text{ kg/s}}}$ and $\underline{\underline{D' = 1.0 \text{ kg/s}}}$

(b) This part of the problem is solved by the McCabe–Thiele method. If the given compositions are converted to mole fractions, then:

$$x_f = 0.44, \quad x_w = 0.024, \quad x_d = 0.974$$

and a mass balance gives for 100 kmol of feed:

$$100 = D + W$$

$$(100 \times 0.44) = 0.974 D + 0.024 W$$

from which $D = 43.8$ and $W = 56.2$ kmol/100 kmol of feed

If $R = 3.5$, then: $L_n/D = 3.5$ and $L_n = 153.3$ kmol

$$V_n = L_n + D \quad \text{and} \quad V_n = 197.1 \text{ kmol}$$

The intercept on the y-axis $= x_d/(R+1) = 0.216$.

As the feed is a cold liquid, the slope of the q-line must be found. Using the given data and employing the method used in earlier problems, q is found to be 1.41 and the slope $= q/(q+1) = 3.44$. This enables the diagram to be completed as shown in

Figure 11r from which it is seen that 10 theoretical plates are required with the feed tray as the fifth from the top.

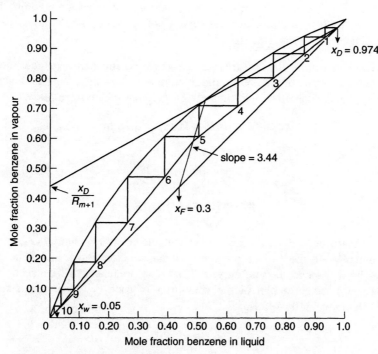

Figure 11r. Equilibrium data for Problem 11.24

(c) The boil-up rate at the bottom of the column $= V_m$.

$$V_m = 238.1 \text{ kmol/100 kmol feed}$$

$$\text{feed rate} = 2.5/(\text{mean mol mass}) = (2.5/86.4) = 0.0289 \text{ kmol/s}$$

Thus: vapour rate $= (238.1/100) \times 0.0289 = 0.069$ kmol/s

The heat load $= (0.069 \times 30) = 2.07$ MW or 2070 kW

The latent heat of steam at 240 kN/m^2 = 2186 kJ/kg (from the Appendix in Volume 2)

Thus: steam required $= (2070/2186) = \underline{\underline{0.95 \text{ kg/s}}}$

(d) At the top of the column the temperature is the boiling-point of essentially pure benzene, that is 353.3 K.

Thus: $C = (1/22.4)(273/353.3) = 0.034$ kmol/m^3

and: $V_n = 197.1$ kmol/100 kmol of feed.

Vapour flow $= (197.1/100) \times 0.0289 = 0.057$ kmol/s

Thus: volumetric flowrate $= (0.057/0.034) = 1.68$ m^3/s

If the vapour velocity is 1.0 m/s, then:

$$\text{the area} = 1.68 \text{ m}^2 \quad \text{and} \quad \text{the diameter} = 1.46 \text{ m}$$

If the diameter is calculated from the velocity at the bottom of the column, the result is a diameter of 1.67 m so that, if the velocity is not to exceed 1 m/s in any part of the column, its diameter must be <u>1.67 m</u>.

(e) The velocity based on the free area (tower area − downcomer area) must not exceed 0.75 m/s. The vapour rate in the bottom of the column is 2.17 m^3/s and, for a single-pass crossflow tray, the free area is approximately 88 per cent of the tower area.

Thus: $\quad A_t = 2.17/(0.75 \times 0.88) = 3.28 \text{ m}^2$

and: $\quad \underline{\underline{D_t = 2.05 \text{ m}}}$

PROBLEM 11.25

For a system that obeys Raoult's law, show that the relative volatility α_{AB} is P_A^0/P_B^0, where P_A^0 and P_B^0 are the vapour pressures of the components **A** and **B** at the given temperature. From vapour pressure curves of benzene, toluene, ethyl benzene and of o-, m- and p-xylenes, obtain a plot of the volatilities of each of the materials relative to m-xylene in the range 340–430 K.

Solution

The volatility of $\mathbf{A} = P_A/x_A$ and the volatility of $\mathbf{B} = P_B/x_B$ and the relative volatility α_{AB} is the ratio of these volatilities,

that is: $\quad \alpha_{AB} = \dfrac{P_A x_B}{x_A P_B}$

For a system that obeys Raoult's law, $P = xP^0$.

Thus: $\quad \alpha_{AB} = \dfrac{(x_A P_A^0)x_B}{x_A(x_B P_B^0)} = \underline{\underline{\dfrac{P_A^0}{P_B^0}}}$

The vapour pressures of the compounds given in the problem are plotted in Figure 11s and are taken from Perry.[1]

For convenience, the vapour pressures are plotted on a logarithmic scale against the reciprocal of the temperature (1/K) given a straight line. The relative volatilities may then be calculated in tabular form as follows:

---

[1] PERRY, R. H., GREEN, D. W., and MALONEY, J. O.: *Perry's Chemical Engineer's Handbook*, 6th edn. (McGraw-Hill, New York, 1987).

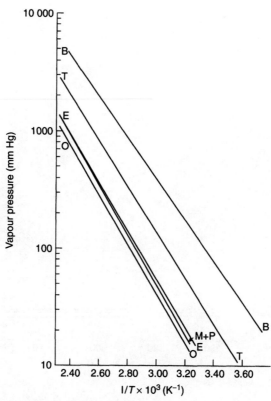

Figure 11s. Vapour pressure data for Problem 11.25

Temperature (K)	$10^3/T$ (1/K)	$(m-x)$	P^0 (mmHg)/ $(p-x)$	$(o-x)$	E	B	T
340	2.94	65	64	56	72	490	180
360	2.78	141	140	118	155	940	360
380	2.63	292	290	240	310	1700	700
400	2.50	554	550	450	570	2900	1250
415	2.41	855	850	700	860	4200	1900
430	2.32	1315	1300	1050	1350	6000	2800

Temperature (K)	α_{pm}	α_{om}	α_{Em}	α_{Bm}	α_{Tm}
340	0.98	0.86	1.11	7.54	2.77
360	0.99	0.84	1.10	6.67	2.55
380	0.99	0.82	1.06	5.82	2.40
400	0.99	0.81	1.03	5.23	2.26
415	0.99	0.82	1.01	4.91	2.22
430	0.99	0.80	1.03	4.56	2.13

These data are plotted in Figure 11t.

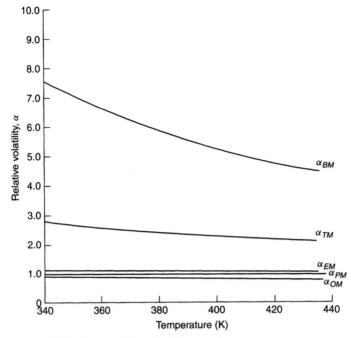

Figure 11t. Relative volatility data for Problem 11.25

PROBLEM 11.26

A still contains a liquor composition of o-xylene 10 per cent, m-xylene 65 per cent, p-xylene 17 per cent, benzene 4 per cent and ethyl benzene 4 per cent. How many plates are required at total reflux to give a product of 80 per cent m-xylene, and 14 per cent p-xylene? The data are given as mass per cent.

Solution

Fenske's equation may be used to find the number of plates at total reflux.

Thus:
$$n + 1 = \frac{\log[(x_A/x_B)_d(x_B/x_A)_s]}{\log \alpha_{AB}}$$
(equation 11.58)

In multicomponent distillation, **A** and **B** are the light and heavy key components respectively. In this problem, the only data given for both top and bottom products are for m- and p-xylene and these will be used with the mean relative volatility calculated in the previous problem. Thus:

$$x_A = 0.8, \quad x_B = 0.14, \quad x_{B_s} = 0.17, \quad x_{A_s} = 0.65$$

$$\alpha_{AB} = (1/0.99) = 1.0101$$

Thus: $n + 1 = \log[(0.8/0.14)(0.17/0.65)]/\log 1.0101$ and $\underline{n = 39 \text{ plates}}$

138

PROBLEM 11.27

The vapour pressures of n-pentane and of n-hexane are:

Pressure (kN/m^2)	1.3	2.6	5.3	8.0	13.3	26.6	53.2	101.3
(mm Hg)	10	20	40	60	100	200	400	760
Temperature (K)								
C_5H_{12}	223.1	233.0	244.0	257.0	260.6	275.1	291.7	309.3
C_6H_{14}	248.2	259.1	270.9	278.6	289.0	304.8	322.8	341.9

The equilibrium data at atmospheric pressure are:

x	0.1	0.2	0.3	0.4	0.5	0.6	0.7	0.8	0.9
y	0.21	0.41	0.54	0.66	0.745	0.82	0.875	0.925	0.975

(a) Determine the relative volatility of pentane to hexane at 273, 293 and 313 K.
(b) A mixture containing 0.52 mole fraction pentane is to be distilled continuously to give a top product of 0.95 mole fraction pentane and a bottom of 0.1 mole fraction pentane. Determine the minimum number of plates, that is the number of plates at total reflux, by the graphical McCabe–Thiele method, and analytically by using the relative volatility method.
(c) Using the conditions in (b), determine the liquid composition on the second plate from the top by Lewis's method, if a reflux ratio of 2 is used.
(d) Using the conditions in (b), determine by the McCabe–Thiele method the total number of plates required, and the position of the feed.

It may be assumed that the feed is all liquid at its boiling-point.

Solution

The vapour pressure data and the equilibrium data are plotted in Figures 11u and 11v.

(a) Using:
$$\alpha_{PH} = P_p^0/P_H^0$$

The following data are obtained:

Temperature (K)	P_p^0	P_H^0	α_{PH}
273	24	6.0	4.0
293	55	16.0	3.44
313	116	36.5	3.18
		Mean =	3.54

(b) The McCabe–Thiele construction is shown in Figure 11v where it is seen that 4 theoretical plates are required at total reflux.

Using Fenske's equation at total reflux:

$$n + 1 = \log[(0.95/0.05)(0.90/0.10)]/\log 3.54 \quad \text{and} \quad \underline{\underline{n = 3.07}}$$

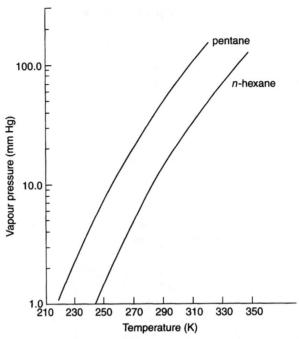

Figure 11u. Vapour pressure data for Problem 11.27

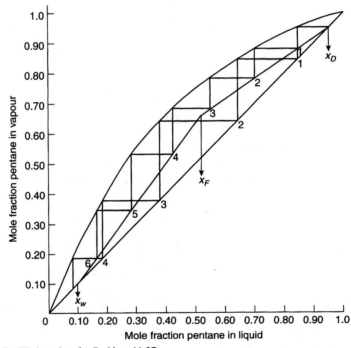

Figure 11v. Equilibrium data for Problem 11.27

The discrepancy here is caused by using a mean value of α although α does in fact vary considerably.

(c) From a mass balance it is found that for 100 kmol of feed and $R = 2$:

$$D = 49.4, \quad W = 50.6, \quad L_n = 98.8, \quad V_n = 148.2$$

Then:
$$y_n = \frac{L_n}{V_n} x_{n+1} + \frac{D}{V_n} x_d \qquad \text{(equation 11.34)}$$

and:
$$y_n = 0.67 x_{n+1} + 0.317$$

The vapour leaving the top plate has the composition of the distillate, that is: $y_t = x_d = 0.95$.

The liquid on the top plate is in equilibrium with this vapour and from the equilibrium curve has a composition $x_t = 0.845$.

The vapour rising to the top tray y_{t-1} is found from the operating line:

$$y_{t-1} = 0.67 \times 0.845 + 0.317 = 0.883$$
$$x_{t-1} = \text{from the equilibrium curve} = 0.707$$
$$y_{t-2} = (0.67 \times 0.707) + 0.317 = 0.790$$

and:
$$x_{t-2} = \underline{\underline{0.56}}$$

(d) From Figure 11v, 6 theoretical plates are required and the feed tray is the third from the top of the column.

PROBLEM 11.28

The vapour pressures of n-pentane and n-hexane are as given in Problem 11.27. Assuming that both Raoult's and Dalton's Laws are obeyed,

(a) Plot the equilibrium curve for a total pressure of 13.3 kN/m^2.
(b) Determine the relative volatility of pentane to hexane as a function of liquid composition for a total pressure of 13.3 kN/m^2.
(c) Would the error caused by assuming the relative volatility constant at its mean value be considerable?
(d) Would it be more advantageous to distil this mixture at a higher pressure?

Solution

(a) The following equations are used where **A** is n-pentane and **B** is n-hexane:

$$x_A = \frac{P - P_B^0}{P_A^0 - P_B^0} \qquad \text{(equation 11.5)}$$

$$y_A = P_A^0 x_A / P \qquad \text{(equation 11.4)}$$

At $P = 13.3$ kN/m^2:

Temperature (K)	P_A^0	P_B^0	x_A	y_A	$\alpha = P_A^0/P_B^0$
260.6	13.3	2.85	1.0	1.0	4.67
265	16.5	3.6	0.752	0.933	4.58
270	21.0	5.0	0.519	0.819	4.20
275	26.0	6.7	0.342	0.669	3.88
280	32.5	8.9	0.186	0.455	3.65
285	40.0	11.0	0.079	0.238	3.64
289	47.0	13.3	0	0	3.53
					Mean $\alpha = \underline{\underline{4.02}}$

These figures are plotted in Figure 11w.

(b) The relative volatility is plotted as a function of liquid composition in Figure 11w.

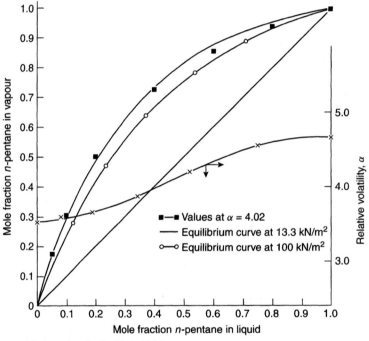

Figure 11w. Equilibrium data for Problem 11.28

(c) If α is taken as 4.02, y_A may be calculated from:

$$y_A = \frac{\alpha x_A}{1 + (\alpha - 1)x_A} \qquad \text{(equation 11.15)}$$

Using equation 11.15, a new equilibrium curve may be calculated as follows:

x_A	0	0.05	0.10	0.20	0.40	0.60	0.80	1.0
y_A	0	0.174	0.308	0.500	0.727	0.857	0.941	1.0

These points are shown in Figure 11w where it may be seen that little error is introduced by the use of this mean value.

(d) If a higher pressure is used, the method used in (a) may be repeated. If $P = 100$ kN/m^2, the temperature range increases to 309–341 K and the new curve is drawn in Figure 11w. Clearly, the higher pressure demands a larger number of plates for the same separation and is not desirable.

PROBLEM 11.29

It is desired to separate a binary mixture by simple distillation. If the feed mixture has a composition of 0.5 mole fraction, calculate the fraction which it is necessary to vaporise in order to obtain:

(a) a product of composition 0.75 mole fraction, when using a continuous process, and
(b) a product whose composition is not less than 0.75 mole fraction at any instant, when using a batch process.

If the product of batch distillation is all collected in a single receiver, what is its mean composition?

It may be assumed that the equilibrium curve is given by:

$$y = 1.2x + 0.3$$

for liquid compositions in the range 0.3–0.8.

Solution

(a) If F = number of kmol of feed of composition x_f,
L = kmol remaining in still with composition x, and
V = kmol of vapour formed with composition y, then:

$$F = V + L \quad \text{and} \quad Fx_f = Vy + Lx$$

For 1 kmol of feed:

$$x_f = Vy + Lx \quad \text{and} \quad y = \frac{x_f}{V} - \frac{L}{V}x$$

This equation is a straight line of slope $-L/V$ which passes through the point (x_f, x_f), so that, if y is known, L/V may be found. This is illustrated in Figure 11x where:

$$-L/V = -5.0$$

As: $\quad F = 1, \quad 1 = V + L$

and: $\quad V = 0.167$ kmol/kmol of feed or <u>16.7 per cent is vaporised</u>

(b) For a batch process it may be shown that:

$$\left(\frac{y - x}{y_0 - x_0}\right) = \left(\frac{S}{S_0}\right)^{m-1} \qquad \text{(equation 11.29)}$$

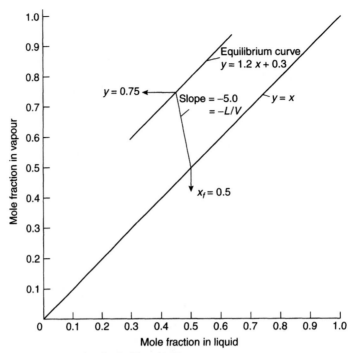

Figure 11x. Graphical construction for Problem 11.29

where S is the number of kmol charged initially = 100 kmol (say), S_0 is the number of kmol remaining, x is the initial still composition = 0.5, y is the initial vapour composition = $(1.2 \times 0.5) + 0.3 = 0.90$, y_0 is the final vapour composition = 0.75 and x_0 is the final liquid composition, is found from:

$$0.75 = 1.2x_0 + 0.3 \quad \text{or} \quad x_0 = 0.375$$

and m is the slope of equilibrium curve = 1.2.

Thus: $$\frac{(0.90 - 0.50)}{(0.75 - 0.375)} = 1.07 = \left(\frac{100}{S_0}\right)^{0.2}$$

and: $$S_0 = 71.3 \text{ kmol}/100 \text{ kmol feed}$$

Therefore: amount vaporised = $\left(\dfrac{100 - 71.3}{100}\right) \times 100 = \underline{\underline{28.7 \text{ per cent}}}$

The distillate composition may be found from a mass balance as follows:

	Total kmol	kmol A	kmol B
Charge	100	50	50
Distillate	28.7	$(50 - 26.7) = 23.3$	$(50 - 44.6) = 5.4$
Residue	71.3	$(0.374 \times 71.3) = 26.7$	$(71.3 - 26.7) = 44.6$

Distillate composition = $(23.3/28.7) \times 100 = \underline{\underline{81.2 \text{ per cent.}}}$

PROBLEM 11.30

A liquor, consisting of phenol and cresols with some xylenol, is separated in a plate column. Given the following data complete the material balance:

Component	Mole per cent		
	Feed	Top	Bottom
C_6H_5OH	35	95.3	5.24
$o\text{-}C_7H_7OH$	15	4.55	—
$m\text{-}C_7H_7OH$	30	0.15	—
C_8H_9OH	20	—	—
Total	100	100	—

Calculate:

(a) the composition on the second plate from the top,
(b) the composition on the second plate from the bottom.

A reflux ratio of 4 is used.

Solution

The mass balance is completed as in Problem 11.24, where it was shown that:

$$x_{wo} = \underline{0.2017}, \quad x_{wm} = \underline{0.4472}, \quad x_{wx} = \underline{0.2987}$$

(a) For 100 kmol feed, from mass balances with $R = 4.0$, the following values are obtained:

$$L_n = 132, \quad V_n = 165, \quad L_n = 232, \quad V_n = 165$$

The equations for the operating lines of each component are obtained from:

$$y_n = \frac{L_n}{V_n}x_{n+1} + \frac{D}{V_n}x_d \quad \text{(equation 11.35)}$$

as: phenols: $y_{np} = 0.8x_{n+1} + 0.191$

o-cresol: $y_{no} = 0.8x_{n+1} + 0.009$

m-cresol: $y_{nm} = 0.8x_{n+1} + 0.0003$

xylenols: $y_{nx} = 0.8x_{n+1}$

The compositions on each plate may then be found by calculating y from the operating line equations and x from $x = \dfrac{y/\alpha}{\Sigma(y/\alpha)}$ to give the following results:

	α	$y_t = x_d$	y_t/α	$x_t = (y_t/\alpha)/0.81$	y_{t-1}	y_{t-1}/α	x_{t-1}
phenols	1.25	0.953	0.762	0.940	0.943	0.762	0.843
o-cresol	1.0	0.0455	0.046	0.057	0.055	0.055	0.061
m-cresol	0.63	0.0015	0.002	0.003	0.002	0.087	0.096
xylenols	0.37	—	—	—	—	—	—
		1.000	$\Sigma = 0.81$	1.000	1.000	0.904	1.000

(b) In a similar way, the following operating lines may be derived for the bottom of the column:

$$\text{phenols:} \quad y_{mp} = 1.406 x_{n+1} - 0.0212$$

$$\text{o-cresol:} \quad y_{mo} = 1.406 x_{n+1} - 0.0819$$

$$\text{m-cresol:} \quad y_{mm} = 1.406 x_{n+1} - 0.1816$$

$$\text{xylenols:} \quad y_{mx} = 1.406 x_{n+1} - 0.1213$$

Thus:

	α	x_s	αx_s	$y_s = \alpha x_s/0.661$	x_1	αx_1	y_1	x_2
phenols	1.25	0.0524	0.066	0.100	0.086	0.108	0.148	0.121
o-cresol	1	0.2017	0.202	0.305	0.275	0.275	0.375	0.324
m-cresol	0.63	0.4472	0.282	0.427	0.433	0.273	0.373	0.394
xylenols	0.37	0.2987	0.111	0.168	0.206	0.076	0.104	0.161
			$\Sigma = 0.661$	1.000	1.000	0.732	1.000	1.000

PROBLEM 11.31

A mixture of 60, 30, and 10 mole per cent benzene, toluene, and xylene respectively is separated by a plate-column to give a top product containing at least 90 mole per cent benzene and a negligible amount of xylene, and a waste containing not more than 60 mole per cent toluene.

Using a reflux ratio of 4, and assuming that the feed is boiling liquid, determine the number of plates required in the column, and the approximate position of the feed.

The relative volatility of benzene to toluene is 2.4 and of xylene to toluene is 0.45, and it may be assumed that these values are constant throughout the column.

Solution

Assuming 100 kmol of feed, the mass balance may be completed to give:

$$D = 60, \quad W = 40 \text{ kmol}$$

and: $\quad x_{dt} = 0.10, \quad x_{wb} = 0.15, \quad x_{wx} = 0.25$

If $R = 4$ and the feed is at its boiling-point then:

$$L_n = 240, \quad V_n = 300, \quad L_m = 340, \quad V_m = 300$$

and the top and bottom operating lines are:

$$y_n = 0.8x_{n+1} + 0.2x_d$$

and:

$$y_m = 1.13x_{n+1} - 0.133x_w$$

A plate-to-plate calculation may be carried out as follows:
In the bottom of the column.

α		x_s	αx_s	y_s	x_1	αx_1	y_1	x_2	αx_2	y_2	x_3
2.4	benzene	0.15	0.360	0.336	0.314	0.754	0.549	0.503	1.207	0.724	0.65
1.0	toluene	0.60	0.600	0.559	0.564	0.564	0.411	0.432	0.432	0.258	0.29
0.45	xylene	0.25	0.113	0.105	0.122	0.055	0.040	0.065	0.029	0.018	0.04
			1.073	1.000	1.000	1.373	1.000	1.000	1.668	1.000	1.00

The composition on the third plate from the bottom corresponds most closely to the feed and above this tray the rectifying equations will be used.

	x_3	αx_3	y_3	x_4	αx_4	y_4
benzene	0.657	1.577	0.832	0.815	1.956	0.917
toluene	0.298	0.298	0.157	0.171	0.171	0.080
xylene	0.045	0.020	0.011	0.014	0.006	0.003
	1.000	1.895	1.000	1.000	2.133	1.000

As the vapour leaving the top plate will be totally condensed to give the product, 4 theoretical plates will be required to meet the given specification.

PROBLEM 11.32

It is desired to concentrate a mixture of ethyl alcohol and water from 40 mole per cent to 70 mole per cent alcohol. A continuous fractionating column, 1.2 m in diameter with 10 plates is available. It is known that the optimum superficial vapour velocity in the column at atmosphere pressure is 1 m/s, giving an overall plate efficiency of 50 per cent.

Assuming that the mixture is fed to the column as a boiling liquid and using a reflux ratio of twice the minimum value possible, determine the location of the feed plate and the rate at which the mixture can be separated.

Equilibria data:

Mole fraction alcohol in liquid	0.1	0.2	0.3	0.4	0.5	0.6	0.7	0.8	0.89
Mole fraction alcohol in vapour	0.43	0.526	0.577	0.615	0.655	0.70	0.754	0.82	0.89

Solution

The equilibrium data are plotted in Figure 11y, where the operating line corresponding to the minimum reflux ratio is drawn from the point (x_d, x_d) through the intersection of the vertical q-line and the equilibrium curve to give an intercept of 0.505.

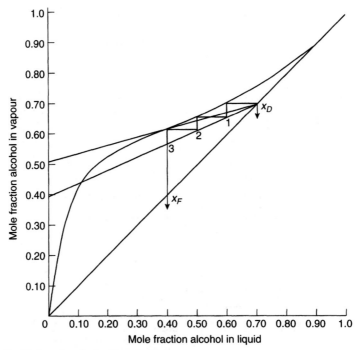

Figure 11y. Equilibrium data for Problem 11.32

Thus: $\qquad x_d/(R_m + 1) = 0.505 \quad \text{and} \quad R_m = 0.386$

The actual value of R is then $(2 \times 0.386) = 0.772$ so that the top operating line may be constructed as shown.

This column contains the equivalent of (10×0.5), that is 5 theoretical plates, so that these may be stepped off from the point (x_d, x_d) to give the feed plate as the third from the top.

The problem as stated gives no bottom-product composition, so that whilst all flow-rates in the top of the column may be calculated, no information about the lower half can be derived. In the absence of these data, the feed rate cannot be determined, though the rate of distillate removal may be calculated as follows.

Mean molecular mass of top product $= (46 \times 0.7 + 18 \times 0.3) = 37.6$ kg/kmol
If the top temperature is assumed to be 353 K, then:

$$C = (1/22.4)(273/353) = 0.0345 \text{ kmol/s}^3$$

If the vapour velocity = 1 m/s, the volumetric vapour flow at the top of the column is:

$$(\pi/4)(1.2)^2 \times 1 = 1.13 \text{ m}^3/\text{s}.$$

Hence: $\qquad V_n = (1.13 \times 0.0345) = 0.039 \text{ kmol/s}$

From the slope of the operating line:

$$L_n/V_n = 0.436$$

Thus: $\qquad L_n = (0.436 \times 0.039) = 0.017 \text{ kmol/s}$

As $R = 0.772$, then:

$$D = (L_n/0.772) = 0.022 \text{ kmol/s} \quad \text{or} \quad \underline{\underline{0.828 \text{ kg/s distillate}}}$$

SECTION 2-12

Absorption of Gases

PROBLEM 12.1

Tests are made on the absorption of carbon dioxide from a carbon dioxide–air mixture in a solution containing 100 kg/m³ of caustic soda, using a 250 mm diameter tower packed to a height of 3 m with 19 mm Raschig rings.

The results obtained at atmospheric pressure were:

Gas rate, $G' = 0.34$ kg/m²s. Liquid rate, $L' = 3.94$ kg/m²s.

The carbon dioxide in the inlet gas was 315 parts per million and the carbon dioxide in the exit gas was 31 parts per million.

What is the value of the overall gas transfer coefficient K_Ga?

Solution

At the bottom of the tower:

$$y_1 = 315 \times 10^{-6}, \quad G' = 0.34 \text{ kg/m}^2\text{s}$$

and: $\quad G'_m = (0.34/29) = 0.0117$ kmol/m²s

At the top of the tower: $y_2 = 31 \times 10^{-6}, \quad x_2 = 0$

and: $\quad L' = 3.94$ kg/m²s

The NaOH, solution contains 100 kg/m³ NaOH.

The mean molecular mass of liquid is:

$$\frac{(100 \times 40) + (900 \times 18)}{1000} = 20.2 \text{ kg/kmol}$$

Thus: $\quad L'_m = (3.94/20.2) = 0.195$ kmol/m²s

For dilute gases, $y = Y$ and a mass balance over the tower gives:

$$G'_m(y_1 - y_2)A = K_Ga P(y - y_e)_{lm} \, ZA$$

It may be assumed that as the solution of NaOH is fairly concentrated, there will be a negligible vapour pressure of CO_2 over the solution, that is <u>all the resistance to transfer lies in the gas phase.</u>

150

Therefore the driving force at the top of the tower $= (y_2 - 0) = 31 \times 10^{-6}$

and: at the bottom of the tower $= (y_1 - 0) = 315 \times 10^{-6}$

The log mean driving force, $(y - y_e)_{\mathrm{lm}} = \dfrac{(315 - 31) \times 10^6}{\ln(315/31)} = 122.5 \times 10^{-6}$

Therefore: $0.0117(315 - 31)10^{-6} = K_G a (101.3 \times 122.5 \times 10^{-6} \times 3)$

from which: $\underline{\underline{K_G a = 8.93 \times 10^{-5} \text{ kmol/m}^3 \text{s (kN/m}^2)}}$

PROBLEM 12.2

An acetone–air mixture containing 0.015 mole fraction of acetone has the mole fraction reduced to 1 per cent of this value by countercurrent absorption with water in a packed tower. The gas flowrate G' is 1 kg/m^{2}s of air and the water flowrate entering is 1.6 kg/m^{2}s. For this system, Henry's law holds and $y_e = 1.75x$, where y_e is the mole fraction of acetone in the vapour in equilibrium with a mole fraction x in the liquid. How many overall transfer units are required?

Solution

See Volume 2, Example 12.3.

PROBLEM 12.3

An oil containing 2.55 mole per cent of a hydrocarbon is stripped by running the oil down a column up which live steam is passed, so that 4 kmol of steam are used/100 kmol of oil stripped. Determine the number of theoretical plates required to reduce the hydrocarbon content to 0.05 mole per cent, assuming that the oil is non-volatile. The vapour–liquid relation of the hydrocarbon in the oil is given by $y_e = 33x$, where y_e is the mole fraction in the vapour and x the mole fraction in the liquid. The temperature is maintained constant by internal heating, so that steam does not condense in the tower.

Solution

If the steam does not condense, $L_m/G_m = (100/4) = 25$.

Inlet oil concentration = 2.55 mole per cent,

$$x_2 = 0.0255 \quad \text{and} \quad X_2 = x_2/(1 - x_2) = 0.0262$$

Exit oil concentration = 0.05 mol per cent and $x_1 = 0.0005$

A mass balance between a plane in the tower, where the concentrations are X and Y, and the bottom of the tower gives:

$$L_m(X - X_1) = G_m(Y - Y_1)$$

$$Y_1 = 0$$

Therefore: $\quad Y = 25X - 25x_1 = 25X - 0.0125$

This is the equation of the operating line and as the equilibrium data are $y_e = 33x$, then:

$$\frac{Y}{1+Y} = \frac{33X}{1+X} \quad \text{or} \quad Y = \frac{33X}{1 - 32X}$$

Using these data, the equilibrium and lines may be drawn as shown in Figure 12a. The number of theoretical plates is then found from a stepping-off procedure as employed for distillation as <u>8 plates.</u>

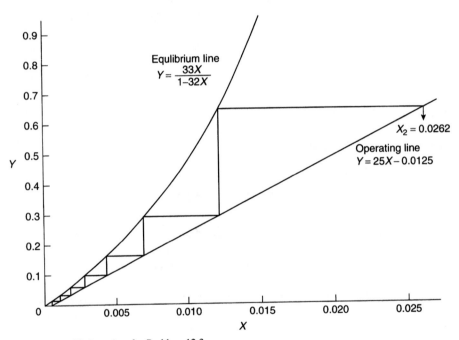

Figure 12a. Equilibrium data for Problem 12.3

PROBLEM 12.4

Gas, from a petroleum distillation column, has its concentration of H_2S reduced from 0.03 kmol H_2S/kmol of inert hydrocarbon gas to 1 per cent of this value, by scrubbing with a triethanolamine-water solvent in a countercurrent tower, operating at 300 K and at atmospheric pressure.

H_2S is soluble in such a solution and the equilibrium relation may be taken as $Y = 2X$, where Y is kmol of H_2S kmol inert gas and X is kmol of H_2S/kmol of solvent.

The solvent enters the tower free of H₂S and leaves containing 0.013 kmol of H_2S/kmol of solvent. If the flow of inert hydrocarbon gas is 0.015 kmol/m²s of tower cross-section and the gas-phase resistance controls the process, calculate:

(a) the height of the absorber necessary, and
(b) the number of transfer units required.

The overall coefficient for absorption $K_G''a$ may be taken as 0.04 kmol/s m³ of tower volume (unit driving force in Y).

Solution

The driving force at the top of column, $(Y_2 - Y_{2e}) = 0.0003$.
The driving force at bottom of column, $(Y_1 - Y_{1e}) = (0.03 - 0.026) = 0.004$.
The logarithmic mean driving force $= (0.004 - 0.0003)/\ln(0.004/0.0003) = 0.00143$.
From equation 12.70: $G_m(Y_1 - Y_2)A = K_G a P(Y - Y_e)_{lm} AZ$

That is: $\quad\quad\quad G_m'(Y_1 - Y_2) = K_G''a(Y - Y_e)_{lm} Z$

Thus: $\quad 0.015(0.03 - 0.0003) = 0.04 \times 0.00143 Z$

and: $\quad\quad\quad\quad\quad\quad Z = (0.000446/0.0000572) = 7.79$ m or 7.8 m

The height of transfer unit, $\mathbf{H}_{OG} = G_m'/K_G''a = (0.015/0.04) = 0.375$ m.
The number of transfer units, $\mathbf{N}_{OG} = (7.79/0.375) = 20.8$ or 21.

PROBLEM 12.5

It is known that the overall liquid transfer coefficient K_La for absorption of SO_2 in water in a column is 0.003 kmol/s m³ (kmol/m³). Obtain an expression for the overall liquid-film coefficient K_La for absorption of NH_3 in water in the same equipment using the same water and gas rates. The diffusivities of SO_2 and NH_3 in air at 273 K are 0.103 and 0.170 cm²/s. SO_2 dissolves in water, and Henry's constant $\mathscr{H}$ is equal to 50 (kN/m²)/(kmol/m³). All the data are expressed for the same temperature.

Solution

See Volume 2, Example 12.1.

PROBLEM 12.6

A packed tower is used for absorbing sulphur dioxide from air by means of a caustic soda solution containing 20 kg/m³ NaOH. At an air flow of 2 kg/m²s, corresponding to a Reynolds number of 5160, the friction factor $R/\rho u^2$ is found to be 0.020.

Calculate the mass transfer coefficient in kg SO_2/s m^2 (kN/m^2) under these conditions if the tower is at atmospheric pressure. At the temperature of absorption, the diffusion coefficient SO_2 is 1.16×10^{-5} m^2/s, the viscosity of the gas is 0.018 mN s/m^2 and the density of the gas stream is 1.154 kg/m^3.

Solution

See Volume 2, Example 12.2.

PROBLEM 12.7

In an absorption tower, ammonia is absorbed from air at atmospheric pressure by acetic acid. The flowrate of 2 kg/m^{2}s in a test corresponds to a Reynolds number of 5100 and hence a friction factor $R/\rho u^2$ of 0.020. At the temperature of absorption the viscosity of the gas stream is 0.018 mN s/m^2, the density is 1.154 kg/m^3 and the diffusion coefficient of ammonia in air is 1.96×10^{-5} m^2/s. Determine the mass transfer coefficient through the gas film in kg/m^{2}s (kN/m^2).

Solution

From equation 12.25: $\left(\dfrac{h_d}{u}\right)\left(\dfrac{P_{Bm}}{P}\right)\left(\dfrac{\mu}{\rho D}\right)^{0.56} = j_d$

and: $\qquad j_d \simeq R/\rho u^2$

Substituting the given data gives:

$$(\mu/\rho D)^{0.56} = 0.88$$

$$\left(\dfrac{h_d}{u}\right)\left(\dfrac{P_{Bm}}{P}\right) = (0.0199/0.88) = 0.0226$$

$$u = G'/\rho = 1.733 \text{ m/s}$$

Therefore: $\quad k_G = \left(\dfrac{h_d}{RT}\right)\left(\dfrac{P_{Bm}}{P}\right) = \left(\dfrac{0.0226 \times 1.733}{8.314 \times 298}\right)$

$$= 1.58 \times 10^{-5} \text{ kmol/m}^2\text{s (kN/m}^2\text{)}$$

and: $\qquad k_G = (1.58 \times 10^{-5} \times 17) = \underline{2.70 \times 10^{-4} \text{ kg/m}^2\text{s (kN/m}^2\text{)}}$

PROBLEM 12.8

Acetone is to be recovered from a 5 per cent acetone–air mixture by scrubbing with water in a packed tower using countercurrent flow. The liquid rate is 0.85 kg/m^{2}s and the gas rate is 0.5 kg/m^{2}s.

The overall absorption coefficient $K_G a$ may be taken as 1.5×10^{-4} kmol/[m^{3}s (kN/m^2) partial pressure difference] and the gas film resistance controls the process.

What height of tower is required tower to remove 98 per cent of the acetone? The equilibrium data for the mixture are:

Mole fraction of acetone in gas	0.0099	0.0196	0.0361	0.0400
Mole fraction of acetone in liquid	0.0076	0.0156	0.0306	0.0333

Solution

At the bottom of the tower: $y_1 = 0.05$

$$G' = (0.95 \times 0.5) \text{ kg/m}^2\text{s} \quad \text{and} \quad G'_m = 0.0164 \text{ kmol/m}^2\text{s}$$

At the top of the tower: $y_2 = (0.02 \times 0.05) = 0.001$,

$$L' = 0.85 \text{ kg/m}^2\text{s} \quad \text{and} \quad L_m = 0.0472 \text{ kmol/m}^2\text{s}$$

The height and number of overall transfer units are defined as $\mathbf{H}_{OG}$ and $\mathbf{N}_{OG}$ by:

$$\mathbf{H}_{OG} = G_m/K_G a P \quad \text{and} \quad \mathbf{N}_{OG} = \int_{y_1}^{y_2} \frac{dy}{y - y_e}$$

(equations 12.80 and 12.77)

Thus: $\mathbf{H}_{OG} = 0.0164/(1.5 \times 10^{-4} \times 101.3) = 1.08$ m

The equilibrium data given are represented by a straight line of slope $m = 1.20$. As shown in Problem 12.12, the equation for $\mathbf{N}_{OG}$ may be integrated directly when the equilibrium line is given by $y_e = mx$ to give:

$$\mathbf{N}_{OG} = \frac{1}{(1 - mG_m/L_m)} \ln\left[\left(1 - \frac{mG'_m}{L'_m}\right)\frac{y_1}{y_2} + \frac{mG'_m}{L'_m}\right]$$

$$m(G'_m/L'_m) = 1.20(0.0164/0.0472) = 0.417$$

$$y_1/y_2 = (0.05/0.001) = 50$$

Thus: $\mathbf{N}_{OG} = \left(\dfrac{1}{1 - 0.417}\right) \ln[(1 - 0.417)50 + 0.417] = 5.80$

The packed height $= \mathbf{N}_{OG} \times \mathbf{H}_{OG}$

$$= (5.80 \times 1.08) = \underline{6.27 \text{ m}}$$

PROBLEM 12.9

Ammonia is to be removed from a 10 per cent ammonia–air mixture by countercurrent scrubbing with water in a packed tower at 293 K so that 99 per cent of the ammonia is removed when working at a total pressure of 101.3 kN/m^2. If the gas rate is 0.95 kg/m^{2}s of tower cross-section and the liquid rate is 0.65 kg/m²s, what is the necessary height of the tower if the absorption coefficient $K_G a = 0.001$ kmol/m³s (kN/m²) partial pressure difference. The equilibrium data are:

Concentration (kmol NH₃/kmol water)	0.021	0.031	0.042	0.053	0.079	0.106	0.150
Partial pressure NH₃ (kN/m²)	1.6	2.4	3.3	4.2	6.7	9.3	15.2

Solution

See Volume 2, Example 12.5.

PROBLEM 12.10

Sulphur dioxide is recovered from a smelter gas containing 3.5 per cent by volume of SO_2, by scrubbing it with water in a countercurrent absorption tower. The gas is fed into the bottom of the tower, and in the exit gas from the top the SO_2 exerts a partial pressure of 1.14 kN/m². The water fed to the top of the tower is free from SO_2, and the exit liquor from the base contains 0.001145 kmol SO_2/kmol water. The process takes place at 293 K, at which the vapour pressure of water is 2.3 kN/m². The water flow rate is 0.43 kmol/s.

If the area of the tower is 1.85 m² and the overall coefficient of absorption for these conditions $K_L'' a$ is 0.19 kmol SO_2/s m³ (kmol of SO_2/kmol H_2O), what is the height of the column required?

The equilibrium data for SO_2 and water at 293 K are:

kmol SO₂/1000 kmol H₂O	0.056	0.14	0.28	0.42	0.56	0.84	1.405
kmol SO₂/1000 kmol inert gas	0.7	1.6	4.3	7.9	11.6	19.4	35.3

Solution

At the top of the column: $P_{SO_2} = 1.14$ kN/m²

That is: $1.14 = 101.3 y_2$ and $y_2 = 0.0113 \simeq Y_2$

At the bottom of the column: $y_1 = 0.035$, that is $Y_1 = 0.036$

$$X_1 = 0.001145$$

$$L_m = 0.43 \text{ kmol/s}$$

The quantity of SO_2 absorbed $= 0.43(0.001145 - 0)$

That is:
$$N_A = 4.94 \times 10^{-4} \text{ kmol } SO_2/s$$
$$N_A = K''_L a (X_e - X)_{lm}$$

The log mean driving force in terms of the liquid phase must now be calculated. Values of X_e corresponding to the gas composition Y may be found from the equilibrium data given (but are not plotted here) as:

When:
$$Y_2 = 0.0113, \quad X_{e2} = 0.54 \times 10^{-3}$$
$$Y_1 = 0.036, \quad X_{e1} = 1.41 \times 10^{-3}$$

Thus: $(X_{e1} - X_1) = (1.41 - 1.145)10^{-3} = 0.265 \times 10^{-3}$ kmol SO_2/kmol H_2O

$(X_{e2} - X_2) = 0.5 \times 10^{-3}$ kmol SO_2/kmol H_2O

Thus: $(X_e - X)_{lm} = \dfrac{(0.54 - 0.265)10^{-3}}{\ln(0.54/0.265)} = 3.86 \times 10^{-4}$ kmol SO_2/kmol H_2O

$4.94 \times 10^{-4} = 0.19 V \times 3.86 \times 10^{-4}$,

from which the packed volume, $V = 6.74$

Thus: packed height $= (6.74/1.35) = \underline{\underline{5.0 \text{ m}}}$

PROBLEM 12.11

Ammonia is removed from a 10 per cent ammonia–air mixture by scrubbing with water in a packed tower, so that 99.9 per cent of the ammonia is removed. What is the required height of tower? The gas enters at 1.2 kg/m²s, the water rate is 0.94 kg/m²s and $K_G a$ is 0.0008 kmol/s m³ (kN/m²).

Solution

The molecular masses of ammonia and air are 17 and 29 kg/kmol respectively. The data in mass per cent must be converted to mole ratios as the inlet gas concentration is high.

Thus:
$$0.10 = \frac{17 y_1}{17 y_1 + 29(1 - y_1)} \quad \text{and} \quad y_1 = 0.159$$

$$Y_1 = \left(\frac{0.159}{1 - 0.159}\right) = 0.189$$

$$Y_2 \simeq y_2 = 0.000159$$

The rates of entering gases are: total $= 1.2$ kg/m²s, ammonia $= 0.12$ kg/m²s, and air $= 1.08$ kg/m²s.

Thus: $G'_m = 0.0372$ kmol/m²s, $\quad L'_m = (0.94/18) = 0.0522$ kmol/m²s

and: $X_2 = 0$ that is ammonia free

The equation of the operating line is found from a mass balance between a plane where the compositions are X and Y and the top of the tower as:

$$0.0372(Y - 0.000159) = 0.0522X$$

or:
$$Y = (1.4X + 0.000159)$$

This equilibrium line is plotted on Figure 12b.

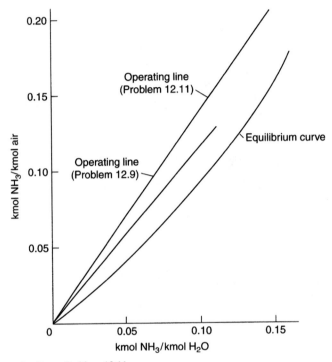

Figure 12b. Operating lines, Problem 12.11

The integral in the following equation may be obtained graphically from Figure 12c as 40.55 using the following data.

$$Z = \frac{G'_m}{k_G a P} \int_{Y_2}^{Y_1} \frac{(1+Y)(1+Y_i)dY}{(Y-Y_i)}$$

Y	Y_i	$(Y - Y_i)$	$(1+Y)(1+Y_i)$	$\dfrac{(1+Y)(1+Y_i)}{(Y-Y_i)}$
0.20	0.152	0.048	3.18	66.3
0.19	0.138	0.052	1.35	26.0
0.15	0.102	0.048	1.27	26.4
0.10	0.063	0.037	1.17	31.6

Y	Y_i	$(Y - Y_i)$	$(1+Y)(1+Y_i)$	$\dfrac{(1+Y)(1+Y_i)}{(Y-Y_i)}$
0.05	0.028	0.022	1.08	49.1
0.04	0.022	0.018	1.06	58.8
0.03	0.016	0.014	1.05	74.7
0.02	0.011	0.009	1.03	114.6
0.01	0.005	0.005	1.015	203.0
0.00015	0.000	0.00015	1.00015	6670.0

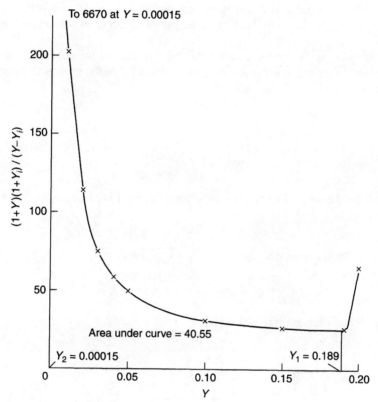

Figure 12c. Evaluation of integral, Problem 12.11

$K_G a$ is approximately equal to $k_G a$ for a very soluble gas so that:

$$Z = \frac{(0.0372 \times 40.55)}{(0.0008 \times 101.3)} = \underline{\underline{18.6 \text{ m}}}$$

It is interesting to note that if $Y = 0.01$ rather than 0.00015, the integral has a value of 8.25 and Z is equal to 3.8 m. Thus 14.8 m of packing is required to remove the last traces of ammonia.

PROBLEM 12.12

A soluble gas is absorbed from a dilute gas–air mixture by countercurrent scrubbing with a solvent in a packed tower. If the liquid fed to the top of the tower contains no solute, show that the number of transfer units required is given by:

$$N = \frac{1}{\left[1 - \dfrac{mG'_m}{L_m}\right]} \ln\left[\left(1 - \frac{mG'_m}{L_m}\right)\frac{y_1}{y_2} + \frac{mG_m}{L_m}\right]$$

where G'_m and L_m are the flowrates of the gas and liquid in kmol/s m^2 tower area, and y_1 and y_2 the mole fractions of the gas at the inlet and outlet of the column. The equilibrium relation between the gas and liquid is represented by a straight line with the equation $y_e = mx$, where y_e is the mole fraction in the gas in equilibrium with mole fraction x in the liquid.

In a given process, it is desired to recover 90 per cent of the solute by using 50 per cent more liquid than the minimum necessary. If the HTU of the proposed tower is 0.6 m, what height of packing will be required?

Solution

By definition:
$$N_{OG} = \int_{y_2}^{y_1} \frac{dy}{y - y_e} \qquad \text{(equation 12.77)}$$

A mass balance between the top and some plane in the tower where the mole fractions are x and y gives:
$$G_m(y - y_2) = L_m(x - x_2)$$

If the inlet liquid is solute free, then:
$$x_2 = 0 \quad \text{and} \quad x = \frac{G'_m}{L'_m}(y - y_2)$$

If the equilibrium data are represented by:
$$y_e = mx$$

then substituting for $y_e = m(G'_m/L'_m)(y - y_2)$ gives:

$$N_{OG} = \int_{y_2}^{y_1} \frac{dy}{y - \dfrac{mG'_m}{L'_m}(y - y_2)}$$

$$= \int_{y_2}^{y_1} \frac{dy}{y\left(1 - \dfrac{mG'_m}{L'_m}\right) + \dfrac{mG'_m}{L'_m}y_2}$$

$$= \left(1 - \frac{mG'_m}{L'_m}\right)^{-1} \ln\left[\left(1 - \frac{mG'_m}{L'_m}\right)\frac{y_1}{y_2} + \frac{mG'_m}{L'_m}\right]$$

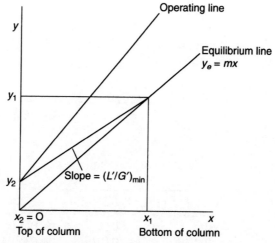

Figure 12d. Graphical construction for Problem 12.12

Referring to Figure 12d:

$$\left(\frac{L'}{G'}\right)_{min} = \left(\frac{y_1 - y_2}{x_1}\right) = \left(\frac{y_1 - y_2}{y_1/m}\right) = m\left(1 - \frac{y_2}{y_1}\right)$$

$$= m\left(1 - \frac{0.1y_1}{y_1}\right) = 0.9\ m$$

If 1.5 $(L'/G')_{min}$ is actually employed, $L'_m/G'_m = (1.5 \times 0.9)\ m = 1.35\ m$

Thus: $\dfrac{mG_m}{L_m} = \dfrac{m}{1.35\ m} = 0.74$

$y_1/y_2 = 10$

Therefore: $\mathbf{N}_{OG} = \dfrac{1}{0.26} \ln[(0.26 \times 10) + 0.74] = 4.64$

$\mathbf{H}_{OG} = 0.6$ m and the height of packing $= (0.6 \times 4.64) = \underline{\underline{2.78\ m}}$

PROBLEM 12.13

A paraffin hydrocarbon of molecular mass 114 kg/kmol at 373 K, is to be separated from a mixture with a non-volatile organic compound of molecular mass 135 kg/kmol by stripping with steam. The liquor contains 8 per cent of the paraffin by mass and this is to be reduced to 0.08 per cent using an upward flow of steam saturated at 373 K. If three times the minimum amount of steam is used, how many theoretical stages will be required? The vapour pressure of the paraffin at 373 K is 53 kN/m² and the process takes place at atmospheric pressure. It may be assumed that the system obeys Raoult's law.

Solution

If Raoult's law applies, the partial pressure $= x$ (vapour pressure)

That is: $P_A = x P_A^0$

$$y = P_A/P \text{ and hence } y_e = x(P_A^0/P) = (53/101.3)x = 0.523x$$

In terms of mole ratios: $\dfrac{Y_e}{1+Y_e} = \dfrac{0.523 X}{1+X}$

Thus the equilibrium curve may be obtained as follows.

X	X/(1+X)	$Y_e/(1+Y_e)$	Y_e
0	0	0	0
0.02	0.0196	0.0103	0.0104
0.04	0.0385	0.020	0.0204
0.06	0.0566	0.0296	0.0305
0.08	0.0741	0.0387	0.0403
0.10	0.0909	0.0475	0.0499
0.12	0.107	0.0560	0.059

This curve is plotted in Figure 12e.

As the inlet gas contains 8 per cent by mass of paraffin, then:

$$X_2 = (8/114)/(92/135) = 0.103$$

and: $\quad X_1 = 0.00103 \quad \text{and} \quad Y_1 = 0$

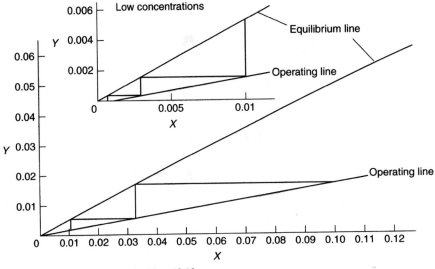

Figure 12e. Equilibrium data for Problem 12.13

The minimum amount is required occurs when the exit streams are in equilibrium, that is when:
$$X_2 = 0.103, \quad Y_{e_2} = 0.0513$$

From an overall mass balance:
$$L_{min}(0.103 - 0.00103) = G_{min}(0.0513 - 0)$$
and:
$$(L/G)_{min} = 0.503 \quad \text{and} \quad (L/G)_{actual} = 0.167$$

Thus the operating line, passing through the point (0.00103, 0) with a slope = 0.167, may be drawn in Figure 12e and the number of theoretical stages is found to be 4.

This problem may also be solved analytically by the use of the absorption factor method. This is illustrated in Problem 12.16.

PROBLEM 12.14

Benzene is to be absorbed from coal gas by means of a wash-oil. The inlet gas contains 3 per cent by volume of benzene, and the exit gas should not contain more than 0.02 per cent benzene by volume. The suggested oil circulation rate is 480 kg oil/100 m³ of inlet gas measured at 273 K and 101.3 kN/m². The wash oil enters the tower solute-free. If the overall height of a transfer unit based on the gas phase is 1.4 m, determine the minimum height of the tower which is required to carry out the absorption. The equilibrium data are:

Benzene in oil (per cent by mass)	0.05	0.01	0.50	1.0	2.0	3.0
Equilibrium partial pressure of benzene in gas (kN/m^2)	0.013	0.033	0.20	0.53	1.33	3.33

Solution

At the top and bottom of the tower respectively:
$$y_2 = 0.0002, \quad x_2 = 0$$
and:
$$y_1 = 0.03, \quad x_1 = \text{exit oil composition}$$

Taking 100 m³ of inlet gas at 273 K and 101.3 kN/m² as the basis of calculation,

Volume of benzene at inlet = $(0.03 \times 100) = 3.0$ m³ in 97.0 m³ of gas.

Volume of benzene at exit = $(0.0002 \times 97) = 0.0194$ m³.

At 273 K and 101.3 kN/m², the kilogramme molecular volume = 22.4 m³.

Thus: $\quad$ kmol of gas = $(97/22.4) = 4.33$ kmol

Volume of benzene absorbed = (3.0 − 0.0194) = 2.9806 m³.

Density of benzene at 273 K and 101.3 kN/m² = (78/22.4) = 3.482 kg/m³.

Thus: mass of benzene absorbed = (2.9806 × 3.482) = 10.38 kg

As the oil rate = 490 kg/100 m³ of gas,

the mass per cent of benzene at the exit = (10.38 × 100)/490 = 2.12 per cent

$$Y_1 = 0.03/(1 - 0.03) = 0.031$$
$$Y_2 \simeq y_2 = 0.0002$$

Thus the operating line may be plotted as shown in Figure 12f. The equilibrium data converted to the appropriate units, are as follows and are plotted in Figure 12f.

Mass per cent benzene	Mass fraction	Equilibrium partial pressure (kN/m²)	$Y_e = P^0/P$
0	0	0	0
0.05	0.0005	0.013	0.00013
0.10	0.001	0.033	0.00033
0.50	0.005	0.20	0.00197
1.0	0.01	0.53	0.00523
2.0	0.02	1.33	0.01313
3.0	0.03	3.33	0.3287

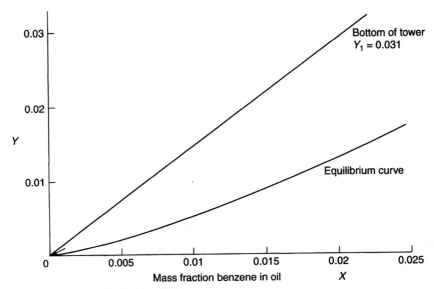

Figure 12f. Operating line for Problem 12.14

$$N_{OG} = \int_{Y_1}^{Y_2} \frac{dY}{Y - Y_e}$$

The value of this integral may be evaluated from the operating and equilibrium line by graphical means for values of Y between 0.0002 and 0.031.

Y	Y_e	$(Y - Y_e)$	$1/(Y - Y_e)$
0.0002	0	0.002	5000
0.0015	0.00033	0.00117	855
0.003	0.007	0.0023	435
0.005	0.0012	0.0038	263
0.0075	0.0021	0.0054	185
0.010	0.0031	0.0069	145
0.015	0.0055	0.0095	105
0.020	0.008	0.012	83
0.025	0.0106	0.0144	69
0.030	0.0137	0.0163	61

From Figure 12g, the area under the curve, $N_{OG} = 8.27$

Thus: height of column $= N_{OG} \times H_{OG} = (8.27 \times 1.4) = \underline{11.6 \text{ m}}$

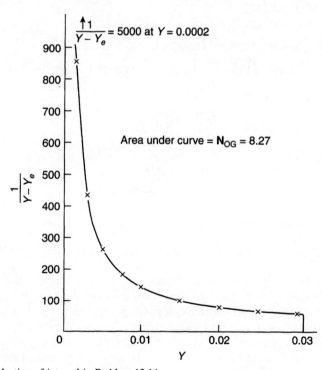

Figure 12g. Evaluation of integral in Problem 12.14

PROBLEM 12.15

Ammonia is to be recovered from a 5 per cent by volume ammonia–air mixture by scrubbing with water in a packed tower. The gas rate is 1.25 m^3/m^{2}s measured at 273 K and 101.3 kN/m^2 and the liquid rate is 1.95 kg/m^{2}s. The temperature of the inlet gas is 298 K and the temperature of the inlet water 293 K. The mass transfer coefficient is $K_G a = 0.113$ kmol/m^{3}s (mole ratio difference) and the total pressure is 101.3 kN/m^2. What is the required height of the tower to remove 95 per cent of the ammonia. The equilibrium data and the heats of solutions are:

Mole fraction in liquid	0.005	0.01	0.015	0.02	0.03
Integral heat of solution (kJ/kmol of solution)	181	363	544	723	1084
Equilibrium partial pressures (kN/m^2)					
at 293 K	0.4	0.77	1.16	1.55	2.33
at 298 K	0.48	0.97	1.43	1.92	2.93
at 303 K	0.61	1.28	1.83	2.47	3.86

Adiabatic conditions may be assumed and heat transfer between phases neglected.

Solution

The data provided are presented in Figure 12h.

The entering gas rate = 12.5 m^3/m^{2}s at 273 K and 101.3 kN/m^2.

Density at 273 K and 101.3 kN/m^2 = $(1/22.4) = 0.0446$ kmol/m^3.

At the bottom of the tower, $y_1 = 0.05$.

Thus: $Y_1 = (0.05/0.95) = 0.0526$

$G'_m = (0.95 \times 1.25 \times 0.0446) = 0.053$ kmol/m^{2}s

$Y_2 = (0.05 \times 0.0526) = 0.00263$

$L'_m = (1.95/18) = 0.108$ kmol/m^{2}s   and $X_2 = 0$

An overall mass balance gives:

$0.108(X_1 - 0) = 0.053(0.0526 - 0.00263)$ and $X_i = 0.0245$

In this problem, the temperature varies throughout the column and the tower will be divided into increments so that by heat and mass balances the terminal conditions over each section may be found. Knowing the compositions and the temperature, the data given may be used in conjunction with the mass coefficient to calculate the height of the chosen increment. Adiabatic conditions will be assumed and, as the sensible heat change of the gas is small, the heat of solution will be used only to raise the temperature of the liquid. The gas temperature will therefore remain constant at 295 K.

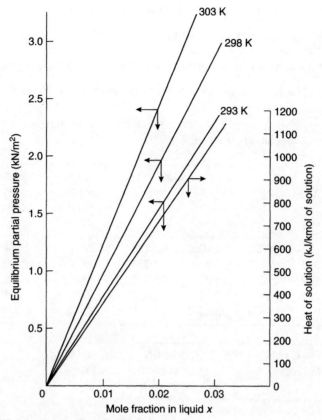

Figure 12h. Equilibrium data, Problem 12.15

Considering conditions at the top of the tower:

$$X_2 = 0, \quad T_1 = 293 \text{ K}, \quad T_g = 295 \text{ K}, \quad Y_2 = 0.00263$$

Choosing an increment such that the exit liquor stream and inlet gas streams have compositions $X = 0.005$ and Y, a mass balance taking 1 m² as a basis gives:

$$L_m(X - 0) = G_m(Y - Y_2)$$

That is: $(0.108 \times 0.005) = 0.053(Y - 0.00263)$ and $Y = 0.0128$

NH$_3$ absorbed in the section $= 0.108 \times 0.005 = 5.4 \times 10^{-4}$ kmol/s.

Heat of solution when $X = 0.005 = 181$ kJ/kmol of solution.

Thus: heat liberated $= (181 \times 0.005) = 19.55$ kW

and: temperature rise $= 19.55/(0.108 \times 18 \times 4.18) = 2.4$ deg K

and the liquid exit temperature $= 295.4$ K

At 295.4 K when $X = 0.005$, $P_e = 0.44$ kN/m², and $Y_e = 0.00434$

At the top of the section, $Y - Y_e = (0.00263 - 0) = 0.00263$.

At the bottom of the section, $Y - Y_e = (0.0128 - 0.00434) = 0.00846$.

Thus: $(Y - Y_e)_{lm} = (0.00846 - 0.00263)/\ln(0.00846/0.00263) = 0.00499$

and: $5.4 \times 10^{-4} = 0.113 \times 1 \times \mathbf{H} \times 0.00499$ (since $A = 1$ m^2) and $\mathbf{H} = 0.958$ m.

In a similar way, further increments may be taken and the heights of each found. A summary of these calculations is as follows.

Increment	Inlet		Outlet		Outlet liquid temperature (K)	Outlet P_e (kN/m^2)	$(Y - Y_e)$ top
	X	Y	X	Y			
1	0	0.00263	0.005	0.0128	295.4	0.44	0.00263
2	0.005	0.0128	0.010	0.023	297.8	0.95	0.00846
3	0.010	0.023	0.015	0.0332	300.2	1.62	0.0136
4	0.015	0.0332	0.020	0.0434	302.6	2.40	0.0172
5	0.020	0.0434	0.0245	0.0526	304.8	3.30	0.0197

Increment	$(Y - Y_e)$ bottom	$(Y - Y_e)_{lm}$	Quantity absorbed (kmol/s)	Height of section (m)
1	0.00846	0.00499	5.4×10^{-4}	0.958
2	0.0136	0.01083	5.4×10^{-4}	0.441
3	0.0172	0.0153	5.4×10^{-4}	0.328
4	0.0197	0.0185	5.4×10^{-4}	0.258
5	0.020	0.0198	5.4×10^{-4}	0.241
				= 2.23 m

Thus the required height of packing = <u>2.23 m</u>.

PROBLEM 12.16

A thirty-plate bubble-cap column is to be used to remove n-pentane from a solvent oil by means of steam stripping. The inlet oil contains 6 kmol of n-pentane/100 kmol of pure oil and it is desired to reduce the solute content of 0.1 kmol/100 kmol of solvent. Assuming isothermal operation and an overall plate efficiency of 30 per cent, what is the specific steam consumption, that is kmol of steam required/kmol of solvent oil treated, and the ratio of the specific and minimum steam consumptions. How many plates would be required if this ratio is 2.0?

The equilibrium relation for the system may be taken as $Y_e = 3.0X$, where Y_e and X are expressed in mole ratios of pentane in the gas and liquid phases respectively.

Solution

See Volume 2, Example 12.6.

PROBLEM 12.17

A mixture of ammonia and air is scrubbed in a plate column with fresh water. If the ammonia concentration is reduced from 5 per cent to 0.01 per cent, and the water and air rates are 0.65 and 0.40 kg/m²s, respectively, how many theoretical plates are required? The equilibrium relationship may be written as $Y = X$, where X is the mole ratio in the liquid phase.

Solution

Assuming that the compositions are given as volume per cent, then:

At the bottom of the tower: $y_1 = 0.05$ and $Y_1 = \dfrac{0.05}{(1 - 0.05)} = 0.0526$.

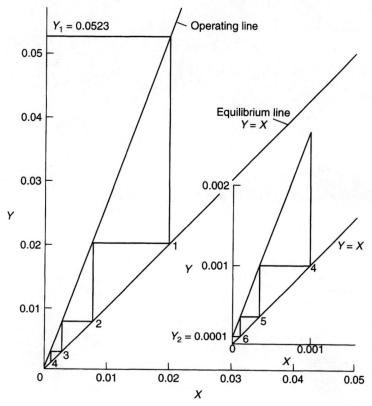

Figure 12i. Graphical construction for Problem 12.17

At the top of the tower: $y_2 = 0.0001 = Y_2$.

$$L'_m = (0.65/18) = 0.036 \text{ kmol/m}^2\text{s}$$
$$G'_m = (0.40/29) = 0.0138 \text{ kmol/m}^2\text{s}$$

A mass balance gives the equation of the operating line as:

$$0.0138(Y - 0.0001) = 0.036(X - 0)$$

or:
$$Y = 2.61 + 0.0001$$

The operating line and equilibrium line are then drawn in and from Figure 12i, 6 theoretical stages are required.

SECTION 2-13

Liquid–Liquid Extraction

PROBLEM 13.1

Tests are made on the extraction of acetic acid from a dilute aqueous solution by means of a ketone in a small spray tower of diameter 46 mm and effective height of 1090 mm with the aqueous phase run into the top of the tower. The ketone enters free from acid at the rate of 0.0014 m^3/s m^2, and leaves with an acid concentration of 0.38 kmol/m^3. The concentration in the aqueous phase falls from 1.19 to 0.82 kmol/m^3.

Calculate the overall extraction coefficient based on the concentrations in the ketone phase, and the height of the corresponding overall transfer unit.

The equilibrium conditions are expressed by:

(Concentration of acid in ketone phase) = 0.548 (Concentration of acid in aqueous phase).

Solution

The solvent flowrate, $L'_E = 0.0014$ m^3/s m^2

and the increase in concentration of the extract stream is given by:

$$(C_{E_2} - C_{E_1}) = (0.38 - 0) = 0.38 \text{ kmol/m}^3$$

At the bottom of the column, $C_{R_1} = 0.82$ kmol/m^3
and the equilibrium concentration is then:

$$C^*_{E_1} = (0.548 \times 0.82) = 0.449 \text{ kmol/m}^3, \quad C_{E_1} = 0$$

Thus: $\quad \Delta C_1 = (C^*_{E_1} - C_{E_1}) = 0.449$ kmol/m^3

At the top of the column:

$$C_{R_2} = 1.19 \text{ kmol/m}^3 \text{ and hence } C^*_{E_2} = (0.548 \times 1.19) = 0.652 \text{ kmol/m}^3$$

$$C_{E_2} = 0.38 \text{ kmol/m}^3$$

Therefore: $\Delta C_2 = (C^*_{E_2} - C_{E_2}) = (0.652 - 0.38) = 0.272$ kmol/m^3

The logarithmic mean driving force, $\Delta C_{\text{lm}} = (0.449 - 0.272)/\ln(0.449/0.272)$

$$= 0.353 \text{ kmol/m}^3$$

The effective height, $Z = 1.09$ m

and in equation 13.30:
$$0.0014(0.38 - 0) = K_E a(0.353 \times 1.09)$$

from which: $\quad K_E a = \underline{0.00138 \text{ s}^{-1}}$

The height of an overall transfer unit based on concentrations in the extract phase is given by equation 13.26:

$$H_{OE} = L'_E / K_E a = (0.0014/0.00138) = \underline{1.02 \text{ m}}$$

PROBLEM 13.2

A laboratory test is carried out into the extraction of acetic acid from dilute aqueous solution, by means of methyl iso-butyl ketone, using a spray tower of 47 mm diameter and 1080 mm high. The aqueous liquor is run into the top of the tower and the ketone enters at the bottom.

The ketone enters at the rate of 0.0022 m³/s m² of tower cross-section. It contains no acetic acid, and leaves with a concentration of 0.21 kmol/m³. The aqueous phase flows at the rate of 0.0013 m³/s m² of tower cross-section, and enters containing 0.68 kmol acid/m³.

Calculate the overall extraction coefficient based on the driving force in the ketone phase. What is the corresponding value of the overall HTU, based on the ketone phase?

Using units of kmol/m³, the equilibrium relationship under these conditions may be taken as:

(Concentration of acid in the ketone phase) = 0.548 (Concentration in the aqueous phase.)

Solution

The increase in concentration of the extract phase, $(C_{E_2} - C_{E_1}) = 0.21$ kmol/m³ and the amount of acid transferred to the ketone phase is given by:

$$L'_E (C_{E_2} - C_{E_1}) = (0.0022 \times 0.21) = 0.000462 \text{ kmol/m}^2\text{s}$$

Making a mass balance by means of equation 13.21 gives:

$$0.000462 = 0.0013(0.68 - C_{R_1}) \quad \text{and} \quad C_{R_1} = 0.325 \text{ kmol/m}^3$$

At the top of the column:

$$C_{R_2} = 0.68 \text{ kmol/m}^3 \text{ and hence } C^*_{E_2} = (0.548 \times 0.68) = 0.373 \text{ kmol/m}^3$$

$$C_{E_2} = 0.21 \text{ kmol/m}^3 \text{ and hence } \Delta C_2 = (0.373 - 0.21) = 0.163 \text{ kmol/m}^3$$

At the bottom of the column:

$$C_{R_1} = 0.325 \text{ kmol/m}^3 \text{ and hence } C^*_{E_1} = (0.548 \times 0.325) = 0.178 \text{ kmol/m}^3$$

$$C_{E_1} = 0 \text{ and hence } \Delta C_1 = (0.178 - 0) = 0.178 \text{ kmol/m}^3$$

The logarithmic mean driving force is then:

$$\Delta C_{lm} = (0.178 - 0.163)/\ln(0.178/0.163) = 0.170 \text{ kmol/m}^3$$

The height $Z = 1.08$ m, and in equation 13.30:

$$0.0022(0.21 - 0) = K_E a(0.170 \times 1.08) \text{ and } K_E a = \underline{0.0025/\text{s}}$$

In equation 13.26: $H_{OE} = (0.0022/0.0025) = \underline{0.88 \text{ m}}$

PROBLEM 13.3

Propionic acid is extracted with water from a dilute solution in benzene, by bubbling the benzene phase into the bottom of a tower to which water is fed at the top. The tower is 1.2 m high and 0.14 m² in area, the drop volume is 0.12 cm³, and the velocity of rise is 12 cm/s. From laboratory tests the value of K_w during the formation of drops is 7.6×10^{-5} kmol/s m² (kmol/m³) and for rising drops $K_w = 4.2 \times 10^{-5}$ kmol/s m² (kmol/m³). What is the value of $K_w a$ for the tower in kmol/sm³ (kmol/m³)?

Solution

Considering drop formation

$$K_W = 6 \times 10^{-5} \text{ kmol/s m}^2 (\text{kmol/m}^3)$$

Droplet volume $= 0.12$ cm³   or 1.2×10^{-7} m³.
Radius of a drop $= [(3 \times 1.2 \times 10^{-7})/4\pi]^{0.33} = 3.08 \times 10^{-3}$ m.

Mean area during formation, as in Problem 13.6 $= 12\pi r^2/5$

$$= 12\pi(3.08 \times 10^{-3})^2/5$$
$$= 7.14 \times 10^{-5} \text{ m}^2$$

Mean time of exposure $= (3t_f/5)$ s, where t_f, the time of formation, may be taken as $t_f = $ (volume of one drop)/(volumetric throughput) $= (1.2 \times 10^{-7}/Q)$, where Q m³/s is the volumetric throughput of the benzene phase.

Thus: mean time of exposure $= (3 \times 1.2 \times 10^{-7})/5Q = (7.2 \times 10^{-8})/Q$ s

and mass transferred $= (6 \times 10^{-5} \times 7.14 \times 10^{-5} \times 7.2 \times 10^{-8})/Q$

$$= (3.24 \times 10^{-16})/Q \text{ kmol/(kmol/m}^3)$$

Considering drop rise

$$K_W = 4.2 \times 10^{-5} \text{ kmol/s m}^2(\text{kmol/m}^3)$$

Residence time = $(1.2/0.12) = 10$ s

Thus: volume in suspension = $10Q$ m^3

and: number of drops rising = $10Q/(1.2 \times 10^{-7}) = (8.3 \times 10^7)Q$

Area of one drop = $4\pi(3.08 \times 10^{-3})^2 = (1.19 \times 10^{-4})$ m^2

and interfacial area available = $(8.3 \times 10^7 Q \times 1.19 \times 10^{-4}) = 9.88Q \times 10^3$ m^2

Thus: mass transferred = $(4.2 \times 10^{-5} \times 10 \times 9.88Q \times 10^3) = 4.15Q$ kmol/(kmol/m^3)

Total mass transferred = $4.15Q + (3.24 \times 10^{-16})/Q$ kmol/(kmol/m^3).

Total residence time = $10 + (1.2 \times 10^{-7}/Q)$ s.

Volume of column = $(1.2 \times 0.14) = 0.168$ m^3.

Therefore: $K_{wa} = (4.15Q + 3.24 \times 10^{-16}/Q)/[0.168(10 + 1.2 \times 10^{-7}/Q)]$

which is approximately equal to $\underline{2.47Q \text{ kmol/s m}^3(\text{kmol/m}^3)}$

Further solution is not possible without information on the volumetric throughput of the benzene phase.

PROBLEM 13.4

A 50 per cent solution of solute **C** in solvent **A** is extracted with a second solvent **B** in a countercurrent multiple contact extraction unit. The mass of **B** is 25 per cent that of the feed solution, and the equilibrium data are:

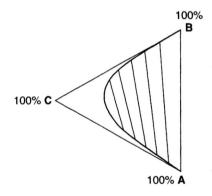

Determine the number of ideal stages required, and the mass and concentration of the first extract if the final raffinate contains 15 per cent of solute **C**.

Solution

See Volume 2, Example 13.2.

PROBLEM 13.5

A solution of 5 per cent acetaldehyde in toluene is to be extracted with water in a five stage co-current unit. If 25 kg water/100 kg feed is used, what is the mass of acetaldehyde extracted and the final concentration? The equilibrium relation is given by:

$$(\text{kg acetaldehyde/kg water}) = 2.20 \, (\text{kg acetaldehyde/kg toluene})$$

Solution

For co-current contact with immiscible solvents where the equilibrium curve is a straight line as in the present case:

$$X_n = [A/(A + Sm)]^n X_f \qquad \text{(equation 13.6)}$$

On the basis of 100 kg feed:

mass of solvent in the feed, $A = 95$ kg,

mass of solvent added, $S = 25$ kg,

slope of the equilibrium line, $m = 2.20$ kg/kg,

number of stages, $n = 5$, and

mass ratio of solute in the feed $X_f = (5/95) = 0.0527$ kg/kg.

Thus: mass ratio of solute in raffinate, $X_n = [95/(95 + 2.2 \times 25)]^5 \times 0.0527$

$$= 0.00538 \text{ kg/kg solvent}$$

Thus, the final solution consists of 0.00538 kg acetaldehyde in 1.00538 kg solution and the concentration $= (100 \times 0.00538)/1.00538 = \underline{0.536 \text{ per cent}}$

With 95 kg toluene in the raffinate, the mass of acetaldehyde $= (0.00538 \times 95) = 0.511$ kg

and the mass of acetaldehyde extracted $= (5.0 - 0.511) = \underline{4.489 \text{ kg}/100 \text{ kg feed}}$

PROBLEM 13.6

If a drop is formed in an immiscible liquid, show that the average surface available during formation of the drop is $12\pi r^2/5$, where r is the final radius of the drop, and that the average time of exposure of the interface is $3t_f/5$, where t_f is the time of formation of the drop.

Solution

If it is assumed that the volumetric input to the drop is constant or,

$$(4\pi r^3/3t) = k, \quad \text{then} \quad r^2 = (3k/4\pi)^{2/3} t^{2/3}$$

The surface area at radius r is given by $a_s = 4\pi r^2$
or substituting for r^2: $\quad a_s = 4\pi (3k/4\pi)^{2/3} t^{2/3}$
The mean area exposed between time 0 and t is then:

$$\bar{a}_s = (1/t) \int_0^t a_s dt$$

$$= (1/t) 4\pi (3k/4\pi)^{2/3} \int_0^t t^{2/3} dt$$

$$= 4\pi (3k/4\pi)^{2/3} (3t^{2/3}/5) = \underline{\underline{12\pi r^2/5}}$$

The area under the curve of a_s as a function of t is:

$$\int_0^t a_s dt \quad \text{or} \quad (12\pi r^2 t/5)$$

and the mean time of exposure, $\bar{t} = (1/a_s) \int_0^t a_s dt = (12\pi r^2 t / 5 a_s)$
When $t = t_f$, the time of formation, $a_s = 4\pi r^2$
and: $\quad \bar{t} = (12\pi r^2 t_f / 20\pi r^2) = \underline{\underline{3t_f/5}}$

PROBLEM 13.7

In the extraction of acetic acid from an aqueous solution with benzene a packed column of height 1.4 m and cross-sectional area 0.0045 m², the concentrations measured at the inlet and the outlet of the column are:

acid concentration in the inlet water phase, $C_{W_2} = 0.69$ kmol/m³.
acid concentration in the outlet water phase, $C_{W_1} = 0.684$ kmol/m³.
flowrate of benzene phase = 5.6×10^{-6} m³/s = 1.24×10^{-3} m³/m²s.
inlet benzene phase concentration, $C_{B_1} = 0.0040$ kmol/m³.
outlet benzene phase concentration, $C_{B_2} = 0.0115$ kmol/m³.
Determine the overall transfer coefficient and the height of the transfer unit.

Solution

The acid transferred to the benzene phase is:

$$5.6 \times 10^{-6} (0.0115 - 0.0040) = 4.2 \times 10^{-8} \text{ kmol/s}$$

The equilibrium relationship for this system is:

$$C_B^* = 0.0247 C_W \quad \text{or} \quad C_{B_1}^* = (0.0247 \times 0.684) = 0.0169 \text{ kmol/m}^3$$

and: $$C_{B_2}^* = (0.0247 \times 0.690) = 0.0171 \text{ kmol/m}^3$$

Thus: driving force at bottom, $\Delta C_1 = (0.0169 - 0.0040) = 0.0129 \text{ kmol/m}^3$

and: driving force at top, $\Delta C_2 = (0.0171 - 0.0115) = 0.0056 \text{ kmol/m}^3$

Logarithmic mean driving force, $\Delta C_{lm} = 0.0087 \text{ kmol/m}^3$.

Therefore:
$$K_B a = (\text{kmol transferred})/(\text{volume of packing} \times \Delta C_{lm})$$
$$= (4.2 \times 10^{-8})/(1.4 \times 0.0045 \times 0.0087)$$
$$= \underline{\underline{7.66 \times 10^{-4}}} \text{ kmol/s m}^3 \text{ (kmol/m}^3\text{)}$$

and: $\mathbf{H}_{OB} = (1.24 \times 10^{-3})/(7.66 \times 10^{-4}) = \underline{1.618 \text{ m}}$

PROBLEM 13.8

It is required to design a spray tower for the extraction of benzoic acid from its solution in benzene.

Tests have been carried out on the rate of extraction of benzoic acid from a dilute solution in benzene to water, in which the benzene phase was bubbled into the base of a 25 mm diameter column and the water fed to the top of the column. The rate of mass transfer was measured during the formation of the bubbles in the water phase and during the rise of the bubbles up the column. For conditions where the drop volume was 0.12 cm^3 and the velocity of rise 12.5 cm/s, the value of K_w for the period of drop formation was 0.000075 kmol/s m^2 (kmol/m^3), and for the period of rise 0.000046 kmol/s m^2 (kmol/s m^3).

If these conditions of drop formation and rise are reproduced in a spray tower of 1.8 m in height and 0.04 m^2 cross-sectional area, what is the transfer coefficient, $K_w a$, kmol/s m^3 (kmol/m^3), where a represents the interfacial area in m^2/unit volume of the column? The benzene phase enters at the flowrate of 38 cm^3/s.

Solution

See Volume 2, Example 13.5.

PROBLEM 13.9

It is proposed to reduce the concentration of acetaldehyde in aqueous solution from 50 per cent to 5 per cent by mass, by extraction with solvent **S** at 293 K. If a countercurrent multiple-contact process is adopted and 0.025 kg/s of the solution is treated with an equal quantity of solvent, determine the number of theoretical stages required and the mass flowrate and concentration of the extract from the first stage.

The equilibrium relationship for this system at 293 K is:

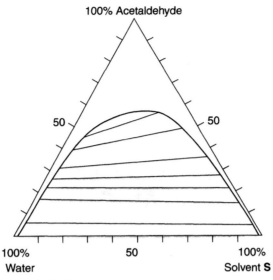

Figure 13a. Equilibrium data for Problem 13.9

Solution

The data are replotted in Figure 13b and the point F, representing the feed, is drawn in at 50 per cent acetaldehyde, 50 per cent water. Similarly, R_n, the raffinate from stage n located on the curve corresponding to 5 per cent acetaldehyde. (This solution then contains 2 per cent S and 93 per cent water.) FS is joined and point M located such that FM = MS, since the ratio of feed solution to solvent is unity. R_nM is projected to meet the equilibrium curve at E_1 and FE_1 and R_nS are projected to meet at P. The tie-line E_1R_1 is drawn in and the line R_1P then meets the curve at E_2. The working is continued in this way and it is found that R_4 is below R_n and hence four theoretical stages are required.

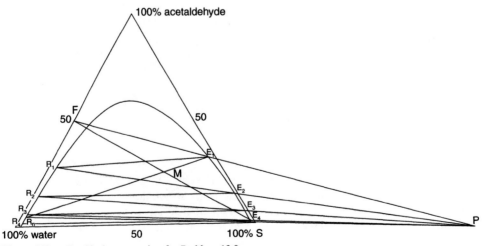

Figure 13b. Graphical construction for Problem 13.9

From Figure 13b, the composition of the extract from stage 1, E_1, is:

3 per cent water, 32 per cent acetaldehyde, and 65 per cent **S**

Making an overall balance, $F + S = R_n + E_1$

Thus: $(0.025 + 0.025) = (R_n + E_1) = 0.050$ kg/s

Making an acetaldehyde balance:

$$(0.50F + 0) = (0.05R_n + 0.32E_1)$$

$$0.05(0.050 - E_1) + 0.32E_1 = (0.50 \times 0.025) \quad \text{and} \quad \underline{\underline{E_1 = 0.037 \text{ kg/s}}}$$

PROBLEM 13.10

160 cm^3/s of a solvent **S** is used to treat 400 cm^3/s of a 10 per cent by mass solution of **A** in **B**, in a three-stage countercurrent multiple contact liquid–liquid extraction plant. What is the composition of the final raffinate?

Using the same total amount of solvent, evenly distributed between the three stages, what would be the composition of the final raffinate, if the equipment were used in a simple multiple-contact arrangement?

Equilibrium data:

kg A/kg **B**	0.05	0.10	0.15
kg A/kg **S**	0.069	0.159	0.258
Densities (kg/m^3)	**A** = 1200,	**B** = 1000,	**S** = 800

Solution

See Volume 2, Example 13.1

PROBLEM 13.11

In order to extract acetic acid from a dilute aqueous solution with isopropyl ether, the two immiscible phases are passed countercurrently through a packed column 3 m in length and 75 mm in diameter. It is found that if 0.5 kg/m^2 of the pure ether is used to extract 0.25 kg/m^{2}s of 4.0 per cent acid by mass, then the ether phase leaves the column with a concentration of 1.0 per cent acid by mass. Calculate:

(a) the number of overall transfer units, based on the raffinate phase, and
(b) the overall extraction coefficient, based on the raffinate phase.

The equilibrium relationship is given by: (kg acid/kg isopropyl ether) = 0.3 (kg acid/kg water).

Solution

See Volume 2, Example 13.5

PROBLEM 13.12

It is proposed to recover material **A** from an aqueous effluent by washing it with a solvent **S** and separating the resulting two phases. The light product phase will contain **A** and the solvent **S** and the heavy phase will contain **A** and water. Show that the most economical solvent rate, W (kg/s) is given by:

$$W = [(F^2 a x_0)/mb)]^{0.5} - F/m$$

where the feedrate is F kg/s water containing x_0 kg A/kg water, the value of **A** in the solvent product phase = £a/kg **A**, the cost of solvent **S** = £b/kg **S** and the equilibrium data are given by:

$$(\text{kg A/kg S})_{\text{product phase}} = m(\text{kg A/kg water})_{\text{water phase}}$$

where a, b and m are constants

Solution

The feed consists of F kg/s water containing Fx_0 kg/s **A**, and this is mixed with W kg/s of solvent **S**. The product consists of a heavy phase containing F kg/s water and, say, Fx kg/s of **A**, and a heavy phase containing W kg/s of **S** and Wy kg/s **A**, where y is the concentration of **A** in **S**, that is kg A/kg S.

The equilibrium relation is of the form: $y = mx$

Making a balance in terms of the solute **A** gives:

$$Fx_0 = Wy + Fx$$

or: $\quad Fx_0 = (Wmx + Fx) = x(mW + F) \quad$ and $\quad x = Fx_0/(mW + F)$

The amount received of **A** recovered = $F(x_0 - x)$ kg/s

which has a value of $Fa(x_0 - x)$ £/s.

Substituting for x, the value of **A** recovered = $Fax_0 - F^2 a x_0/(mW + F)$ £/s

The cost involved is that of the solvent used, Wb £/s.

Taking the profit P as the value of **A** recovered less the cost of solvent, all other costs being equal, then:

$$P = Fax_0[1 - F/(mW + F)] - Wb \text{ £/s}$$

Differentiating: $\quad dP/dW = F^2 a x_0 m/(mW + F)^2 - b$

Putting the differential equal to zero for maximum profit, then:

$$F^2 a x_0 m = b(mW + F)^2$$

and: $\quad\quad\quad\quad \underline{\underline{W = (F^2 a x_0/mb)^{0.5} - F/m}}$

SECTION 2-14

Evaporation

PROBLEM 14.1

A single-effect evaporator is used to concentrate 7 kg/s of a solution from 10 to 50 per cent of solids. Steam is available at 205 kN/m² and evaporation takes place at 13.5 kN/m². If the overall heat transfer coefficient is 3 kW/m² K, calculate the heating surface required and the amount of steam used if the feed to the evaporator is at 294 K and the condensate leaves the heating space at 352.7 K. The specific heat capacity of a 10 per cent solution is 3.76 kJ/kg K, the specific heat capacity of a 50 per cent solution is 3.14 kJ/kg K.

Solution

From the Steam Tables in Volume 2, assuming the steam is dry and saturated at 205 kN/m², the steam temperature = 394 K and the enthalpy = 2530 kJ/kg.

At 13.5 kN/m² water boils at 325 K and in the absence of data as to the boiling-point rise, this will be taken as the temperature of evaporation, assuming an aqueous solution. The total enthalpy of steam at 325 K is 2594 kJ/kg.

The feed containing 10 per cent solids has to be heated therefore from 294 to 325 K at which the evaporation takes place.

In the feed, the mass of dry solids $= (7 \times 10)/100 = 0.7$ kg/s and for x kg/s of water in the product:

$$(0.7 \times 100)/(0.7 + x) = 50 \quad \text{and} \quad x = 0.7 \text{ kg/s}$$

Thus: water to be evaporated $= (7.0 - 0.7) - 0.7 = 5.6$ kg/s

Summarising:

	Solids (kg/s)	Liquid (kg/s)	Total (kg/s)
Feed	0.7	6.3	7.0
Product	0.7	0.7	1.4
Evaporation		5.6	5.6

Using a datum of 273 K:

Heat entering with feed $= (7.0 \times 3.76)(294 - 273) = 552.7$ kW.

Heat leaving with product $= (1.4 \times 3.14)(325 - 273) = 228.6$ kW.

Heat leaving with evaporated water $= (5.6 \times 2594) = 14{,}526$ kW.

Thus: heat transferred from steam $= (14{,}526 + 228.6) - 552.7 = 14{,}202$ kW.

The condensed steam leaves at 352.7 K at which the enthalpy is:

$$= 4.18(352.7 - 273) = 333.2 \text{ kJ/kg}$$

Thus the heat transferred from 1 kg steam $= (2530 - 333.2) = 2196.8$ kJ/kg and the steam required $= (14{,}202/2196.8) = \underline{\underline{6.47 \text{ kg/s}}}$.

The temperature driving force will be taken as the difference between the temperature of the condensing steam and that of the evaporating water as the preheating of the solution and subcooling of the condensate represent but a small proportion of the total heat load, that is $\Delta T = (394 - 325) = 69$ deg K.

Thus from equation 14.1:

$$A = Q/U\Delta T$$
$$= 14{,}202/(3 \times 69) = \underline{\underline{68.6 \text{ m}^2}}$$

PROBLEM 14.2

A solution containing 10 per cent of caustic soda is to be concentrated to a 35 per cent solution at the rate of 180,000 kg/day during a year of 300 working days. A suitable single-effect evaporator for this purpose, neglecting the condensing plant, costs £1600 and for a multiple-effect evaporator the cost may be taken as £1600N, where N is the number of effects.

Boiler steam may be purchased at £0.2/1000 kg and the vapour produced may be assumed to be $0.85N$ kg/kg of boiler steam. Assuming that interest on capital, depreciation, and other fixed charges amount to 45 per cent of the capital involved per annum, and that the cost of labour is constant and independent of the number of effects employed, determine the number of effects which, based on the data given, will give the maximum economy.

Solution

The first step is to prepare a mass balance.

Mass of caustic soda in 180,000 kg/day of feed $= (180{,}000 \times 10)/100 = 18{,}000$ kg/day and mass of water in feed $= (180{,}000 - 18{,}000) = 162{,}000$ kg/day.

For x kg water in product/day:

$$(18{,}000 \times 100)/(18{,}000 + x) = 35$$

and water in product $x = 33{,}430$ kg/day.

Therefore: evaporation $= (162{,}000 - 33{,}430) = 128{,}570$ kg/day

Summarising:

	Solids (kg/day)	Liquid (kg/day)	Total (kg/day)
Feed	18,000	162,000	180,000
Product	18,000	33,430	51,430
Evaporation		128,570	128,570

The evaporation in one year is $(128{,}570 \times 300) = 3.86 \times 10^7$ kg

The boiler steam required $= (1/0.85N)$ kg/kg of vapour produced

or: $\qquad (3.86 \times 10^7)/0.85N = 4.54 \times 10^7/N$ kg/year

$\qquad$ Thus the annual cost of steam $= (0.2 \times 4.54 \times 10^7/N)/1000$

$\qquad\qquad = 9076/N$ £/year

The capital cost of the installation $= £1600N$ and the annual capital charges $= (1600N \times 45)/100 = £702N$/year.

The labour cost is independent of the number of effects and hence the total annual cost C is made up of the capital charges plus the cost of steam, or:

$$C = (720N + 9076/N) \text{ £/year}$$

Thus: $\qquad dC/dN = (720 - 9076/N^2)$

In order to minimise the costs:

$\qquad dC/dN = 0 \quad$ or $\quad (9076/N^2) = 720$ from which $N = 3.55$

Thus N must be either 3 or 4 effects.

When $N = 3$, $\qquad C = (720 \times 3) + (9076/3) = 5185$ £/year

When $N = 4$, $\qquad C = (720 \times 4) + (9076/4) = 5149$ £/year

and hence <u>4 effects</u> would be specified.

PROBLEM 14.3

Saturated steam leaves an evaporator at atmospheric pressure and is compressed by means of saturated steam at 1135 kN/m² in a steam jet to a pressure of 135 kN/m². If 1 kg of the high pressure steam compresses 1.6 kg of the evaporated atmospheric steam, what is the efficiency of the compressor?

Solution

See Volume 2, Example 14.3.

PROBLEM 14.4

A single effect evaporator operates at 13 kN/m². What will be the heating surface necessary to concentrate 1.25 kg/s of 10 per cent caustic soda to 41 per cent, assuming a value of U of 1.25 kW/m² K, using steam at 390 K? The heating surface is 1.2 m below the liquid level.

The boiling-point rise of the solution is 30 deg K, the feed temperature is 291 K, the specific heat capacity of the feed is 4.0 kJ/kg deg K, the specific heat capacity of the product is 3.26 kJ/kg deg K and the density of the boiling liquid is 1390 kg/m³.

Solution

Making a mass balance:

In 1.25 kg/s feed, mass of caustic soda $= (1.25 \times 10)/100 = 0.125$ kg/s

For x kg/s water in the product:

$(100 \times 0.125)/(x + 0.125) = 41.0$ from which $x = 0.180$ kg/s.

Thus:

	Solids (kg/s)	Liquid (kg/s)	Total (kg/s)
Feed	0.125	1.125	1.250
Product	0.125	0.180	0.305
Evaporation		0.945	0.945

At a pressure of 13 kN/m², from the Steam Tables in Volume 2, water boils at 324 K. Thus at the surface of the liquid the temperature will be $(324 + 30) = 354$ K. The pressure due to the hydrostatic head of liquid $= (1.2 \times 1390 \times 9.81)/1000 = 16.4$ kN/m² and hence the pressure at the heating surface $= (16.4 + 13) = 29.4$ kN/m² at which pressure the temperature of saturated steam $= 341$ K.

Thus the temperature at which the liquid boils at the heating surface is 371 K, at which the enthalpy of steam $= 2672$ kJ/kg.

Thus: temperature difference, $\Delta T = (390 - 371) = 19$ deg K

The heat load Q is the heat in the vapour plus the enthalpy of the product minus the enthalpy of the feed, or:

$Q = (0.945 \times 2672) + [0.305 \times 3.26(371 - 273)] - [1.250 \times 4.0(291 - 273)]$

assuming the product is withdrawn at 371 K.

Therefore: $Q = (2525 + 97.4 - 90) = 2532.4$ kW

Thus in equation 14.1:

$A = 2532.4/(1.25 \times 19) = \underline{106.6 \text{ m}^2}$

PROBLEM 14.5

Distilled water is produced from sea water by evaporation in a single-effect evaporator working on the vapour compression system. The vapour produced is compressed by a mechanical compressor of 50 per cent efficiency, and then returned to the calandria of the evaporator. Extra steam, dry and saturated at 650 kN/m^2, is bled into the steam space through a throttling valve. The distilled water is withdrawn as condensate from the steam space. 50 per cent of the sea water is evaporated in the plant. The energy supplied in addition to that necessary to compress the vapour may be assumed to appear as superheat in the vapour. Calculate the quantity of extra steam required in kg/s. The production rate of distillate is 0.125 kg/s, the pressure in the vapour space is 101.3 kN/m^2, the temperature difference from steam to liquor is 8 deg K, the boiling-point rise of sea water is 1.1 deg K and the specific heat capacity of sea water is 4.18 kJ/kg K.

The sea water enters the evaporator at 344 K from an external heater.

Solution

See Volume 2, Example 14.4.

PROBLEM 14.6

It is claimed that a jet booster requires 0.06 kg/s of dry and saturated steam at 700 kN/m^2 to compress 0.125 kg/s of dry and saturated vapour from 3.5 kN/m to 14.0 kN/m^2. Is this claim reasonable?

Solution

With the nomenclature used in Volume 2, Example 14.3, $H_1 = 2765$ kJ/kg and, assuming isentropic expansion to 3.5 kN/m^2, from the entropy—enthalpy chart:

$$H_2 = 2015 \text{ kJ/kg}$$

Making an enthalpy balance across the system, noting that the enthalpy of saturated steam at 3.5 kN/m^2 is 2540 kJ/kg then:

$$(0.06 \times 2765) + (0.125 \times 2540) = (0.06 + 0.125)H_4$$

and:
$$H_4 = 2612 \text{ kJ/kg}$$

Assuming isentropic compression from 3.5 kN/m^2 to 14.0 kN/m^2, then $H_3 = 2420$ kJ/kg (again using the chart).

Using the equation for efficiency given in Example 14.3:

$$\eta = (0.06 + 0.125)(2612 - 2420)/[0.06(2765 - 2015)] = \underline{\underline{0.79}}$$

As stated in the solution to Example 14.3, with good design, overall efficiencies may approach 0.75–0.80 and the claim here is therefore reasonable.

In Example 14.3 and Problem 14.6 use has been made of entropy—enthalpy diagrams. The change in enthalpy due to isentropic compression or expansion may also be calculated, however, using equations 8.30 and 8.32 in Volume 1.

PROBLEM 14.7

A forward-feed double-effect evaporator, having 10 m² of heating surface in each effect, is used to concentrate 0.4 kg/s of caustic soda solution from 10 to 50 per cent by mass. During a particular run, when the feed is at 328 K, the pressures in the two calandrias are 375 and 180 kN/m² respectively, and the condenser operates at 15 kN/m². For these conditions, calculate:

(a) the load on the condenser;
(b) the steam economy and
(c) the overall heat transfer coefficient in each effect.

Would there be any advantages in using backward feed in this case? Heat losses to the surroundings are negligible.

Physical properties of caustic soda solutions:

Solids (per cent by mass)	Boiling-point rise (deg K)	Specific heat capacity (kJ/kg K)	Heat of dilution (kJ/kg)
10	1.6	3.85	0
20	6.1	3.72	2.3
30	15.0	3.64	9.3
50	41.6	3.22	220

Solution

A mass balance may be made as follows:

	Solids (kg/s)	Liquor (kg/s)	Total (kg/s)
Feed	0.04	0.36	0.40
Product	0.04	0.04	0.08
Evaporation	—	0.32	0.32

Thus: $(D_1 + D_2) = 0.32$ kg/s

From the given data:

at 375 kN/m², $T_0 = 414.5$ K, $\lambda_0 = 2141$ kJ/kg
at 180 kN/m², $T_1 = 390$ K, $\lambda_1 = 2211$ kJ/kg
at 15 kN/m², $T_2 = 328$ K, $\lambda_2 = 2370$ kJ/kg

Making a heat balance around stage 1:

$$D_0\lambda_0 = WC_{p1}(T_1' - T_f) + Wh_{d1} + D_1\lambda_1 \quad \text{(i)}$$

where C_{p1} is the mean specific heat capacity between T_1' and T_f; $T_1' = T_1$ + the boiling-point rise in stage 1; and h_{d1} is the difference in the heat of dilution between the concentration in the first effect and the feed concentration.

Similarly, around stage 2:

$$D_1\lambda_1 + (W - D_1)C_{p2}(T_1' - T_2') = D_2\lambda_2 + (W - D_1)h_{d2} \quad \text{(ii)}$$

Values of D_1 and D_2 are now selected such that a balance is obtained in equation (ii)

Trying $D_1 = 0.17$ kg/s, $D_2 = 0.15$ kg/s.

Thus: concentration of solids in the first effect $= 0.04/[0.04 + (0.36 - 0.17)]$
$$= 0.174 \text{ kg/kg solution}$$

at which, the boiling-point rise = 5.0 deg K

the specific heat capacity = 3.75 kJ/kg K

and the heat of dilution = 1.6 kJ/kg

Thus:
$$T_1' = (390 + 5.0) = 395 \text{ K}$$
$$T_2' = (328 + 41.6) = 369.6 \text{ K}$$
$$C_{p2} = (3.75 + 3.22)/2 = 3.49 \text{ kJ/kg K}$$

The heat of dilution to be provided in the second effect $= (220 - 1.6) = 218.4$ kJ/kg

Thus in equation (ii):

$$(2211 \times 0.17) + (0.4 - 0.17)3.49(395 - 369.6) = (0.15 \times 2370) + (0.4 - 0.17)218.$$

or: $\qquad 396.3 = 405.7$

which is close enough for the purposes of this calculation, and hence the load on the condenser, $\underline{\underline{D_2 = 0.15 \text{ kg/s}}}$.

For the first effect:
$$C_{p1} = (3.85 + 3.75)/2 = 3.80 \text{ kJ/kg K}$$
$$h_1 = (1.6 - 0) = 1.6 \text{ kJ/kg}$$

and in equation (i):

$2141 D_0 = (0.4 \times 3.80)(395 - 328) + (0.4 \times 1.6) + (0.17 \times 2211)$ and $D_0 = 0.23$ kg/s

and the economy is $(0.36/0.23) = \underline{1.57}$

The overall heat transfer coefficient in the first effect is given by:

$$U_1 = D_0\lambda_0/A_1\Delta T_1$$
$$= (0.23 \times 2141)/[10(414.5 - 395)] = \underline{\underline{2.53 \text{ kW/m}^2 \text{ K}}}$$

and for the second effect:

$$U_2 = (0.17 \times 2211)/10(390 - 369.6) = \underline{1.84 \text{ kW/m}^2 \text{ K}}$$

With a *backward feed* arrangement, the concentration of solids in the first effect would be 50 per cent, which gives a boiling-point rise of 41.6 deg K. The vapour passing to the second effect must condense at 390 K in the calandria of the second effect to give a pressure there of 180 kN/m². Thus the temperature of the liquor in the first effect would be $(390 + 41.6) = 431.6$ K, which is higher than the feed steam temperature, 414.5 K, and thus there is no temperature gradient. There is therefore no advantage in using backward feed.

PROBLEM 14.8

A 12 per cent glycerol—water mixture produced as a secondary product in a continuous process plant flows from the reactor at 4.5 MN/m² and at 525 K. Suggest, with preliminary calculations, a method of concentration to 75 per cent glycerol in a plant where no low-pressure steam is available.

Solution

Making a mass balance on the basis of 1 kg feed:

	Glycerol (kg)	Water (kg)	Total (kg)
Feed	0.12	0.88	1.00
Product	0.12	0.04	0.16
Evaporation	—	0.84	0.84

The total evaporation is 0.84 kg water/kg feed.

The first possibility is to take account of the fact that the feed is at high pressure which could be reduced to, say, atmospheric pressure and the water removed by flash evaporation. For this to be possible, the heat content of the feed must be at least equal to the latent heat of the water evaporated.

Assuming evaporation at 101.3 kN/m², that is at 373 K, and a specific heat capacity for a 12 per cent glycerol solution of 4.0 kJ/kg K, then:

$$\text{heat in feed} = (1.0 \times 4.0)(525 - 375) = 608 \text{ kJ}$$

$$\text{heat in water evaporated} = (0.84 \times 2256) = 1895 \text{ kJ}$$

and hence only $(608/2256) = 0.27$ kg water could be evaporated by this means, giving a solution containing $(100 \times 0.12)/[0.12 + (0.88 - 0.27)] = 16.5$ per cent glycerol.

It is therefore necessary to provide an additional source of heat. Although low-pressure steam is not available, presumably a high-pressure supply (say 1135 kN/m²) exists, and vapour recompression using a steam-jet ejector could be considered.

Assuming that the discharge pressure from the ejector, that is in the steam-chest, is 170 kN/m² at which the latent heat, $\lambda_0 = 2216$ kN/m², then a heat balance across the unit gives:

$$D_0\lambda_0 = WC_p(T_1 - T_f) + D_1\lambda_1$$

Thus: $2216 D_0 = (1.0 \times 4.0)(373 - 525) + (0.84 \times 2256)$ and $D_0 = 0.58$ kg.

Using Figure 14.12b in Volume 2, with a pressure of entrained vapour of 101.3 kN/m², a live-steam pressure of 1135 kN/m² and a discharge pressure of 170 kN/m², then 1.5 kg live steam is required/kg of entrained vapour.

Thus, if x kg is the amount of entrained vapour, then:

$$(1 + 1.5x) = 0.58 \quad \text{and} \quad x = 0.23 \text{ kg}.$$

The proposal is therefore to condense $(0.84 - 0.23) = 0.61$ kg of the water evaporated and to entrain 0.23 kg with $(1.5 \times 0.23) = 0.35$ kg of steam at 1135 kN/m² in an ejector to provide 0.58 kg of steam at 170 kN/m² which is then fed to the calandria.

These values represent only one solution to the problem and variation of the calandria and live-steam pressures may result in even lower requirements of high-pressure steam.

PROBLEM 14.9

A forward-feed double-effect standard vertical evaporator with equal heating areas in each effect is fed with 5 kg/s of a liquor of specific heat capacity of 4.18 kJ/kg K, and with no boiling-point rise, so that 50 per cent of the feed liquor is evaporated. The overall heat transfer coefficient in the second effect is 75 per cent of that in the first effect. Steam is fed at 395 K and the boiling-point in the second effect is 373 K. The feed is heated to its boiling point by an external heater in the first effect.

It is decided to bleed off 0.25 kg/s of vapour from the vapour line to the second effect for use in another process. If the feed is still heated to the boiling-point of the first effect by external means, what will be the change in the steam consumption of the evaporator unit?

For the purposes of calculation, the latent heat of the vapours and of the live steam may be taken as 2230 kJ/kg.

Solution

The total evaporation, $(D_1 + D_2) = (5.0/2) = 2.5$ kg/s.

In equation 14.7:

$$U_1 A_1 \Delta T_1 = U_2 A_2 \Delta T_2$$

Since: $A_1 = A_2$ and $U_2 = 0.75 U_1$

Therefore: $\Delta T_1 = 0.75 \Delta T_2$

Also: $\sum \Delta T = \Delta T_1 + \Delta T_2 = (395 - 373) = 22$ deg K and solving between these two equations gives:

$$\Delta T_1 = 9.5 \text{ deg K} \quad \text{and} \quad \Delta T_2 = 12.5 \text{ deg K}$$

$$\text{For steam to the first effect, } T_0 = 395 \text{ K}$$

For steam to the second effect, $T_1 = (395 - 9.5) = 385.5$ K and $T_2 = 373$ K.
The latent heat, λ is 2230 kJ/kg in each case.
Making a heat balance across the first effect then:

$$D_0 \lambda_0 = W C_p (T_1 - T_f) + D_1 \lambda_1$$

or: $\quad 2230 D_0 = (5.0 \times 4.18)(385.5 - 385.5) + 2230 D_1 \quad \text{and} \quad D_0 = D_1.$

Making a heat balance across the second effect then:

$$D_1 \lambda_1 + (W - D_1) C_p (T_1 - T_2) = D_2 \lambda_2$$

or: $\quad 2230 D_1 + (5.0 - D_1) 4.18 (385.5 - 273) = 2230 D_2$

and: $\quad 2178 D_1 = 2230 D_2 - 261.3$

But: $\quad D_2 = (2.5 - D_1)$

Therefore: $\quad D_1 = 1.21$ kg/s and the steam consumption, $\underline{\underline{D_0 = 1.21 \text{ kg/s}}}$

Considering the case where 0.25 kg/s is bled from the steam line to the calandria of the second effect, a heat balance across the first effect gives:

$$D_0 = D_1 \text{ as before.}$$

Making a heat balance across the second effect, gives:

$$(D_1 - 0.25)\lambda_1 + (W - D_1) C_p (T_1 - T_2) = D_2 \lambda_2$$

Thus: $\quad 2230(D_1 - 0.25) + (5.0 - D_1) 4.18 (385.5 - 373) = 2230 D_2$

and: $\quad 177.7 D_1 = 2230 D_2 + 296.2$

Substituting $(2.5 - D_1)$ for D_2, then:

$$D_1 = 1.33 \text{ kg/s and the steam consumption, } \underline{\underline{D_0 = 1.33 \text{ kg/s}}}$$

The change in steam consumption is therefore an increase of $\underline{\underline{0.12 \text{ kg/s}}}$.

PROBLEM 14.10

A liquor containing 15 per cent solids is concentrated to 55 per cent solids in a double-effect evaporator operating at a pressure of 18 kN/m^2 in the second effect. No crystals are formed. The feedrate is 2.5 kg/s at a temperature of 375 K with a specific heat capacity of 3.75 kJ/kg K. The boiling-point rise of the concentrated liquor is 6 deg K and the pressure of the steam fed to the first effect is 240 kN/m^2. The overall heat transfer coefficients in the first and second effects are 1.8 and 0.63 kW/m² K, respectively. If the heat transfer area is to be the same in each effect, what areas should be specified?

Solution

Making a mass balance based on a flow of feed of 2.5 kg/s, gives:

	Solids (kg/s)	Liquor (kg/s)	Total (kg/s)
Feed	0.375	2.125	2.50
Product	0.375	0.307	0.682
Evaporation	—	1.818	1.818

Thus:
$$(D_1 + D_2) = 1.818 \text{ kg/s} \tag{i}$$

At 18 kN/m² pressure, $T_2 = 331$ K and T_2', the temperature of the liquor in the second effect, allowing for the boiling-point rise, $= (331 + 6) = 337$ K.

At 240 kN/m² pressure, $T_0 = 399$ K, then:

$$\Delta T_1 + \Delta T_2 = (399 - 337) = 62 \text{ deg K} \tag{ii}$$

From equation 14.8:
$$U_1 \Delta T_1 = U_2 \Delta T_2$$

Substituting $U_1 = 1.8$ and $U_2 = 0.63$ kW/m² K, and combining with equation (ii), gives:

$$\Delta T_1 = 16 \text{ deg K} \quad \text{and} \quad \Delta T_2 = 46 \text{ deg K}$$

Thus $T_1 = (399 - 16) = 383$ K and hence the feed enters slightly cooler than the temperature of the liquor in the first effect. Hence ΔT_1 will be slightly greater and the following values will be assumed:

$$\Delta T_1 = 17 \text{ deg K}, \quad \Delta T_2 = 45 \text{ deg K}$$

Thus, for steam to 1: $T_0 = 399$ K, $\lambda_0 = 2185$ kJ/kg

For steam to 2: $T_1 = 382$ K, $\lambda_1 = 2232$ kJ/kg

and: $T_2 = 331$ K, $\lambda_2 = 2363$ kJ/kg

Making a heat balance around each effect:

(1) $\quad D_0 \lambda_0 = W C_p (T_1 - T_f) + D_1 \lambda_1$

$\quad\quad$ or $\quad 2185 D_0 = (2.5 \times 3.75)(382 - 375) + 2232 D_1 \tag{iii}$

(2) $\quad D_1 \lambda_1 + (W - D_1) C_p (T_1 - T_2') = D_2 \lambda_2$

$\quad\quad$ or $\quad 2232 D_1 + (2.5 - D_1) 3.75 (382 - 337) = 2363 D_2 \tag{iv}$

Solving equations (i), (iii), and (iv) simultaneously:

$$D_0 = 0.924 \text{ kg/s}, \quad D_1 = 0.875 \text{ kg/s}, \quad \text{and} \quad D_2 = 0.943 \text{ kg/s}.$$

The areas are given by:

$$A_1 = D_0\lambda_0/U_1\Delta T_1 = (0.924 \times 2185)/(1.8 \times 17) = 66.1 \text{ m}^2$$
$$A_2 = D_1\lambda_1/U_2\Delta T_2 = (0.875 \times 2232)/(0.63 \times 45) = 68.8 \text{ m}^2$$

which are sufficiently close to justify the assumed values of ΔT_1 and ΔT_2. A total area of 134.9 m² is required and hence an area of, say, $\underline{67.5 \text{ m}^2}$ would be specified for each effect.

PROBLEM 14.11

Liquor containing 5 per cent solids is fed at 340 K to a four-effect evaporator. Forward feed is used to give a product containing 28.5 per cent solids. Do the following figures indicate normal operation? If not, why not?

Effect	1	2	3	4
Solids entering (per cent)	5.0	6.6	9.1	13.1
Temperature in steam chest (K)	382	374	367	357.5
Temperature of boiling solution (K)	369.5	364.5	359.6	336.6

Solution

Examination of the data indicates one obvious point in that the temperatures in the steam chests in effects 2 and 3 are higher than the temperatures of the boiling solution in the previous effects. The explanation for this is not clear although a steam leak in the previous effect is a possibility. Further calculations may be made as follows, starting with a mass balance on the basis of 1 kg feed.

	Solid (kg)	Liquor (kg)	Total (kg)	
Feed	0.05	0.950	1.00	
Product from 1	0.05	0.708	0.758	$D_1 = (0.950 - 0.708) = 0.242$ kg
Product from 2	0.05	0.500	0.550	$D_2 = (0.708 - 0.500) = 0.208$ kg
Product from 3	0.05	0.332	0.382	$D_3 = (0.500 - 0.332) = 0.168$ kg
Product from 4	0.05	0.125	0.175	$D_4 = (0.332 - 0.125) = 0.207$ kg

and the total evaporation $= (0.242 + 0.208 + 0.168 + 0.207) = 0.825$ kg.

The steam fed to the plant is obtained by a heat balance across stage 1, given:

$$D_0\lambda_0 = WC_p(T_1 - T_f) + D_1\lambda_1$$

Taking C_p as 4.18 kJ/kg K and λ_0 and λ_1 as 2300 kJ/kg in all effects,

$$2300 D_0 = (1.0 \times 4.18)(369.5 - 340) + (2300 \times 0.242) \quad \text{and} \quad D_0 = 0.296 \text{ kg}$$

The overall coefficient in each effect assuming equal areas, A m^2, is:

$$U_1 = D_0 \lambda_0 / A \Delta T_1 = (0.296 \times 2300)/(12.5 A) = (54.5/A) \text{ kW/m}^2 \text{ K}$$
$$U_2 = D_1 \lambda_1 / A \Delta T_2 = (0.242 \times 2300)/(9.5 A) = (58.6/A) \text{ kW/m}^2 \text{ K}$$
$$U_3 = D_2 \lambda_2 / A \Delta T_3 = (0.208 \times 2300)/(6.6 A) = (72.5/A) \text{ kW/m}^2 \text{ K}$$
$$U_4 = D_3 \lambda_3 / A \Delta T_4 = (0.168 \times 2300)/(20.9 A) = (18.5/A) \text{ kW/m}^2 \text{ K}$$

These results are surprising in that a reduction in U is normally obtained with a decrease in boiling temperature. On this basis U_3 is high, which may indicate a change in boiling mechanism although ΔT_3 is reasonable. Even more important is the very low value of U in effect 4. This must surely indicate that part of the area is inoperative, possibly due to the deposition of crystals from the highly concentrated liquor.

PROBLEM 14.12

1.25 kg/s of a solution is concentrated from 10 to 50 per cent solids in a triple-effect evaporator using steam at 393 K, and a vacuum such that the boiling point in the last effect is 325 K. If the feed is initially at 297 K and backward feed is used, what is the steam consumption, the temperature distribution in the system and the heat transfer area in each effect, each effect being identical?

For the purpose of calculation, it may be assumed that the specific heat capacity is 4.18 kJ/kg K, that there is no boiling point rise, and that the latent heat of vaporisation is constant at 2330 kJ/kg over the temperature range in the system. The overall heat transfer coefficients may be taken as 2.5, 2.0 and 1.6 kW/m^2 K in the first, second and third effects, respectively.

Solution

Making a mass balance:

	Solid (kg/s)	Liquor (kg/s)	Total (kg/s)
Feed	0.125	1.125	1.250
Product	0.125	0.125	0.250
Evaporation	—	1.0	1.0

Thus:
$$D_1 + D_2 + D_3 = 1.0 \text{ kg/s} \quad \text{(i)}$$

$$\sum \Delta T = (T_0 - T_3) = (393 - 325) = 68 \text{ deg K} \quad \text{(ii)}$$

From equation 14.8:
$$2.5 \Delta T_1 = 2.0 \Delta T_2 = 1.6 \Delta T_3 \quad \text{(iii)}$$

and from equations (ii) and (iii):

$$\Delta T_1 = 18 \text{ deg K}, \quad \Delta T_2 = 22 \text{ deg K}, \quad \text{and} \quad \Delta T_3 = 28 \text{ deg K}$$

Modifying the figures to take account of the effect of the feed temperature, it will be assumed that:

$$\Delta T_1 = 19 \text{ deg K}, \quad \Delta T_2 = 24 \text{ deg K}, \quad \text{and} \quad \Delta T_3 = 25 \text{ deg K}$$

The temperatures in each effect and the corresponding latent heats are then:

$$T_0 = 393 \text{ K}, \quad \lambda_0 = 2202 \text{ kJ/kg}$$
$$T_1 = 374 \text{ K}, \quad \lambda_1 = 2254 \text{ kJ/kg}$$
$$T_2 = 350 \text{ K}, \quad \lambda_2 = 2315 \text{ kJ/kg}$$
$$T_3 = 325 \text{ K}, \quad \lambda_3 = 2376 \text{ kJ/kg}$$

Making a heat balance for each effect:

(1) $\quad D_0\lambda_0 = (W - D_3 - D_2)C_p(T_1 - T_2) + D_1\lambda_1$

$\quad\quad$ or $\quad 2202 D_0 = (1.25 - D_2 - D_3)4.18(374 - 350) + 2254 D_1$ $\quad$ (iv)

(2) $\quad D_1\lambda_1 = (W - D_3)C_p(T_2 - T_3) + D_2\lambda_2$

$\quad\quad$ or $\quad 2254 D_1 = (1.25 - D_3)4.18(350 - 325) + 2315 D_2$ $\quad$ (v)

(3) $\quad D_2\lambda_2 = WC_p(T_3 - T_f) + D_3\lambda_3$

$\quad\quad$ or $\quad 2315 D_2 = (1.25 \times 4.18)(325 - 297) + 2376 D_3$ $\quad$ (vi)

Solving equations (i), (iv), (v), and (vi) simultaneously:

$$D_0 = 0.432 \text{ kg/s}, \quad D_1 = 0.393 \text{ kg/s}, \quad D_2 = 0.339 \text{ kg/s}, \quad \text{and} \quad D_3 = 0.268 \text{ kg/s}.$$

The areas of transfer surface are:

$$A_1 = D_0\lambda_0/U_1\Delta T_1 = (0.432 \times 2202)/(2.5 \times 19) = 20.0 \text{ m}^2$$
$$A_2 = D_1\lambda_1/U_2\Delta T_2 = (0.393 \times 2254)/(2.0 \times 24) = 18.5 \text{ m}^2$$
$$A_3 = D_2\lambda_2/U_3\Delta T_3 = (0.268 \times 2315)/(1.6 \times 25) = 15.5 \text{ m}^2$$

which are probably sufficiently close for design purposes; the mean area being $\underline{18.0 \text{ m}^2}$. The steam consumption is therefore, $D_0 = \underline{0.432 \text{ kg/s}}$

The temperatures in each effect are: (1) 374 K, (2) 350 K, and (3) 325 K.

PROBLEM 14.13

A liquid with no appreciable elevation of boiling-point is concentrated in a triple-effect evaporator. If the temperature of the steam to the first effect is 395 K and vacuum is applied to the third effect so that the boiling-point is 325 K, what are the approximate boiling-points in the three effects? The overall transfer coefficients may be taken as 3.1, 2.3, and 1.1 kW/m² K in the three effects respectively.

Solution

For equal thermal loads in each effect, that is $Q_1 = Q_2 = Q_3$, then:

$$U_1 A_1 \Delta T_1 = U_2 A_2 \Delta T_2 = U_3 A_3 \Delta T_3 \quad \text{(equation 14.7)}$$

or, for equal areas in each effect:

$$U_1 \Delta T_1 = U_2 \Delta T_2 = U_3 \Delta T_3 \quad \text{(equation 14.8)}$$

In this case:

$$3.1 \Delta T_1 = 2.3 \Delta T_2 = 1.1 \Delta T_3$$

Thus: $\Delta T_1 = 0.742 \Delta T_2$ and $\Delta T_3 = 2.091 \Delta T_2$

$$\sum \Delta T = \Delta T_1 + \Delta T_2 + \Delta T_3 = (395 - 325) = 70 \text{ deg K}$$

Thus: $0.742 \Delta T_2 + \Delta T_2 + 2.091 \Delta T_2 = 70$ deg K and $\Delta T_2 = 18.3$ deg K

and: $\Delta T_1 = 13.5$ deg K, $\Delta T_3 = 38.2$ deg K

The temperatures in each effect are therefore:

$$T_1 = (395 - 13.5) = \underline{\underline{381.5 \text{ K}}}$$

$$T_2 = (381.5 - 18.3) = \underline{\underline{363.2 \text{ K}}}, \quad \text{and} \quad T_3 = (363.2 - 38.2) = \underline{\underline{325 \text{ K}}}$$

PROBLEM 14.14

A three-stage evaporator is fed with 1.25 kg/s of a liquor which is concentrated from 10 to 40 per cent solids. The heat transfer coefficients may be taken as 3.1, 2.5, and 1.7 kW/m² K in each effect respectively. Calculate the required steam flowrate at 170 kN/m² and the temperature distribution in the three effects, if:

(a) if the feed is at 294 K, and
(b) if the feed is at 355 K.

Forward feed is used in each case, and the values of U are the same for the two systems. The boiling-point in the third effect is 325 K, and the liquor has no boiling-point rise.

Solution

(a) In the absence of any data to the contrary, the specific heat capacity will be taken as 4.18 kJ/kg K.

Making a mass balance:

	Solids (kg/s)	Liquor (kg/s)	Total (kg/s)
Feed	0.125	1.125	1.250
Product	0.125	0.188	0.313
Evaporation	—	0.937	0.937

Thus: $(D_1 + D_2 + D_3) = 0.937$ kg/s

For steam at 170 kN/m^2: $T_0 = 388$ K

Therefore: $\sum \Delta T = (388 - 325) = 63$ deg K

From equation 14.8: $3.1\Delta T_1 = 2.5\Delta T_2 = 1.7\Delta T_3$

and hence: $\Delta T_1 = 15.5$ deg K, $\Delta T_2 = 19$ deg K, and $\Delta T_3 = 28.5$ deg K

Allowing for the cold feed (294 K), it will be assumed that:

$\Delta T_1 = 20$ deg K, $\Delta T_2 = 17$ deg K, and $\Delta T_3 = 26$ deg K

and hence:
$T_0 = 388$ K, $\lambda_0 = 2216$ kJ/kg
$T_1 = 368$ K, $\lambda_1 = 2270$ kJ/kg
$T_2 = 351$ K, $\lambda_2 = 2312$ kJ/kg
$T_3 = 325$ K, $\lambda_3 = 2376$ kJ/kg

Making a heat balance across each effect:

(1) $2216D_0 = (1.25 \times 4.18)(368 - 294) + 2270D_1$ or $D_0 = (0.175 + 1.024D_1)$

(2) $2270D_1 + (1.25 - D_1)4.18(368 - 351) = 2312D_2$ or $D_2 = (0.951D_1 + 0.038)$

(3) $2312D_2 + (1.25 - D_1 - D_2)4.18(351 - 325) = 2376D_3$

or: $D_3 = (0.927D_2 - 0.046D_1 + 0.057)$

Noting that $D_1 + D_2 + D_3 = 0.937$ kg/s, these equations may be solved to give:

$D_0 = 0.472$ kg/s, $D_1 = 0.290$ kg/s, $D_2 = 0.313$ kg/s and $D_3 = 0.334$ kg/s

The area of each effect is then:

$A_1 = D_0\lambda_0/U_1\Delta T_1 = (0.472 \times 2216)/(3.1 \times 20) = 16.9$ m^2

$A_2 = D_1\lambda_1/U_2\Delta T_2 = (0.290 \times 2270)/(2.5 \times 17) = 15.5$ m^2

$A_3 = D_2\lambda_2/U_3\Delta T_3 = (0.334 \times 2312)/(1.7 \times 26) = 17.4$ m^2

These are probably sufficiently close for a first approximation, and hence the steam consumption, $\underline{\underline{D_0 = 0.472 \text{ kg/s}}}$

and the temperatures in each effect are:

$$(1)\ 368\ K,\quad (2)\ 351\ K,\quad (3)\ 325\ K$$

(b) In this case the feed is much hotter and hence less modification to the estimated values of ΔT will be required. It is assumed that:

$$\Delta T_1 = 17\ \text{deg K},\quad \Delta T_2 = 18\ \text{deg K},\quad \text{and}\quad \Delta T_3 = 28\ \text{deg K}$$

and hence:
$$T_0 = 388\ K,\quad \lambda_0 = 2216\ kJ/kg$$
$$T_1 = 371\ K,\quad \lambda_1 = 2262\ kJ/kg$$
$$T_2 = 353\ K,\quad \lambda_2 = 2308\ kJ/kg$$
$$T_3 = 325\ K,\quad \lambda_3 = 2376\ kJ/kg$$

The heat balances are now:

(1) $2216D_0 = (1.25 \times 4.18)(371 - 355) + 2262D_1$ or $D_0 = (0.038 + 1.021D_1)$

(2) $2262D_1 + (1.25 - D_1)4.18(371 - 353) = 2308D_2$ or $D_2 = (0.948D_1 + 0.041)$

(3) $2308D_2 + (1.25 - D_1 - D_2)4.18(353 - 325) = 2376D_3$

or $D_3 = (0.922D_2 - 0.049D_1 + 0.062)$

Again:
$$(D_1 + D_2 + D_3) = 0.937\ kg/s$$

and: $D_0 = 0.331$ kg/s, $D_1 = 0.287$ kg/s, $D_2 = 0.313$ kg/s, and $D_3 = 0.337$ kg/s

Thus:
$$A_1 = (0.331 \times 2216)/(3.1 \times 17) = 13.9\ m^2$$
$$A_2 = (0.287 \times 2262)/(2.5 \times 18) = 14.4\ m^2$$
$$A_3 = (0.313 \times 2308)/(1.7 \times 28) = 15.1\ m^2$$

These are close enough for design purposes and hence:

the steam consumption $D_0 = 0.331$ kg/s

and the temperatures in each effect are: (1) 371 K, (2) 353 K, (3) 325 K

PROBLEM 14.15

An evaporator operating on the thermo-recompression principle employs a steam ejector to maintain atmospheric pressure over the boiling liquid. The ejector uses 0.14 kg/s of steam at 650 kN/m², and is superheated by 100 K and the pressure in the steam chest is 205 kN/m². A condenser removes surplus vapour from the atmospheric pressure line. What is the capacity and economy of the system and how could the economy be improved?

The feed enters the evaporator at 295 K and the concentrated liquor is withdrawn at the rate of 0.025 kg/s. The concentrated liquor exhibits a boiling-point rise of 10 deg K. Heat losses to the surroundings are negligible. The nozzle efficiency is 0.95, the efficiency of momentum transfer is 0.80 and the efficiency of compression is 0.90.

Solution

See Volume 2, Example 14.5.

PROBLEM 14.16

A single-effect evaporator is used to concentrate 0.075 kg/s of a 10 per cent caustic soda liquor to 30 per cent. The unit employs forced circulation in which the liquor is pumped through the vertical tubes of the calandria which are 32 mm o.d. by 28 mm i.d. and 1.2 m long. Steam is supplied at 394 K, dry and saturated, and the boiling-point rise of the 30 per cent solution is 15 deg K. If the overall heat transfer coefficient is 1.75 kW/m² K, how many tubes should be used, and what material of construction would be specified for the evaporator? The latent heat of vaporisation under these conditions is 2270 kJ/kg.

Solution

Making a mass balance:

	Solids (kg/s)	Liquor (kg/s)	Total (kg/s)
Feed	0.0075	0.0675	0.0750
Product	0.0075	0.0175	0.0250
Evaporation	—	0.0500	0.0500

The temperature of boiling liquor in the tubes, assuming atmospheric pressure, $T_1' = (373 + 15) = 388$ K. In the absence of any other data it will be assumed that the solution enters at 373 K and the specific heat capacity will be taken as 4.18 kJ/kg K.

A heat balance then gives:

$$D_0 \lambda_0 = W C_p (T_1' - T_f) + D_1 \lambda_1$$
$$= (0.0750 \times 4.18)(388 - 373) + (0.050 \times 2270) = 118.2 \text{ kW}$$

$$\Delta T_1 = (394 - 388) = 6 \text{ deg K}$$

and the area: $A_1 = D_0 \lambda_0 / U_1 \Delta T_1 = 118.2/(1.75 \times 6) = 11.25 \text{ m}^2.$

The tube o.d. is 0.032 m and the outside area per unit length of tubing is given by:

$$(\pi \times 0.032) = 0.101 \text{ m}^2/\text{m}$$

Thus: total length of tubing required $= (11.25/0.101) = 112$ m

and number of tubes required $= (112/1.2) = \underline{\underline{93}}$

Mild steel does not cope with caustic soda solutions, and stainless steel has limitations at higher temperatures. Aluminium bronze, copper, nickel, and nickel—copper alloys may be used, together with neoprene and butyl rubber, though from the cost viewpoint and the need for a good conductivity, graphite tubes would probably be specified.

PROBLEM 14.17

A steam-jet booster compresses 0.1 kg/s of dry and saturated vapour from 3.4 kN/m² to 13.4 kN/m². The high-pressure steam consumption is 0.05 kg/s at 690 kN/m². (a) What must be the condition of the high pressure steam for the booster discharge to be superheated by 20 deg K? (b) What is the overall efficiency of the booster if the compression efficiency is 100 per cent?

Solution

(a) Considering the outlet stream at 13.4 kN/m² and 20 deg K superheat, this has an enthalpy, $H_4 = 2638$ kJ/kg.

The enthalpy of the entrained vapours, $H_3' = 2540$ kJ/kg assuming that these are dry and saturated.

If H_1 is the enthalpy of the high pressure steam, then an enthalpy balance gives:

$$0.05 H_1 + (0.1 \times 2540) = (0.15 \times 2638) \quad \text{and} \quad H_1 = 2834 \text{ kJ/kg}$$

At 690 kN/m², this corresponds to a temperature of 453 K.

At 690 kN/m², steam is saturated at 438 K, and the high pressure steam must be superheated by <u>15 deg K</u>.

(b) Assuming the high pressure steam is expanded isentropically from 690 kN/m² (and 453 K) to 3.4 kN/m², its enthalpy H_2 (assuming 100 per cent efficiency) = 2045 kJ/kg.

If the outlet stream is expanded isentropically from 3.4 kN/m² to 13.4 kN/m² then $H_3 = 2435$ kJ/kg and the efficiency is given by:

$$(0.1/0.05) = [(2834 - 2045)/(2638 - 2435)]\eta - 1$$

from which: $\eta = \underline{0.77}$

PROBLEM 14.18

A triple-effect backward-feed evaporator concentrates 5 kg/s of liquor from 10 per cent to 50 per cent solids. Steam is available at 375 kN/m² and the condenser operates at 13.5 kN/m². What is the area required in each effect, assumed identical, and the economy of the unit?

The specific heat capacity is 4.18 kJ/kg K at all concentrations and that there is no boiling-point rise. The overall heat transfer coefficients are 2.3, 2.0 and 1.7 kW/m² K respectively in the three effects, and the feed enters the third effect at 300 K.

Solution

As a variation, this problem will be solved using Storrow's Method (Volume 2 page 786). A mass balance gives the total evaporation as follows:

	Solids (kg/s)	Liquor (kg/s)	Total (kg/s)
Feed	0.50	4.50	5.0
Product	0.50	0.50	1.0
Evaporation	—	4.0	4.0

For steam at 375 kN/m^2, $T_0 = 414$ K

For steam at 13.5 kN/m^2, $T_3 = 325$ K

Thus: $\sum \Delta T = (414 - 325) = 89$ deg K

For equal heat transfer rates in each effect:

$$2.3\Delta T_1 = 2.0\Delta T_2 = 1.7\Delta T_3 \qquad \text{(equation 14.8)}$$

and hence: $\Delta T_1 = 26$ deg K, $\Delta T_2 = 29$ deg K, $\Delta T_3 = 34$ deg K

Modifying the values to take account of the feed temperature, it will be assumed that:

$$\Delta T_1 = 27 \text{ deg K}, \quad \Delta T_2 = 30 \text{ deg K}, \quad \text{and} \quad \Delta T_3 = 32 \text{ deg K}$$

and hence: $T_1 = 387$ K, $T_2 = 357$ K,   and $T_3 = 325$ K

As a first approximation, equal evaporation in each effect will be assumed, or:

$$D_1 = D_2 = D_3 = 1.33 \text{ kg/s}$$

With backward feed, the liquor has to be heated to its boiling-point as it enters each effect.

Heat required to raise the feed to the second effect to its boiling-point is given by:

$$(4.0 - 1.33)4.18(357 - 325) = 357.1 \text{ kW}$$

Heat required to raise the feed to the first effect to its boiling-point is

$$[4.0 - (2 \times 1.33)]4.18(387 - 357) = 168.1 \text{ kW}$$

At $T_0 = 414$ K: $\lambda_0 = 2140$ kJ/kg

At $T_3 = 325$ K: $\lambda_3 = 2376$ kJ/kg

A mean value of 2258 kJ/kg which will be taken as the value of the latent heat in all three effects.

The relation between the heat transferred in each effect and in the condenser is given by:

$$(Q_1 - 168.1) = Q_2 = (Q_3 + 357.1) = (Q_c + 357.1) + [5.0 \times 4.18(325 - 300)]$$
$$= (Q_c + 879.6)$$

The total evaporation is: $(Q_2 + Q_3 + Q_c)/2258 = 4.0$

or: $\quad Q_2 + (Q_2 - 357.1) + (Q_2 - 879.6) = 9032$

Therefore: $\quad Q_2 = 3423 = (A\Delta T_2 \times 2.0)$

$$Q_3 = 3066 = (A\Delta T_3 \times 1.7)$$
$$Q_1 = 3591 = (A\Delta T_1 \times 2.3)$$

Thus:

$$A\Delta T_1 = 1561 \text{ m}^2 \text{ K}, \; A\Delta T_2 = 1712 \text{ m}^2 \text{ K}, \; \text{and} \; A\Delta T_3 = 1804 \text{ m}^2 \text{ K}$$
$$(\Delta T_1 + \Delta T_2 + \Delta T_3) = 89 \text{ deg K}$$

Values of ΔT_1, ΔT_2, and ΔT_3 deg K are now chosen by trial and error to give equal value of A m² in each effect as follows:

ΔT_1	A_1	ΔT_2	A_2	ΔT_3	A_3
27	57.8	30	57.1	32	56.4
27.5	56.8	30.5	56.1	31	58.1
27.25	57.3	30.25	56.6	31.5	57.3

These areas are approximately equal and the assumed values of ΔT are acceptable

The economy = $4.0/(3591/2258) = \underline{\underline{2.52}}$

and the area to be specified for each effect = $\underline{\underline{57 \text{ m}^2}}$

PROBLEM 14.19

A double-effect climbing-film evaporator is connected so that the feed passes through two preheaters, one heated by vapour from the first effect and the other by vapour from the second effect. The condensate from the first effect is passed into the steam space of the second effect. The temperature of the feed is initially 289 K, 348 K after the first heater and 383 K after the second heater. The vapour temperature in the first effect is 398 K and in the second effect 373 K. The feed flowrate is 0.25 kg/s and the steam is dry and saturated at 413 K. What is the economy of the unit if the evaporation rate is 0.125 kg/s?

Solution

A heat balance across the first effect gives:

$D_0 \lambda_0 = W C_p (T_1 - T_f) + D_1 \lambda$ where T_f is the temperature of the feed leaving the second preheater, 383 K.

When $T_0 = 413$ K, $\quad \lambda_0 = 2140$ kJ/kg

When $T_1 = 398$ K, $\quad \lambda_1 = 2190$ kJ/kg

Taking $C_p = 4.18$ kJ/kg K throughout, then:

$$2140 D_0 = (0.25 \times 4.18)(398 - 383) + 2190 D_1$$

Thus: $\quad D_0 = (0.0073 + 0.023 D_1)$ kg/s

Assuming that the first preheater is heated by steam from the first effect, then the heat transferred in this unit is:

$$(0.25 \times 4.18)(348 - 289) = 61.7 \text{ kW}$$

and hence the steam condensed in the preheater $= (61.7/2190) = 0.028$ kg/s.

Therefore the flowrate of vapour from the first effect which is condensed in the steam space of the second effect $= (D_1 - 0.028)$ kg/s.

A heat balance on the second effect is thus:

$$(D_1 - 0.028)\lambda_1 + (W - D_1)C_p(T_1 - T_2) = D_2 \lambda_2$$

Since the total evaporation $= 0.125$ kg/s, $D_2 = (0.125 - D_1)$

and: $\quad (D_1 - 0.028)2190 + (0.25 - D_1)4.18(398 - 373) = (0.125 - D_1)2256$

Therefore: $\quad D_1 = 0.073$ kg/s

Thus: $\quad D_0 = 0.0073 + (1.023 \times 0.073) = 0.082$ kg/s

and the economy is: $\quad (0.125/0.082) = \underline{\underline{1.5 \text{ kg/kg}}}$

PROBLEM 14.20

A triple-effect evaporator is fed with 5 kg/s of a liquor containing 15 per cent solids. The concentration in the last effect, which operates at 13.5 kN/m^2, is 60 per cent solids. If the overall heat transfer coefficients in the three effects are 2.5, 2.0, and 1.1 kW/m^2 K, respectively, and the steam is fed at 388 K to the first effect, determine the temperature distribution and the area of heating surface required in each effect? The calandrias are identical. What is the economy and what is the heat load on the condenser?

The feed temperature is 294 K and the specific heat capacity of all liquors is 4.18 kJ/kg K.

If the unit is run as a backward-feed system, the coefficients are 2.3, 2.0, and 1.6 kW/m^2 K respectively. Determine the new temperatures, the heat economy, and the heating surface required under these conditions.

Solution

(a) *Forward feed* A mass balance gives:

	Solids (kg/s)	Liquor (kg/s)	Total (kg/s)
Feed	0.75	4.25	5.0
Product	0.75	0.50	1.25
Evaporation	—	3.75	3.75

For steam saturated at 13.5 kN/m^2:

$$T_3 = 325 \text{ K} \quad \text{and} \quad \lambda_3 = 2375 \text{ kJ/kg}$$
$$T_0 = 388 \text{ K} \quad \text{and} \quad \lambda_0 = 2216 \text{ kJ/kg}$$

Thus: $\sum \Delta T = (388 - 325) = 63 \text{ deg K}$

For equal heat transfer rates in each effect:

$$U_1 \Delta T_1 = U_2 \Delta T_2 = U_3 \Delta T_3 \quad \text{(equation 14.8)}$$

Thus: $2.5 \Delta T_1 = 2.0 \Delta T_2 = 1.1 \Delta T_3$

and: $\Delta T_1 = 14 \text{ deg K}, \quad \Delta T_2 = 17.5 \text{ deg K}, \quad \text{and} \quad \Delta T_3 = 31.5 \text{ deg K}$

Allowing for the cold feed it will be assumed that:

$$\Delta T_1 = 18 \text{ deg K}, \quad \Delta T_2 = 16 \text{ deg K}, \quad \text{and} \quad \Delta T_3 = 29 \text{ deg K}$$

and hence:
$T_0 = 388 \text{ K}, \quad \lambda_0 = 2216 \text{ kJ/kg}$
$T_1 = 370 \text{ K}, \quad \lambda_1 = 2266 \text{ kJ/kg}$
$T_2 = 354 \text{ K}, \quad \lambda_2 = 2305 \text{ kJ/kg}$
$T_3 = 325 \text{ K}, \quad \lambda_3 = 2375 \text{ kJ/kg}$

Making heat balances over each effect:

(1) $D_0 \lambda_0 = W C_p (T_1 - T_f) + D_1 \lambda_1$ or $2216 D_0 = (5.0 \times 4.18)(370 - 294) + 2266 D_1$

(2) $D_1 \lambda_1 + (W - D_1) C_p (T_1 - T_2) = D_2 \lambda_2$

or $2266 D_1 + (5.0 - D_1) 4.18 (370 - 354) = 2305 D_2$

(3) $D_2 \lambda_2 + (W - D_1 - D_2) C_p (T_2 - T_3) = D_3 \lambda_3$

or $2305 D_2 + (5.0 - D_1 - D_2) 4.18 (354 - 325) = 2375 D_3$

Also: $(D_1 + D_2 + D_3) = 3.75 \text{ kg/s}$

Solving simultaneously:

$D_0 = 1.90 \text{ kg/s}, \quad D_1 = 1.16 \text{ kg/s}, \quad D_2 = 1.25 \text{ kg/s}, \quad \text{and} \quad D_3 = 1.35 \text{ kg/s}$

The areas are now given by:

$$A_1 = D_0 \lambda_0 / U_1 \Delta T_1 = (1.90 \times 2216)/(2.5 \times 18) = 93.6 \text{ m}^2$$
$$A_2 = D_1 \lambda_1 / U_2 \Delta T_2 = (1.16 \times 2266)/(2.0 \times 16) = 82.1 \text{ m}^2$$
$$A_3 = D_2 \lambda_2 / U_3 \Delta T_3 = (1.25 \times 2305)/(1.1 \times 29) = 90.3 \text{ m}^2$$

It is apparent that the modification to the temperature difference made to take account of the effect of the cold feed has been incorrect and it is now assumed that:

$$\Delta T_1 = 19 \text{ deg K}, \quad \Delta T_2 = 15 \text{ deg K}, \quad \text{and} \quad \Delta T_3 = 29 \text{ deg K}$$

Thus:
$$T_0 = 388 \text{ K} \quad \text{and} \quad \lambda_0 = 2216 \text{ kJ/kg}$$
$$T_1 = 369 \text{ K} \quad \text{and} \quad \lambda_1 = 2267 \text{ kJ/kg}$$
$$T_2 = 354 \text{ K} \quad \text{and} \quad \lambda_2 = 2305 \text{ kJ/kg}$$
$$T_3 = 325 \text{ K} \quad \text{and} \quad \lambda_3 = 2375 \text{ kJ/kg}$$

The heat balances now become:

(1) $2216 D_0 = (5.0 \times 4.18)(369 - 294) + 2267 D_1$

(2) $2267 D_1 + (5.0 - D_1)4.18(369 - 354) = 2305 D_2$

(3) $2305 D_2 + (5.0 - D_1 - D_2)4.18(354 - 325) = 2375 D_3$

and solving:

$$D_0 = 1.89 \text{ kg/s}, \quad D_1 = 1.16 \text{ kg/s}, \quad D_2 = 1.25 \text{ kg/s}, \quad \text{and} \quad D_3 = 1.34 \text{ kg/s}$$

Hence:
$$A_1 = (1.89 \times 2216)/(2.5 \times 19) = 88.2 \text{ m}^2$$
$$A_2 = (1.16 \times 2267)/(2.0 \times 15) = 87.8 \text{ m}^2$$
$$A_3 = (1.25 \times 2305)/(1.1 \times 29) = 90.3 \text{ m}^2$$

giving much closer values for the three areas.

The temperature distribution is now:

(1) 369 K, (2) 354 K, (3) 325 K

The area in each effect should be about $\underline{\underline{89 \text{ m}^2}}$.

The economy is $(3.75/1.89) = \underline{\underline{2.0}}$ and the heat load on the condenser is:

$$D_3 \lambda_3 = (1.34 \times 2375) = \underline{\underline{31.8 \text{ kW}}}$$

(b) *Backward feed*

As before: $(D_1 + D_2 + D_3) = 3.75 \text{ kg/s}$

and: $\sum \Delta T = (388 - 325) = 63 \text{ deg k}$

In this case: $2.3 \Delta T_1 = 2.0 \Delta T_2 = 1.6 \Delta T_3$

and: $\Delta T_1 = 17.5 \text{ deg K}, \quad \Delta T_2 = 20.0 \text{ deg K}, \quad \text{and} \quad \Delta T_3 = 25.5 \text{ deg K}$

Modifying for the effect of the cold feed, it is assumed that:

$$\Delta T_1 = 18.5 \text{ deg K}, \quad \Delta T_2 = 20.5 \text{ deg K}, \quad \text{and} \quad \Delta T_3 = 24 \text{ deg K}$$

Thus:
$$T_0 = 388 \text{ K}, \quad \lambda_0 = 2216 \text{ kJ/kg}$$
$$T_1 = 369.5 \text{ K}, \quad \lambda_1 = 2266 \text{ kJ/kg}$$
$$T_2 = 349 \text{ K}, \quad \lambda_2 = 2318 \text{ kJ/kg}$$
$$T_3 = 325 \text{ K}, \quad \lambda_3 = 2375 \text{ kJ/kg}$$

The heat balances become:

(1) $D_0\lambda_0 = (W - D_3 - D_2)C_p(T_1 - T_2) + D_1\lambda_1$

or $2216 D_0 = (5.0 - D_3 - D_2)4.18(369.5 - 349) + 2266 D_1$

(2) $D_1\lambda_1 = (W - D_3)C_p(T_2 - T_3) + D_2\lambda_2$

or $2266 D_1 = (5.0 - D_3)4.18(349 - 325) + 2318 D_2$

(3) $D_2\lambda_2 = WC_p(T_3 - T_f) + D_3\lambda_3$ or $2318 D_2 = (5.0 \times 4.18)(325 - 294) + 2375 D_3$

Solving:

$$D_0 = 1.62 \text{ kg/s}, \quad D_1 = 1.49 \text{ kg/s}, \quad D_2 = 1.28 \text{ kg/s}, \quad \text{and} \quad D_3 = 0.98 \text{ kg/s}$$

The areas are therefore:

$$A_1 = (1.62 \times 2216)/(2.3 \times 18.5) = 84.4 \text{ m}^2$$
$$A_2 = (1.49 \times 2266)/(2.0 \times 20.5) = 82.4 \text{ m}^2$$
$$A_3 = (1.28 \times 2318)/(1.6 \times 24) = 77.3 \text{ m}^2$$

which are probably close enough for design purposes.

Thus, the temperatures in each effect are now:

(1) 369.5 K, (2) 349 K, and (3) 325 K

The economy is $(3.75/1.62) = \underline{2.3}$, and the area required in each effect is approximately $\underline{81 \text{ m}^2}$.

PROBLEM 14.21

A double-effect forward-feed evaporator is required to give a product which contains 50 per cent by mass of solids. Each effect has 10 m² of heating surface and the heat transfer coefficients are 2.8 and 1.7 kW/m² K in the first and second effects respectively. Dry and saturated steam is available at 375 kN/m² and the condenser operates at 13.5 kN/m². The concentrated solution exhibits a boiling-point rise of 3 deg K. What is the maximum permissible feed rate if the feed contains 10 per cent solids and is at 310 K? The latent heat is 2330 kJ/kg and the specific heat capacity is 4.18 kJ/kg under all the above conditions.

Solution

Making a mass balance on the basis of W kg/s feed:

	Solids (kg/s)	Liquor (kg/s)	Total (kg/s)
Feed	0.10W	0.90W	W
Product	0.10W	0.10W	0.20W
Evaporation	—	0.80W	0.80W

Thus: $(D_1 + D_2) = 0.8W$ kg/s

At 375 kN/m²: $T_0 = 413$ K

At 13.5 kN/m²: $T_2 = 325$ K

and the temperature of the boiling liquor in the second effect, $T_2' = (325 + 3) = 328$ K.

Therefore: $\sum \Delta T = (413 - 328) = 85$ deg K

At this stage, values may be assumed for ΔT_1 and ΔT_2 and heat balances made until equal areas are obtained. There are sufficient data, however, to enable an exact solution to be obtained as follows.

Making heat balances:

1st effect

$$D_0 \lambda_0 = W C_p (T_1 - T_f) + D_1 \lambda_1$$

or: $U_1 A (T_0 - T_1) = W(T_1 - T_f) + U_2 A(T_1 - T_2')$

Thus: $(2.8 \times 10)(413 - T_1) = (W \times 4.18)(T_1 - 310) + (1.7 \times 10)(T_1 - 328)$

and: $W = (17{,}140 - 45T_1)/(4.18T_1 - 1296)$

2nd effect

$$D_1 \lambda_1 = (W - D_1) C_p (T_1 - T_2') + D_2 \lambda_2$$

But: $D_1 = (1.7 \times 10)(T_1 - 328)/2330 = (0.0073T_1 - 2.393)$

and: $D_2 = (0.8W - D_1) = (0.8W - 0.0073T_1 - 2.393)$

Thus: $2330(0.0073T_1 - 2.393) =$
$(W - 0.0073T_1 + 2.393)4.18(T_1 - 328) + (0.8W - 0.0073T_1 - 2.393)2330$

and: $W = (14T_1 - 7872 + 0.0305T_1^2)/(4.18T_1 + 493)$

Equations (i) and (ii) are plotted in Figure 14a and the two curves coincide when $T_1 = 375$ K and $W = 0.83$ kg/s.

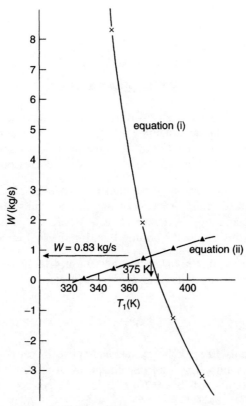

Figure 14a. Graphical construction for Problem 14.21

PROBLEM 14.22

For the concentration of fruit juice by evaporation, it is proposed to use a falling-film evaporator and to incorporate a heat-pump cycle with ammonia as the medium. The ammonia in vapour form enters the evaporator at 312 K and the water is evaporated from the juices at 287 K. The ammonia in the vapour—liquid mixture enters the condenser at 278 K and the vapour then passes to the compressor. It is estimated that the work required to compress the ammonia is 150 kJ/kg of ammonia and that 2.28 kg of ammonia is cycled/kg water evaporated. The following proposals are made for driving the compressor:

(a) To use a diesel engine drive taking 0.4 kg of fuel/MJ. The calorific value of the fuel is 42 MJ/kg, and the cost £0.02/kg.
(b) To pass steam, costing £0.01/10 kg, through a turbine which operates at 70 per cent isentropic efficiency, between 700 and 101.3 kN/m^2.

Explain by means of a diagram how this plant will work, illustrating all necessary major items of equipment. Which method for driving the compressor is to be preferred?

Solution

See Volume 2, Example 14.6.

PROBLEM 14.23

A double-effect forward-feed evaporator is required to give a product consisting of 30 per cent crystals and a mother liquor containing 40 per cent by mass of dissolved solids. Heat transfer coefficients are 2.8 and 1.7 kW/m² K in the first and second effects respectively. Dry saturated steam is supplied at 375 kN/m² and the condenser operates at 13.5 kN/m².

(a) What area of heating surface is required in each effect, assuming they are both identical, if the feed rate is 0.6 kg/s of liquor, containing 20 per cent by mass of dissolved solids, and the feed temperature is 313 K?
(b) What is the pressure above the boiling liquid in the first effect?

The specific heat capacity may be taken as constant at 4.18 kJ/kg K, and the effects of boiling point rise and of hydrostatic head may be neglected.

Solution

The final product contains 30 per cent crystals and hence 70 per cent solution containing 40 per cent dissolved solids. The total percentage of dissolved and undissolved solids in the final product is $0.30 + (0.40 \times 0.70) = 0.58$ or 58 per cent, and the mass balance becomes:

	Solids (kg/s)	Liquor (kg/s)	Total (kg/s)
Feed	0.12	0.48	0.60
Product	0.12	0.087	0.207
Evaporation	—	$(D_1 + D_2) = 0.393$	0.393

At 375 kN/m²: $T_0 = 413$ K and at 13.5 kN/m², $T_2 = 325$ K

Thus: $\sum \Delta T = (413 - 325) = 88$ deg K

For equal heat transfer areas in each effect: $2.8 \Delta T_1 = 1.7 \Delta T_2$ (equation 14.8)

and: $\Delta T_1 = 33$ deg K   and $\Delta T_2 = 55$ deg K

Since the feed is cold, it will be assumed that:

$$\Delta T_1 = 40 \text{ deg K} \quad \text{and} \quad \Delta T_2 = 48 \text{ deg K}$$

and hence: $T_0 = 413$ K, $\lambda_0 = 2140$ kJ/kg

$T_1 = 373$ K, $\lambda_1 = 2257$ kJ/kg

$T_2 = 325$ K, $\lambda_1 = 2375$ kJ/kg

Taking the specific heat capacity of the liquor as constant at 4.18 kJ/kg K at all times, then the heat balance over each effect becomes:

(1) $D_0 \lambda_0 = W C_p(T_1 - T_f) + D_1 \lambda_1$ or $2140 D_0 = (0.6 \times 4.18)(373 - 313) + 2257 D_1$

(2) $D_1 \lambda_1 + (W - D_1) C_p (T_1 - T_2) = D_2 \lambda_2$

or $2257 D_1 + (0.6 - D_1) 4.18 (373 - 325) = 2375 D_2$

Solving: $D_0 = 0.263$ kg/s, $D_1 = 0.184$ kg/s, and $D_2 = 0.209$ kg/s

The areas are now given by:

$$A_1 = D_0 \lambda_0 / U_1 \Delta T_1 = (0.263 \times 2140)/(2.8 \times 40) = 5.04 \text{ m}^2$$

$$A_2 = D_1 \lambda_1 / U_2 \Delta T_2 = (0.184 \times 2257)/(1.7 \times 48) = 5.08 \text{ m}^2$$

which show close agreement.

Thus: the area of heating to be specified = $\underline{5.06 \text{ m}^2}$ or approximately 5 m² in each effect.

The temperature of liquor boiling in the first effect, assuming no boiling-point rise, is 373 K at which temperature steam is saturated at 101.3 kN/m².

The pressure in the first effect is, therefore, atmospheric.

PROBLEM 14.24

1.9 kg/s of a liquid containing 10 per cent by mass of dissolved solids is fed at 338 K to a forward-feed double-effect evaporator. The product consists of 25 per cent by mass of solids and a mother liquor containing 25 per cent by mass of dissolved solids. The steam fed to the first effect is dry and saturated at 240 kN/m² and the pressure in the second effect is 20 kN/m². The specific heat capacity of the solid may be taken as 2.5 kJ/kg K, both in solid form and in solution, and the heat of solution may be neglected. The mother liquor exhibits a boiling point rise of 6 deg K. If the two effects are identical, what area is required if the heat transfer coefficients in the first and second effects are 1.7 and 1.1 kW/m² K respectively?

Solution

The percentage by mass of dissolved and undissolved solids in the final product = $(0.25 \times 75) + 25 = 43.8$ per cent and hence the mass balance becomes:

	Solids (kg/s)	Liquor (kg/s)	Total (kg/s)
Feed	0.19	1.71	1.90
Product	0.19	0.244	0.434
Evaporation	—	$(D_1 + D_2) = 1.466$	1.466

At 240 kN/m²: $T_0 = 399$ K At 20 kN/m², $T_2 = 333$ K

Thus: $T_2' = (333 + 6) = 339$ K

and, allowing for the boiling-point rise in the first effect:

$$\Delta T_1 + \Delta T_2 = (399 - 339) - 6 = 54 \text{ deg K}$$

From equation 14.8: $1.7\Delta T_1 = 1.1\Delta T_2$

and: $\Delta T_1 = 21$ deg K and $\Delta T_2 = 33$ deg K

Modifying these values to allow for the cold feed, it will be assumed that:

$$\Delta T_1 = 23 \text{ deg K} \quad \text{and} \quad \Delta T_2 = 31 \text{ deg K}$$

Assuming that the liquor exhibits a 6 deg K boiling-point rise at all concentrations, then, with T_1' as the temperature of boiling liquor in the first effect and T_2' that in the second effect:

$T_0 = 399$ K at which $\lambda_0 = 2185$ kJ/kg

$T_1' = (399 - 23) = 376$ K

$T_1 = (376 - 6) = 370$ K at which $\lambda_1 = 2266$ kJ/kg

$T_2' = 339$ K

$T_2 = 333$ K at which $\lambda_2 = 2258$ K

Making a heat balance over each effect:

(1) $D_0\lambda_0 = WC_p(T_1' - T_f) + D_1\lambda_1$ or $2185D_0 = (1.90 \times 2.5)(376 - 338) + 2266D_1$

(2) $D_1\lambda_1 + (W - D_1)C_p(T_1' - T_2') = D_2\lambda_2$

or: $2358D_2 = (1.90 - D_1)2.5(376 - 339) + 2266D_1$

Solving: $D_0 = 0.833$ kg/s, $D_1 = 0.724$ kg/s, and $D_2 = 0.742$ kg/s

The areas are then given by:

$$A_1 = D_0\lambda_0/U_1(T_0 - T_1') = 0.833 \times 2185/[1.7(399 - 376)] = 46.7 \text{ m}^2$$

$$A_2 = D_1\lambda_1/U_2(T_1 - T_2') = 0.724 \times 2266/[1.1(370 - 339)] = 48.0 \text{ m}^2$$

which are close enough for design purposes.

The area to be specified for each effect is approximately $\underline{\underline{47.5 \text{ m}^2}}$.

PROBLEM 14.25

2.5 kg/s of a solution at 288 K containing 10 per cent of dissolved solids is fed to a forward-feed double-effect evaporator, operating at 14 kN/m² in the last effect. If the product is to consist of a liquid containing 50 per cent by mass of dissolved solids and dry saturated steam is fed to the steam coils, what should be the pressure of the steam? The surface in each effect is 50 m² and the coefficients for heat transfer in the first and second effects are 2.8 and 1.7 kW/m² K respectively. It may be assumed that the concentrated solution exhibits a boiling-point rise of 5 deg K, that the latent heat has a constant value of 2260 kJ/kg and that the specific heat capacity of the liquid stream is constant at 3.75 kJ/kg K.

Solution

A mass balance gives:

	Solids (kg/s)	Liquor (kg/s)	Total (kg/s)
Feed	0.25	2.25	2.50
Product	0.25	0.25	0.50
Evaporation	—	2.0	2.0

Thus: $(D_1 + D_2) = 2.0$ kg/s

At 14.0 kN/m²: $T_2 = 326$ K

and allowing for the boiling-point rise in the second effect, $T_2' = (326 + 5) = 331$ K.

Writing a heat balance for each effect, then:

(1) $D_0 \lambda_0 = W(T_1 - T_f) + D_1 \lambda_1$ or $2260 D_0 = (2.5 \times 3.75)(T_1 - 288) + 2260 D_1$ (i)

(2) $D_1 \lambda_1 + (W - D_1)(T_1 - T_2') = D_2 \lambda_2$

or: $2260 D_1 + (2.5 - D_1)3.75(T_1 - 331) = 2260 D_2$

since $D_2 = (2.0 - D_1)$,

Thus: $4520(1 - D_1) = (9.375 - 3.75 D_1)(T_1 - 331)$ (ii)

But: $D_1 \lambda_1 = U_2 A_2 (T_1 - T_2')$

Therefore: $2260 D_1 = (1.7 \times 50)(T_1 - 331)$ or $T_1 = (26.6 D_1 + 331)$

Substituting in equation (ii) for T_1:

$$(D_1^2 - 48.26 D_1 + 45.31) = 0$$

and: $D_1 = 47.30$ kg/s (which is clearly impossible) or 0.96 kg/s

Thus: $T_1 = (26.6 \times 0.96) + 331 = 356.5$ K

211

In equation (i):

$$2260 D_0 = (2.5 \times 3.75)(356.5 - 288) + (2260 \times 0.96) \quad \text{and} \quad D_0 = 1.24$$

But: $\quad D_0 \lambda_0 = U_1 A_1 (T_0 - T_1)$

or: $\quad (1.24 \times 2260) = (2.8 \times 50)(T_0 - 356.5) \quad \text{and} \quad T_0 = 376.5 \text{ K}$

Steam is dry and saturated at 376.5 K at a pressure of $\underline{\underline{115 \text{ kN/m}^2}}$.

PROBLEM 14.26

A salt solution at 293 K is fed at the rate of 6.3 kg/s to a forward-feed triple-effect evaporator and is concentrated from 2 per cent to 10 per cent of solids. Saturated steam at 170 kN/m² is introduced into the calandria of the first effect and a pressure of 34 kN/m² is maintained in the last effect. If the heat transfer coefficients in the three effects are 1.7, 1.4 and 1.1 kW/m² K respectively and the specific heat capacity of the liquid is approximately 4 kJ/kg K, what area is required if each effect is identical? Condensate may be assumed to leave at the vapour temperature at each stage, and the effects of boiling point rise may be neglected. The latent heat of vaporisation may be taken as constant throughout.

Solution

The mass balance is as follows:

	Solids (kg/s)	Liquor (kg/s)	Total (kg/s)
Feed	0.126	6.174	6.30
Product	0.126	1.134	1.26
Evaporation	—	5.04	5.04

Thus: $\quad (D_1 + D_2 + D_3) = 5.04 \text{ kg/s}$

At 170 kN/m²: $\quad T_0 = 388 \text{ K} \quad \text{and} \quad \lambda_0 = 2216 \text{ kJ/kg}$

At 34 kN/m²: $\quad T_3 = 345 \text{ K} \quad \text{and} \quad \lambda_3 = 2328 \text{ kJ/kg}$

Thus the latent heat will be taken as 2270 kJ/kg throughout and:

$$\sum \Delta T = (388 - 345) = 43 \text{ deg K}$$

From equation 14.8:
$$1.7 \Delta T_1 = 1.4 \Delta T_2 = 1.1 \Delta T_3$$

and hence: $\quad \Delta T_1 = 11.5 \text{ deg K}, \quad \Delta T_2 = 14 \text{ deg K}, \quad \text{and} \quad \Delta T_3 = 17.5 \text{ deg K}$

Modifying these values for the cold feed, it will be assumed that:

$$\Delta T_1 = 15 \text{ deg K}, \quad \Delta T_2 = 12 \text{ deg K}, \quad \text{and} \quad \Delta T_3 = 16 \text{ deg K}$$

Thus: $T_0 = 388$ K, $T_1 = 373$ K, $T_2 = 361$ K, and $T_3 = 345$ K

$$\lambda_0 = \lambda_1 = \lambda_2 = \lambda_3 = 2270 \text{ kJ/kg}$$

The heat balance over each effect is now:

(1) $D_0 \lambda_0 = W C_p (T_1 - T_f) + D_1 \lambda_1$ or $2270 D_0 = (6.3 \times 4)(373 - 293) + 2270 D_1$

(2) $D_1 \lambda_1 + (W - D_1) C_p (T_1 - T_2) = D_2 \lambda_2$

or: $2270 D_1 + (6.3 - D_1) 4 (373 - 361) = 2270 D_2$

(3) $D_2 \lambda_2 + (W - D_1 - D_2) C_p (T_1 - T_2) = D_3 \lambda_3$

or: $2270 D_2 + (6.3 - D_1 - D_2) 4 (361 - 345) = 2270 D_3$

Solving:

$$D_0 = 2.48 \text{ kg/s}, \quad D_1 = 1.59 \text{ kg/s}, \quad D_2 = 1.69 \text{ kg/s}, \quad D_3 = 1.78 \text{ kg/s}$$

The areas are given by:

$$A_1 = D_0 \lambda_0 / U_1 \Delta T_1 = (2270 \times 2.48)/(1.7 \times 15) = 220.8 \text{ m}^2$$
$$A_2 = D_1 \lambda_1 / U_2 \Delta T_2 = (2270 \times 1.59)/(1.4 \times 12) = 214.8 \text{ m}^2$$
$$A_3 = D_2 \lambda_2 / U_3 \Delta T_3 = (2270 \times 1.69)/(1.1 \times 16) = 217.9 \text{ m}^2$$

which are close enough for design purposes.

The area specified for each stage is therefore $\underline{218 \text{ m}^2}$.

PROBLEM 14.27

A single-effect evaporator with a heating surface area of 10 m² is used to concentrate a NaOH solution flowing at 0.38 kg/s from 10 per cent to 33.3 per cent. The feed enters at 338 K and its specific heat capacity is 3.2 kJ/kg K. The pressure in the vapour space is 13.5 kN/m² and 0.3 kg/s of steam is used from a supply at 375 K. Calculate:

(a) The apparent overall heat transfer coefficient.
(b) The coefficient corrected for boiling point rise of dissolved solids.
(c) The corrected coefficient if the depth of liquid is 1.5 m.

Solution

Mass of solids in feed $= (0.38 \times 10/100) = 0.038$ kg/s.

Mass flow of product $= (0.038 \times 100/33.3) = 0.114$ kg/s.

Thus: evaporation, $D_1 = (0.38 - 0.114) = 0.266$ kg/s

At a pressure of 13.5 kN/m², steam is saturated at 325 K and $\lambda_1 = 2376$ kJ/kg.

At 375 K, steam is saturated at 413 K and $\lambda_0 = 2140$ kJ/kg.

(a) Ignoring any boiling-point rise, it may be assumed that the temperature of the boiling liquor, $T_1 = 325$ K.

Thus:
$$\Delta T_1 = (375 - 325) = 50 \text{ deg K}$$
$$U_1 = D_0\lambda_0/A_1\Delta T_1$$
$$= (0.3 \times 2140)/(10 \times 50) = \underline{1.28 \text{ kW/m}^2 \text{ K}}$$

(b) Allowing for a boiling-point rise, the temperature of the boiling liquor in the effect, T_1' may be calculated from a heat balance:

$$D_0\lambda_0 = WC_p(T_1' - T_f) + D_1\lambda_1$$

or: $(0.3 \times 2140) = (0.38 \times 3.2)(T_1' - 338) + (0.266 \times 2376)$

and: $T_1' = 346$ K

$$\Delta T_1 = (375 - 346) = 29 \text{ deg K}$$

and: $U_1 = (0.3 \times 2140)/(10 \times 29) = \underline{2.21 \text{ kW/m}^2 \text{ K}}$

(c) Taking the density of the fluid as 1000 kg/m³, the pressure due to a height of liquid of 1.5 m = $(1.5 \times 1000 \times 9.81) = 14{,}715$ N/m² or 14.7 kN/m².

The pressure outside the tubes is therefore $(13.5 + 14.7) = 28.2$ kN/m² at which pressure, water boils at 341 K.

Thus: $\Delta T_1 = (375 - 341) = 34 \text{ deg K}$

and: $U_1 = (0.3 \times 2140)/(10 \times 34) = \underline{1.89 \text{ kW/m}^2 \text{ K}}$

A value of the boiling liquor temperature $T_1' = 346$ K obtained in (b) by heat balance must take into account the effects of hydrostatic head *and* of boiling-point rise. The true boiling-point rise is $(346 - 341) = 5$ deg K.

Thus:
$$T_1' = (325 + 5) = 330 \text{ K}$$
$$T_1' = (375 - 330) = 45 \text{ K}$$

and: $U_1 = (0.3 \times 2140)/(10 \times 45) = \underline{1.43 \text{ kN/m}^2 \text{ K}}$.

PROBLEM 14.28

An evaporator, working at atmospheric pressure, is used to concentrate a solution from 5 per cent to 20 per cent solids at the rate of 1.25 kg/s. The solution, which has a specific heat capacity of 4.18 kJ/kg K, is fed to the evaporator at 295 K and boils at 380 K. Dry saturated steam at 240 kN/m² is fed to the calandria, and the condensate leaves at the temperature of the condensing stream. If the heat transfer coefficient is 2.3 kW/m² K, what is the required area of heat transfer surface and how much steam is required? The latent heat of vaporisation of the solution may be taken as being the same as that of water.

Solution

A material balance gives:

	Solids (kg/s)	Liquor (kg/s)	Total (kg/s)
Feed	0.0625	1.1875	1.2500
Product	0.0625	0.2500	0.3125
Evaporation	—	0.9375	$0.9375 = D_1$

At 240 kN/m^2, $T_0 = 399$ K and $\lambda_0 = 2185$ kJ/kg

At a pressure of 101.3 kN/m^2, $\lambda_1 = 2257$ kJ/kg

Making a heat balance across the unit:

$$D_1\lambda_1 + WC_p(T_1 - T_f) = D_0\lambda_0$$

or: $(2257 \times 0.9375) + (1.25 \times 4.18)(380 - 295) = 2185 D_0$ and $\underline{D_0 = 1.17 \text{ kg/s}}$

The heat transfer area is given by:

$$A = D_0\lambda_0/UT_1$$
$$= (1.17 \times 2185)/[2.3(399 - 380)] = \underline{58.5 \text{ m}^2}$$

SECTION 2-15

Crystallisation

PROBLEM 15.1

A saturated solution containing 1500 kg of potassium chloride at 360 K is cooled in an open tank to 290 K. If the density of the solution is 1200 kg/m^3, the solubility of potassium chloride/100 parts of water by mass is 53.55 at 360 K and 34.5 at 290 K calculate:

(a) the capacity of the tank required, and
(b) the mass of crystals obtained, neglecting any loss of water by evaporation.

Solution

At 360 K, 1500 kg KCl will be dissolved in $(1500 \times 100)/53.55 = 2801$ kg water.

The total mass of the solution $= (1500 + 2801) = 4301$ kg.

The density of the solution $= (1.2 \times 1000) = 1200$ kg/m^3 and hence the capacity of the tank $= (4301/1200) = \underline{3.58 \text{ m}^3}$.

At 290 K, the mass of KCl dissolved in 2801 kg water $= (2801 \times 34.5)/100 = 966$ kg

Thus: mass of crystals which has come out of solution $= (1500 - 966) = \underline{\underline{534 \text{ kg}}}$

PROBLEM 15.2

Explain how fractional crystallisation may be applied to a mixture of sodium chloride and sodium nitrate, given the following data. At 290 K, the solubility of sodium chloride is 36 kg/100 kg water and of sodium nitrate 88 kg/100 kg water. Whilst at this temperature, a saturated solution comprising both salts will contain 25 kg sodium chloride and 59 kg sodium nitrate/100 parts of water. At 357 K these values, again per 100 kg of water, are 40 and 176, and 17 and 160 kg respectively.

Solution

See Volume 2, Example 15.9.

PROBLEM 15.3

10 Mg of a solution containing 0.3 kg Na_2CO_3/kg solution is cooled slowly to 293 K to form crystals of $Na_2CO_3.10H_2O$. What is the yield of crystals if the solubility of Na_2CO_3 at 293 K is 21.5 kg/100 kg water and during cooling 3 per cent of the original solution is lost by evaporation?

Solution

The initial concentration of the solution = 0.3 kg/kg solution

or: $\quad c_1 = 0.3/(1 - 0.3) = 0.428$ kg/kg water.

The final concentration of the solution, $c_2 = (21.5/100) = 0.215$ kg/kg water.

The feed of 10 Mg of solution contains $(10 \times 0.3) = 3$ Mg of anhydrous salt and $(10 - 3) = 7$ Mg of water.

Thus: the initial mass of solvent in the liquid, $w_1 = (7 \times 1000) = 7000$ kg.

3 per cent of the original solution or $(10 \times 1000 \times 3)/100 = 300$ kg is evaporated.

Thus the mass of solvent evaporated/mass of solvent in initial solution is given by:

$$E = [300/(10 \times 1000)] = 0.03 \text{ kg/kg solution.}$$

The final mass of solvent in the liquid, $w_2 = (7000 - 300) = 6700$ kg.

The molecular mass of $Na_2CO_3 = 106$ kg/kmol and the molecular mass of $Na_2CO_3.10H_2O = 286.2$ kg/kmol and hence the molecular mass of hydrate/molecular mass of anhydrous salt is given by:

$$R = (286.2/106) = 2.7$$

Therefore from equation 15.22: $y = Rw_1[c_1 - c_2(1 - E)]/[1 - c_2(R - 1)]$

and by substituting, the yield is:

$$y = (2.7 \times 7000)[0.428 - 0.215(1 - 0.03)]/[1 - 0.215(2.7 - 1.0)]$$

$$= \underline{\underline{6536 \text{ kg}}}.$$

PROBLEM 15.4

The heat required when 1 kmol of $MgSO_4.7H_2O$ is dissolved isothermally at 291 K in a large mass of water is 13.3 MJ. What is the heat of crystallisation per unit mass of the salt?

Solution

The molecular mass of $MgSO_4.7H_2O = 246.5$ kg/kmol

Thus: $\quad$ heat of crystallisation $= (13.3 \times 1000)/246.5 = \underline{\underline{53.9 \text{ kJ/kg}}}$

PROBLEM 15.5

A solution of 500 kg of Na_2SO_4 in 2500 kg water is cooled from 333 K to 283 K in an agitated mild steel vessel of mass 750 kg. At 283 K, the solubility of the anhydrous salt is 8.9 kg/100 kg water and the stable crystalline phase is $Na_2SO_4.10H_2O$. At 291 K, the heat of solution is -78.5 MJ/kmol and the specific heat capacities of the solution and mild steel are 3.6 and 0.5 kJ/kg deg K respectively. If, during cooling, 2 per cent of the water initially present is lost by evaporation, estimate the heat which must be removed.

Solution

It is assumed that the heat of crystallisation = $-$(the heat of solution)

$$= 78.5 \text{ MJ/kmol}$$

or: $(78.5 \times 1000)/322 = 244$ kJ/kg

where 322 kg/kmol is the molecular mass of the hydrate.

Molecular mass of $Na_2SO_4 = 142$ kg/kmol, and $R = (322/142) = 2.27$

The latent heat of vaporisation of water will be taken as 2395 kJ/kmol.

The initial concentration of the solution, $c_1 = (500/2500) = 0.2$ kg/kg water and the final concentration of the solution, $c_2 = (8.9/100) = 0.089$ kg/kg water at 283 K.

The initial mass of water, $w_1 = 2500$ kg

and the evaporation is 2 per cent of the initial water, or $E = 0.02$ kg/kg water.

Thus: In equation 15.22, the yield is:

$$y = (2.27 \times 2500)[0.2 - 0.089(1 - 0.02)]/[1 - 0.089(2.27 - 1)] = 723 \text{ kg}.$$

Therefore: The heat of crystallisation = $(723 \times 244) = 176{,}412$ kJ.

Heat removed from the solution, assuming crystallisation takes place at 283 K, is then:

$$= (500 + 2500)3.6(333 - 283) = 540{,}000 \text{ kJ}$$

Heat removed from the mild steel vessel = $(750 \times 0.5)(333 - 283) = 18{,}750$ kJ

and total heat to be removed = $(176{,}412 + 540{,}000 + 18{,}750) = 735{,}162$ kJ

But: $[(2500 \times 2)/100]2395 = 119{,}750$ kJ is lost by evaporation.

Thus: net heat to be removed = $(735{,}162 - 119{,}750) \approx \underline{\underline{615{,}000 \text{ kJ}}}$

PROBLEM 15.6

A batch of 1500 kg of saturated potassium chloride solution is cooled from 360 K to 290 K in an unagitated tank. If the solubilities of KCl are 53 and 34 kg/100 kg water at 360 K and 290 K respectively and water losses due to evaporation may be neglected, what is the yield of crystals?

Solution

The initial concentration of the solution is $(53/100) = 0.53$ kg/kg water

or: $\qquad c_1 = 0.53/(1 + 0.53) = 0.346$ kg/kg solution.

Mass of potassium chloride in the original batch $= (1500 \times 0.346) = 520$ kg
and hence, mass of water in the original batch, $w_1 = (1500 - 520) = 980$ kg.
The concentration of potassium chloride in the final solution is:

$$c_2 = (34/100) = 0.34 \text{ kg/kg water.}$$

With no hydrate formation and negligible evaporation, $R = 1.0$ and $E = 0$ respectively and in equation 15.22, the yield is:

$$y = (1.0 \times 980)[0.53 - 0.34(1 - 0)]/[1 - 0.34(1 - 1)] = \underline{\underline{186 \text{ kg}}}$$

PROBLEM 15.7

Glauber's salt, $Na_2SO_4 \cdot 10H_2O$, is to be produced in a Swenson–Walker crystalliser by cooling to 290 K a solution of anhydrous Na_2SO_4 which saturates between 300 K and 290 K. If cooling water enters and leaves the unit at 280 K and 290 K respectively and evaporation is negligible, how many sections of crystalliser, each 3 m long, will be required to process 0.25 kg/s of the product? The solubilities of anhydrous Na_2SO_4 in water are 40 and 14 kg/100 kg water at 300 K and 290 K respectively, the mean heat capacity of the liquor is 3.8 kJ/kg K and the heat of crystallisation is 230 kJ/kg. For the crystalliser, the available heat transfer area is 3 m^2/m length, the overall coefficient of heat transfer is 0.15 kW/m^2 K and the molecular masses are $Na_2SO_4 \cdot 10H_2O = 322$ kg/kmol and $Na_2SO_4 = 142$ kg/kmol.

Solution

The ratio of the molecular mass of the hydrate to that of the anhydrous salt,
$R = (322/142) = 2.27$ and the evaporation, E may be neglected.
The concentration of salt in the feed $= (40/100) = 0.40$ kg/kg water

and: $\qquad c_1 = 0.40/(1 - 0.40) = 0.286$ kg/kg solution.

Similarly: $\qquad c_2 = (14/100) = 0.140$ kg/kg solution.

In 1 kg of feed, the mass of salt $= 0.286$ kg

and the water present, $\quad w_1 = (1 - 0.286) = 0.714$ kg/kg feed.

In equation 15.22, the yield is:

$$y = (2.27 \times 0.714)[0.40 = 0.14(1 - 0)]/[1 - 0.14(2.27 - 1)] = 0.573 \text{ kg/kg feed.}$$

In order to produce 0.25 kg/s of crystals, a feed rate of:

$$(1 \times 0.25)/0.573 = 0.487 \text{ kg/s is required.}$$

The heat to be removed from the solution is given by:

$$(0.487 \times 3.8)(300 - 290) = 18.5 \text{ kW (assuming crystals are formed at 290 K)}$$

The heat of crystallisation is given by:

$$(0.25 \times 330) = 57.5 \text{ kW}$$

and the total heat to be removed $= (18.5 + 57.5) = 76.0$ kW.

The logarithmic mean temperature difference, assuming counter current flow is:

$$\Delta T_m = [(300 - 290) - (290 - 280)]/\ln[(300 - 290)/(290 - 280)] = 10 \text{ deg K}$$

and with an overall coefficient of heat transfer of 0.15 kW/m² deg K, the area required is:

$$A = Q/U\Delta T_m$$
$$= 76.0/(0.15 \times 10) = 50.67 \text{ m}^2$$

Area per unit length of crystallise section $= 3 \text{ m}^2/\text{m}$
Hence: Total length of unit required $= (50.67/3) = 16.9$ m
and 6 sections each of 3 m length would be specified.

PROBLEM 15.8

What is the evaporation rate and yield of the sodium acetate hydrate $CH_3COONa.3H_2O$ from a continuous evaporative crystalliser operating at 1 kN/m² when it is fed with 1 kg/s of a 50 per cent by mass aqueous solution of sodium acetate hydrate at 350 K? The boiling point elevation of the solution is 10 deg K and the heat of crystallisation is 150 kJ/kg. The mean heat capacity of the solution is 3.5 kJ/kg K and, at 1 kN/m², water boils at 280 K at which temperature the latent heat of vaporisation is 2.482 MJ/kg. Over the range 270–305 K, the solubility of sodium acetate hydrate in water s at $T(K)$ is given approximately by:

$$s = 0.61T - 132.4 \text{ kg/100 kg water}$$

Molecular masses: $CH_3COONa.3H_2O = 136$ kg/kmol, $CH_3COONa = 82$ kg/kmol.

Solution

Allowing for the boiling point elevation, the temperature of the solution at equilibrium is:

$$(280 + 10) = 290 \text{ K}$$

The concentration of the feed solution is $(50/100) = 0.5$ kg/kg solution and the initial concentration of solution is:

$$c_1 = 0.5/(1 - 0.5) = 1.0 \text{ kg/kg water.}$$

Using the given relationship for the solubility, at 290 K, the final concentration of solution is:
$$c_2 = [(0.61 \times 290) - 132.4]/100 = 0.445 \text{ kg/kg water.}$$

The heat of crystallisation, $q_c = 150$ kJ/kg
and the ratio of the molecular masses, $R = (136/82) = 1.66$.
The evaporation rate is given by equation 15.23 as:

$$E = [q_c R(c_1 - c_2) + C_p(T_1 - T_2)(1 + c_1)(1 - c_2(R - 1))]$$
$$/[L(1 - c_2(R - 1)) - q_c R c_2]$$

or: $E = [150 \times 1.66(1.0 - 0.445) + 3.5(350 - 290)(1 + 1.0)(1 - 0.445(1.66 - 1))]$
$/[2482(1 - 0.445(1.66 - 1)) - (150 \times 1.66 \times 0.445)]$
$= 0.265$ kg/kg

The actual feed is 1 kg/s of solution containing 0.5 kg/s of salt
and 0.5 kg/s of water $= w_1$

Therefore: Actual evaporation rate $= (0.265 \times 0.5) = \underline{\underline{0.132 \text{ kg/s}}}$

The yield is then given by equation 15.22:

$$y = R w_1 [c_1 - c_2(1 - E)]/[1 - c_2(R - 1)]$$
$$= (1.66 \times 0.5)(1.0 - 0.445(1 - 0.265))/(1 - 0.445(1.66 - 1))$$
$$= \underline{\underline{0.791 \text{ kg/s}}}$$

SECTION 2-16

Drying

PROBLEM 16.1

A wet solid is dried from 35 to 10 per cent moisture under constant drying conditions in 18 ks (5 h). If the equilibrium moisture content is 4 per cent and the critical moisture content is 14 per cent, how long will it take to dry to 6 per cent moisture under the same conditions?

Solution

See Volume 2, Example 16.1.

PROBLEM 16.2

Strips of a material 10 mm thick are dried under constant drying conditions from 28 per cent to 13 per cent moisture in 25 ks. If the equilibrium moisture content is 7 per cent, what is the time taken to dry 60 mm planks from 22 to 10 per cent moisture under the same conditions, assuming no loss from the edges? All moisture contents are expressed on the wet basis. The relation between E, the ratio of the average free moisture content at time t to the initial free moisture content, and the parameter f is given by:

E	1	0.64	0.49	0.38	0.295	0.22	0.14
f	0	0.1	0.2	0.3	0.5	0.6	0.7

where $f = kt/l^2$, k is a constant, t is the time in ks and $2l$ is the thickness of the sheet of material in mm.

Solution

See Volume 2, Example 16.2.

PROBLEM 16.3

A granular material containing 40 per cent moisture is fed to a countercurrent rotary dryer at 295 K and is withdrawn at 305 K containing 5 per cent moisture. The air supplied, which contains 0.006 kg water vapour/kg of dry air, enters at 385 K and leaves at 310 K. The dryer handles 0.125 kg/s wet stock.

Assuming that radiation losses amount to 20 kJ/kg of dry air used, determine the mass flow of dry air supplied to the dryer and the humidity of the outlet air. The latent heat of water vapour at 295 K = 2449 kJ/kg, the specific heat capacity of dried material = 0.88 kJ/kg K, the specific heat capacity of dry air = 1.00 kJ/kg K, and the specific heat capacity of water vapour = 2.01 kJ/kg K.

Solution

See Volume 2, Example 16.3.

PROBLEM 16.4

1 Mg of dry mass of a non-porous solid is dried under constant drying conditions in an air stream flowing at 0.75 m/s. The area of surface drying is 55 m^2. If the initial rate of drying is 0.3 g/m^{2}s, how long will it take to dry the material from 0.15 to 0.025 kg water/kg dry solid? The critical moisture content of the material may be taken as 0.125 kg water/kg dry solid. If the air velocity were increased to 4.0 m/s, what would be the anticipated saving in time if the process were surface-evaporation controlled?

Solution

During the constant rate period, that is whilst the moisture content falls from 0.15 to 0.125 kg/kg, the rate of drying is:

$$(dw/dt)/A = (0.3/1000) = 0.0003 \text{ kg/m}^2\text{s}$$

At the start of the falling rate period, $w = w_c = 0.125$ kg/kg

and: $\quad (dw/dt)/A = m(w_c - w_e)$

or: $\quad 0.0003 = m(0.125 - 0.025)$

and: $\quad m = 0.003 \text{ kg/m}^2\text{s kg dry solid}$

$$= (0.003/1000) = 3.0 \times 10^{-6} \text{ kg/m}^2\text{s Mg dry solid}$$

The total drying time is given by equation 16.14:

$$t = (1/mA)[\ln(f_c/f) + (f_1 - f_c)/f_c]$$

where: $\quad f = (0.025 - 0) = 0.025$ kg/kg (taking w_e as zero)

$\quad\quad\quad\; f_c = (0.125 - 0) = 0.125$ kg/kg

$\quad\quad\quad\; f_1 = (0.15 - 0) = 0.150$ kg/kg

Thus: $\quad t = [1/(3.0 \times 10^{-6} \times 55)][\ln(0.125/0.025) + (0.150 - 0.125)/0.125]$

$\quad\quad\quad\;\; = 10{,}960 \text{ s} \quad \text{or} \quad \underline{10.96 \text{ ks}} \text{ (3 h)}$

As a first approximation it may be assumed that the rate of evaporation is proportional to the air velocity raised to the power of 0.8. For the second case, m may then be calculated as:

$$m = (3.0 \times 10^{-6})(4.0/0.75)^{0.8} = (1.15 \times 10^{-5}) \text{ kg water/m}^2\text{s Mg dry solid}$$

The time of drying is then:

$$t = [1/(1.15 \times 10^{-5} \times 55)](1.609 + 0.20) = 2860 \text{ s or } 2.86 \text{ ks}$$

and the time saved is therefore: $(10.96 - 2.86) = \underline{\underline{8.10 \text{ ks}}}$ (2.25 h)

PROBLEM 16.5

A 100 kg batch of granular solids containing 30 per cent of moisture is to be dried in a tray dryer to 15.5 per cent moisture by passing a current of air at 350 K tangentially across its surface at the velocity of 1.8 m/s. If the constant rate of drying under these conditions is 0.7 g/s m^2 and the critical moisture content is 15 per cent, calculate the approximate drying time. It may be assumed that the drying surface is 0.03 m^2/kg dry mass.

Solution

See Volume 2, Example 16.4.

PROBLEM 16.6

A flow of 0.35 kg/s of a solid is to be dried from 15 per cent to 0.5 per cent moisture on a dry basis. The mean specific heat capacity of the solids is 2.2 kJ/kg deg K. It is proposed that a co-current adiabatic dryer should be used with the solids entering at 300 K and, because of the heat sensitive nature of the solids, leaving at 325 K. Hot air is available at 400 K with a humidity of 0.01 kg/kg dry air and the maximum allowable mass velocity of the air is 0.95 kg/m^{2}s. What diameter and length should be specified for the proposed dryer?

Solution

See Volume 2, Example 16.5.

PROBLEM 16.7

0.126 kg/s of a solid product containing 4 per cent water is produced in a dryer from a wet feed containing 42 per cent water on a wet basis. Ambient air at 294 K and of 40 per cent relative humidity is heated to 366 K in a preheater before entering the dryer from which it leaves at 60 per cent relative humidity. Assuming that the dryer operates adiabatically, what must be the flowrate of air to the preheater and how much heat must be added to the preheater? How will these values be affected if the air enters the dryer at 340 K and sufficient heat is supplied within the dryer so that the air again leaves at 340 K with a relative humidity of 60 per cent?

Solution

From the humidity chart, Figure 13.4 in Volume 1, air at 294 K and of 40 per cent relative humidity has a humidity of 0.006 kg/kg. This remains unchanged on heating to 366 K. At the dryer inlet, the wet bulb temperature of the air is 306 K. In the dryer, the cooling takes place along the adiabatic cooling line until 60 per cent relative humidity is reached.
At this point:

$$\text{the humidity} = 0.028 \text{ kg/kg} \quad \text{and}$$
$$\text{the dry bulb temperature} = 312 \text{ K}.$$

The water picked up by 1 kg of dry air = $(0.028 - 0.006) = 0.22$ kg water/kg dry air.

The wet feed contains:

$$42 \text{ kg water}/100 \text{ kg feed} \quad \text{or} \quad 42/(100 - 42) = 0.725 \text{ kg water/kg dry solids}$$

The product contains:

$$4 \text{ kg water}/100 \text{ kg product} \quad \text{or} \quad 4/(100 - 4) = 0.0417 \text{ kg water/kg dry solids}.$$

Thus:
$$\text{water evaporated} = (0.725 - 0.0417) = 0.6833 \text{ kg/kg dry solids}$$

The throughput of dry solids is:

$$0.126(100 - 4)/100 = 0.121 \text{ kg/s}$$

and the water evaporated is:

$$(0.121 \times 0.6833) = 0.0825 \text{ kg/s}$$

The required air throughput is then:

$$0.0825/(0.078 - 0.006) = 3.76 \text{ kg/s}$$

At 294 K, from Figure 13.4 in Volume 1:

$$\text{specific volume of the air} = 0.84 \text{ m}^3/\text{kg}$$

and the volume of air required is:

$$(0.84 \times 3.76) = \underline{\underline{3.16 \text{ m}^3/\text{s}}}$$

At a humidity of 0.006 kg water/kg dry air, from Figure 13.4, the humid heat is 1.02 kJ/kg deg K and hence the heat required in the preheater is:

$$3.76 \times 1.02(366 - 294) = \underline{276 \text{ kW}}$$

In the second case, air both enters and leaves at 340 K, the outlet humidity is 0.111 kg water/kg dry air and the water picked up by the air is:

$$(0.111 - 0.006) = 0.105 \text{ kg/kg dry air}$$

The air requirements are then:

$$(0.0825/0.105) = 0.786 \text{ kg/s} \quad \text{or} \quad (0.786 \times 0.84) = \underline{\underline{0.66 \text{ m}^3/\text{s}}}$$

The heat to be added in the preheater is:

$$0.66 \times 1.02(340 - 294) = \underline{31.0 \text{ kW}}$$

The heat to be supplied within the dryer is that required to heat the water to 340 K plus its latent heat at 340 K.

That is: $\qquad 0.0825[4.18(340 - 290) + 2345] = 211 \text{ kW}$

Thus: $\qquad$ Total heat to be supplied $= (81 + 211) = \underline{242 \text{ kW}}$.

PROBLEM 16.8

A wet solid is dried from 40 to 8 per cent moisture in 20 ks. If the critical and the equilibrium moisture contents are 15 and 4 per cent respectively, how long will it take to dry the solid to 5 per cent moisture under the some drying conditions? All moisture contents are on a dry basis.

Solution

For the first drying operation:

$$w_1 = 0.40 \text{ kg/kg}, \quad w = 0.08 \text{ kg/kg}, \quad w_c = 0.15 \text{ kg/kg} \quad \text{and} \quad w_e = 0.04 \text{ kg/kg}$$

Thus: $\qquad f_1 = (w_1 - w_e) = (0.40 - 0.04) = 0.36 \text{ kg/kg}$

$\qquad\qquad f_c = (w_c - w_e) = (0.15 - 0.04) = 0.11 \text{ kg/kg}$

$\qquad\qquad f = (w - w_e) = (0.08 - 0.04) = 0.04 \text{ kg/kg}.$

From equation 16.14, the total drying time is:

$$t = (1/mA)[(f_1 - f_c)/f_c + \ln(f_c/f)]$$

or: $\qquad 20 = (1/mA)[(0.36 - 0.11)/0.11 + \ln(0.11/0.04)]$

and: $\qquad mA = 0.05(2.27 + 1.012) = 0.164 \text{ kg/ks}$

For the second drying operation:

$w_1 = 0.40$ kg/kg, $w = 0.05$ kg/kg, $w_c = 0.15$ kg/kg, and $w_e = 0.04$ kg/kg

Thus:
$$f_1 = (w_1 - w_e) = (0.40 - 0.04) = 0.36 \text{ kg/kg}$$
$$f_e = (w_c - w_e) = (0.15 - 0.04) = 0.11 \text{ kg/kg}$$
$$f = (w - w_e) = (0.05 - 0.04) = 0.01 \text{ kg/kg}$$

The total drying time is then:
$$t = (1/0.164)[(0.36 - 0.11)/0.11 + \ln(0.11/0.01)]$$
$$= 6.098(2.273 + 2.398)$$
$$= \underline{\underline{28.48 \text{ ks (7.9 h)}}}$$

PROBLEM 16.9

A solid is to be dried from 1 kg water/kg dry solids to 0.01 kg water/kg dry solids in a tray dryer consisting of a single tier of 50 trays, each 0.02 m deep and 0.7 m square completely filled with wet material. The mean air temperature is 350 K and the relative humidity across the trays may be taken as constant at 10 per cent. The mean air velocity is 2.0 m/s and the convective coefficient of heat transfer is given by:

$$h_c = 14.3 G'^{0.8} \quad \text{W/m}^2 \text{ deg K}$$

where G' is the mass velocity of the air in kg/m²s. The critical and equilibrium moisture contents of the solid are 0.3 and 0 kg water/kg dry solids respectively and the bulk density of the dry solid is 6000 kg/m³. Assuming that the drying is by convection from the top surface of the trays only, what is the drying time?

Solution

From Figure 13.4 in Volume 1, the specific volume of moist air at 350 K and 10 per cent relative humidity is 1.06 m³/kg.

Thus:
$$\text{mass velocity, } G' = (2.0/1.06) = 1.88 \text{ kg/m}^2\text{s}$$

and:
$$\text{convective heat transfer coefficient, } h_c = (14.3 \times 1.88^{0.8})$$
$$= 23.8 \text{ W/m}^2 \text{ K} \quad \text{or} \quad 0.0238 \text{ kW/m}^2 \text{ K}.$$

If it may be assumed that the temperature of the surface is equal to the wet bulb temperature of the air which, from Figure 13.4, is 317 K, then:

$$\text{mean temperature driving force, } \Delta T = (350 - 317) = 33 \text{ deg K}.$$

The area of the top surface of the trays is:

$$A = (50 \times 0.7^2) = 24.5 \text{ m}^2$$

From steam tables in the Appendix of Volume 2, the latent heat of vaporisation of water at 317 K is $\lambda = 2395$ kJ/kg.

The drying rate during the constant drying period is then given by:

$$W = h_c A \Delta T / \lambda \quad \text{(equation 16.9)}$$

or:
$$W = (0.0238 \times 24.5 \times 33)/2359$$
$$= 0.00816 \text{ kg/s}$$

The total volume of dry material is:

$$(50 \times 0.7^2 \times 0.02) = 0.49 \text{ m}^3$$

and the mass of dry material is:

$$(0.49 \times 6000) = 2940 \text{ kg}$$

The rate of drying during the constant rate period is then:

$$R_c = 0.00816/(2940 \times 24.5) = 1.133 \times 10^{-7} \text{ kg water/m}^2\text{s kg dry solid}$$

In this problem:

$$w_1 = 1.0 \text{ kg/kg}, \quad w = 0.01 \text{ kg/kg}, \quad w_c = 0.3 \text{ kg/kg} \quad \text{and} \quad w_e = 0$$

Thus:
$$f_1 = (w_1 - w_e) = (1.0 - 0) = 1.0 \text{ kg/kg}$$
$$f_c = (w_c - w_e) = (0.3 - 0) = 0.3 \text{ kg/kg}$$

and:
$$f = (w - w_e) = (0.01 - 0) = 0.01 \text{ kg/kg}$$

From equation 16.13:

$$R_c = mf_c$$

or:
$$1.133 \times 10^{-7} = m \times 0.3$$

and:
$$m = 3.78 \times 10^{-7}$$

Thus, in equation 16.14, the total drying time is:

$$t = [1/(3.78 \times 10^{-7} \times 24.5)][(1.0 - 0.3)/0.3 + \ln(0.3/0.01)]$$
$$= 1.081 \times 10^5 (2.33 + 3.40)$$
$$= 6.19 \times 10^5 \text{ s} \quad \text{or} \quad \underline{\underline{619 \text{ ks (172 h)}}}$$

PROBLEM 16.10

Skeins of a synthetic fibre are dried from 46 per cent to 8.5 per cent moisture on a wet basis in a 10 m long tunnel dryer by a countercurrent flow of hot air. The air mass velocity, G' is 1.36 kg/m²s and the inlet conditions are 355 K and a humidity of 0.03 kg moisture/kg dry air. The air temperature is maintained at 355 K throughout the dryer by internal heating and, at the outlet, the humidity of the air is 0.08 kg moisture/kg dry air. The equilibrium moisture content is given by:

$$w_e = 0.25 \text{ (per cent relative humidity)}$$

and the drying rate by:

$$R = 1.34 \times 10^{-4} G'^{1.47}(w - w_e)(\mathscr{H}_w - \mathscr{H}) \text{ kg/s kg dry fibres}$$

where $\mathscr{H}$ is the humidity of dry air and $\mathscr{H}_w$ the saturation humidity at the wet bulb temperature. Data relating w, $\mathscr{H}$ and $\mathscr{H}_w$ are as follows:

w (kg/kg dry fibre)	$\mathscr{H}$ (kg/kg dry air)	$\mathscr{H}_w$ (kg/kg dry air)	relative humidity (per cent)
0.852	0.080	0.095	22.4
0.80	0.0767	0.092	21.5
0.60	0.0635	0.079	18.2
0.40	0.0503	0.068	14.6
0.20	0.0371	0.055	11.1
0.093	0.030	0.049	9.0

At what velocity should the skeins be passed through the dryer?

Solution

For $G' = 1.36$ kg/m²s, the rate of drying is given by:

$$R = 1.34 \times 10^{-4} \times 1.36^{1.47}(w - w_e)(\mathscr{H}_w - \mathscr{H})$$
$$= -2.11 \times 10^{-4}(w_e - w)(\mathscr{H}_w - \mathscr{H}) \text{ kg/s kg dry fibre}$$

The working is now laid out in tabular form, noting that an inlet moisture content of 46.0 per cent on a wet basis is equivalent to:

$$46.0/(100 - 46.0) = 0.852 \text{ kg/kg dry fibre.}$$

and an outlet moisture content of 8.5 per cent on a wet basis is equivalent to:

$$8.5/(100 - 8.5) = 0.093 \text{ kg/kg dry fibre.}$$

w (kg/kg dry fibre)	0.852	0.80	0.60	0.40	0.20	0.093
relative humidity (per cent)	22.4	21.5	18.2	14.6	11.1	9.0
w_e (kg/kg dry fibre) (= 0.25 RH/100)	0.056	0.054	0.046	0.037	0.028	0.023
$(w - w_e)$ (kg/kg dry fibre)	0.794	0.746	0.537	0.363	0.172	0.070

$\mathscr{H}_w$ (kg/kg dry air)	0.095	0.092	0.079	0.068	0.055	0.049
$\mathscr{H}$ (kg/kg dry air)	0.0800	0.0767	0.0635	0.0503	0.0371	0.0300
$(\mathscr{H}_w - \mathscr{H})$ (kg/kg dry air)	0.0150	0.0153	0.0155	0.0177	0.0179	0.0190
R (kg/s kg dry fibre)	−0.0112	−0.0103	−0.0076	−0.0043	−0.0020	−0.0008
$1/R$ (kg dry fibre s/kg)	−89.0	−97.0	−132	−233	−500	−1250

Because R/w is equal to the drying time in s, the area under a plot of $1/R$ and w is the drying time. Such a plot is shown in Figure 16a from which the area between $w = 0.852$ and $w = 0.093$ kg/kg is equivalent to 203 s.

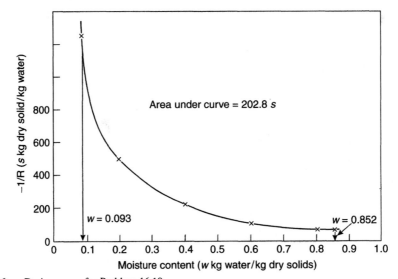

Figure 16a. Drying curve for Problem 16.10

Thus:

the velocity at which material should be passed through the dryers

$$= (10/203) = 0.0494 \quad \text{or} \quad \underline{\underline{0.05 \text{ m/s}}}$$

SECTION 2-17

Adsorption

PROBLEM 17.1

Spherical particles 15 nm in diameter and of density 2290 kg/m³ are pressed together to form a pellet. The following equilibrium data were obtained for the sorption of nitrogen at 77 K. Obtain estimates of the surface area of the pellet from the adsorption isotherm and compare the estimates with the geometric surface. The density of liquid nitrogen at 77 K is 808 kg/m³.

P/P^0	0.1	0.2	0.3	0.4	0.5	0.6	0.7	0.8	0.9
m³ liquid $N_2 \times 10^6$/kg solid	66.7	75.2	83.9	93.4	108.4	130.0	150.2	202.0	348.0

where P is the pressure of the sorbate and P^0 is its vapour pressure at 77 K.

Solution

See Volume 2, Example 17.1.

PROBLEM 17.2

In a volume of 1 m³ of a mixture of air and acetone vapour, the temperature is 303 K and the total pressure is 100 kN/m². If the relative saturation of the air by acetone is 40 per cent, what mass of activated carbon must be added to the space so that at equilibrium the value is reduced to 5 per cent at 303 K?

If 1.6 kg carbon is added, what is relative saturation of the equilibrium mixture assuming the temperature to be unchanged? The vapour pressure of acetone at 303 K is 37.9 kN/m² and the adsorption equilibrium data for acetone on carbon at 303 K are:

Partial pressure acetone $\times 10^{-2}$ (N/m²)	0	5	10	30	50	90
x_r (kg acetone/kg carbon)	0	0.14	0.19	0.27	0.31	0.35

Solution

The data are plotted in Figure 17a.

The final partial pressure of acetone $= (0.05 \times 37.9 \times 1000)$

$$= 1895 \text{ N/m}^2$$

From the isotherm, the acetone in the carbon at equilibrium = 0.23 kg/kg carbon.

The mass of acetone in the air initially = $(1/22.4)(273/303)(0.4 \times 37,900 \times 58/10^5)$ kg

where the molecular mass of acetone = 58 kg/kmol.

Thus the acetone removed from the air is:

$$= (1/22.4)(273/303)(37,900 \times 58/10^5)(0.4 - 0.05)$$
$$= 0.3095 \text{ kg.}$$

Thus the mass of carbon to be added = $(0.3095/0.23)$

$$= \underline{\underline{1.35 \text{ kg}}}$$

If the mass of carbon added is 1.6 kg, then:

$$1.6x = (1/22.4)(273/303)(37,900 \times 58/10^5)(0.4 - y)$$

and: $\quad x = 0.221 - 0.553y = 0.221 - 1.458 \times 10^{-3} P.$

From Figure 17a, the two curves intersect at $x = \underline{\underline{0.203 \text{ kg/kg}}}$

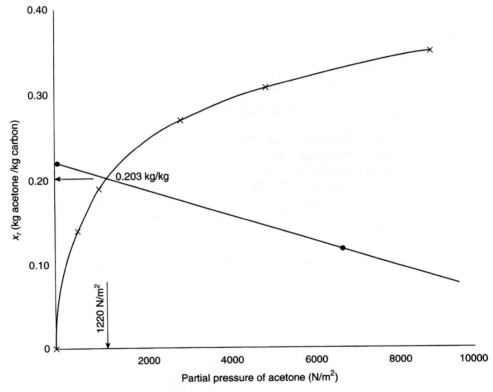

Figure 17a. Adsorption data for Problem 17.2

and:
$$P = 1220 \text{ N/m}^2$$
$$= 100(1220/37,900)$$
$$= \underline{\underline{3.2 \text{ per cent relative saturation.}}}$$

PROBLEM 17.3

A solvent contaminated with 0.03 kmol/m³ of a fatty acid is to be purified by passing it through a fixed bed of activated carbon which adsorbs the acid but not the solvent. If the operation is essentially isothermal and equilibrium is maintained between the liquid and the solid, calculate the length of a bed of 0.15 m in diameter which will allow operation for one hour when the fluid is fed at 1×10^{-4} m³/s. The bed is free of adsorbate initially and the intergranular voidage is 0.4. Use an equilibrium, fixed-bed theory to obtain the length for three types of isotherm:

(a) $C_s = 10\ C$.
(b) $C_s = 3.0\ C^{0.3}$ (use the mean slope).
(c) $C_s = 10^4\ C^2$ (the breakthrough concentration is 0.003 kmol/m³).

C and C_s refer to concentrations in kmol/m³ in the gas phase and the absorbent, respectively.

Solution

See Volume 2, Example 17.3.

SECTION 2-18

Ion Exchange

PROBLEM 18.1

A single pellet of resin is exposed to a flow of solution and the temperature is maintained constant. The take-up of exchanged ion is followed automatically and the following results are obtained:

t (min)	2	4	10	20	40	60	120
x_r (kg/kg)	0.091	0.097	0.105	0.113	0.125	0.128	0.132

On the assumption that the resistance to mass transfer in the external film is negligible, predict the values of x_r, the mass of sorbed phase per unit mass of resin as a function of time t, for a pellet of twice the radius.

Solution

See Volume 2, Example 18.1.

SECTION 2-19

Chromatographic Separations

PROBLEM 19.1

Describe the principle of separation involved in elution chromatography and derive the retention equation:

$$t_R = t_M \left[1 + \frac{(1-\varepsilon)}{\varepsilon} K \right]$$

Solution

The principles of separation by elution chromatography are considered in Volume 2, Sections 19.1 and 19.2. The derivation of the retention equation is given by equations 19.1–19.8 in Section 19.2.2 of Volume 2.

PROBLEM 19.2

In chemical analysis, chromatography may permit the separation of more than a hundred components from a mixture in a single run. Explain why chromatography can provide such a large separating power. In production chromatography, the complete separation of a mixture containing more than a few components is likely to involve two or three columns for optimum economic performance. Why is this?

Solution

In chemical analysis by the elution technique, only very small samples are injected into the column. The resolution is then given by equation 19.14 and can be very high for most components if the stationary phase is suitably chosen. The equation governs the resolution regardless of the relative amounts of adjacent eluted components. In production chromatography using the elution method, the injected feed bands are much larger in both height (solute concentration) and width (duration of the injection). Equation 19.14 then no longer applies. Components present in high concentrations will give asymmetrical, broader bands than adjacent components of smaller concentration, which may then appear only as unresolved shoulders on the broader bands of their larger neighbours. To optimise the overall separation performance and achieve high throughout and product purity (Section 19.5.4), it may then be best to use one column for the main separation and further columns to complete the resolution of the closest adjacent pairs. The alternative, of resolving all components on one column, might require a much longer column which would not be optimum for other component pairs already resolved on a short column. In the selective adsorption of biological macromolecules more than one type of chromatographic column is often used for a different reason, namely to take advantage of the different types of chromatography such as ion-exchange, affinity and size exclusion, as described in Section 19.6.2.

PROBLEM 19.3

By using the chromatogram shown in Figure 19.3, of Volume 2, show that $k'_1 = 3.65$, $k'_2 = 4.83$, $\alpha = 1.32$, $R_s = 1.26$ and $N = 500$. Show also that, if $\varepsilon = 0.8$ and $L = 1.0$ m, then $K_1 = 14.6$, $K_2 = 19.3$ and $H = 2.0$ mm. Calculate the ratio of plate height to particle diameter to confirm that the column is inefficient, as might be anticipated from the wide bands shown in Figure 19.3. It may be assumed that the particle size is that of a typical GC column, as given in Table 19.3 in Volume 2.

Solution

See Volume 2, Example 19.1

PROBLEM 19.4

Suggest one or more types of chromatography to separate each of the following mixtures:

(a) α- and β-pinenes
(b) blood serum proteins
(c) hexane isomers
(d) purification of cefonicid, a synthetic β-lactam antibiotic.

Solution

Examination of Table 19.2 in Volume 2 suggests that the following types of chromatograph might be suitable for separating the various mixtures:

(a) Gas — liquid (GLC) or reverse phase (RP-BPC) chromatography
(b) Ion — exchange (IEC) or affinity (AC) chromatography
(c) Gas chromatography (GLC or GSC)
(d) Reverse phase (RP-BPC) chromatography

It may be noted that these techniques are not exclusive and reference may be made to Section 19.6.2.

SECTION 3-1

Reactor Design — General Principles

PROBLEM 1.1

A preliminary assessment of a process for the hydrodealkylation of toluene is to be made. The reaction involved is:

$$C_6H_5 \cdot CH_3 + H_2 \rightleftharpoons C_6H_6 + CH_4$$

The feed to the reactor will consist of hydrogen and toluene in the ratio $2H_2 : 1C_6H_5 \cdot CH_3$.

(a) Show that with this feed and an outlet temperature of 900 K, the maximum conversion attainable, that is the equilibrium conversion, is 0.996 based on toluene. The equilibrium constant of the reaction at 900 K is $K_p = 227$.

(b) Calculate the temperature rise which would occur with this feed if the reactor were operated adiabatically and the products were withdrawn at equilibrium. For reaction at 900 K, $-\Delta H = 50{,}000$ kJ/kmol.

(c) If the maximum permissible temperature rise for this process is 100 deg K suggest a suitable design for the reactor.

Specific heat capacities at 900 K (kJ/kmol K): $C_6H_6 = 198$, $C_6H_5CH_3 = 240$, $CH_4 = 67$, $H_2 = 30$.

Solution

(a) If, on the basis of 1 kmol toluene fed to the reactor, α kmol react, then a material balance is:

	In (kmol)	Out (kmol)	Out (mole fraction)	Out (partial pressure)
$C_6H_5 \cdot CH_3$	1	$(1-\alpha)$	$(1-\alpha)/3$	$(1-\alpha)P/3$
H_2	2	$(2-\alpha)$	$(2-\alpha)/3$	$(2-\alpha)P/3$
C_6H_6	0	α	$\alpha/3$	$\alpha P/3$
CH_4	0	α	$\alpha/3$	$\alpha P/3$
Total	3	3	1	P

If equilibrium is reached at 900 K at which $K_p = 227$, then:

$$K_p = (P_{C_6H_6} \times P_{CH_4})/(P_{C_6H_5 \cdot CH_3} \times P_{H_2})$$
$$= \alpha^2/(1-\alpha)(2-\alpha) = 227$$

Thus: $(1-\alpha)(2-\alpha) = \alpha^2/227 = 0.00441\alpha^2$

or: $0.996\alpha^2 - 3\alpha + 2 = 0$

Because α cannot exceed 1, only one root of this quadratic equation is acceptable and:

$$\underline{\underline{\alpha = 0.996}}$$

It may be noted that, if a feed ratio of $1C_6H_5 \cdot CH_3$: $1H_2$ had been used, the conversion at equilibrium would have been only 0.938, illustrating the advantage of using an excess of hydrogen, the less expensive reactant, rather than the stoichiometric proportion.

(b) For the adiabatic reaction, a thermodynamically equivalent path may be considered such that the reactants enter at a temperature T_{in} and the heat produced by the reaction is used to heat the product mixture to 900 K.

On the basis of 1 kmol toluene fed to the reactor and neglecting any changes in the heat capacity and enthalpy of reaction with temperature, then:

Heat released in the reaction = increase in sensible heat of the gas

or: $\alpha(-\Delta H) = \sum n_i(C_{pi}(900 - T_{in}))$

Thus: $(0.996 \times 50{,}000) = [(0.996 \times 198) + (0.996 \times 67) + ((1 - 0.996) \times 240)$
$+ ((1 - 0.996) \times 30)](900 - T_{in})$

$49{,}800 = 295(900 - T_{in})$

and: $(900 - T_{in}) = \underline{\underline{169 \text{ deg K}}}$.

This is the temperature rise which would occur if the reactor were designed and operated as a single adiabatic stage.

(c) If the maximum permissible temperature rise in the process is 100 deg K, then a suitable design for the reactor would be a two-stage arrangement with inter-stage cooling, where:

1st stage: inlet = 800 K

outlet = (800 + 100) = 900 K.

2nd stage: inlet = 900 − (169 − 100) = 831 K

outlet = 900 K.

although the inlet temperatures would be adjusted in order to optimise the process.

PROBLEM 1.2

In a process for the production of hydrogen required for the manufacture of ammonia, natural gas is to be reformed with steam according to the reactions:

$$CH_4 + H_2O \rightleftharpoons CO + 3H_2, \quad K_p(\text{at } 1173\ K) = 1.43 \times 10^{13}\ N^2/m^4$$

$$CO + H_2O \rightleftharpoons CO_2 + H_2, \quad K_p(\text{at } 1173\ K) = 0.784$$

The natural gas is mixed with steam in the mole ratio $1CH_4 : 5H_2O$ and passed into a catalytic reactor which operates at a pressure of 3 MN/m² (30 bar). The gases leave the reactor virtually at equilibrium at 1173 K.

(a) Show that for every 1 mole of CH_4 entering the reactor, 0.950 mole reacts, and 0.44 mole of CO_2 is formed.

(b) Explain why other reactions such as:

$$CH_4 + 2H_2O \rightleftharpoons CO_2 + 4H_2$$

need not be considered.

(c) By considering the reaction:

$$2CO \rightleftharpoons CO_2 + C$$

for which $K_p = P_{CO_2}/P_{CO}^2 = 2.76 \times 10^{-7}\ m^2/N$ at 1173 K, show that carbon deposition on the catalyst is unlikely to occur under the operating conditions.

(d) What will be the effect on the composition of the exit gas of increasing the total pressure in the reformer? Why, for ammonia manufacture, is the reformer operated at 3 MN/m² (30 bar) instead of at a considerably lower pressure? The reforming step is followed by a shift conversion:

$$CO + H_2O \rightleftharpoons CO_2 + H_2,$$

absorption of the CO_2, and ammonia synthesis according to the reaction:

$$N_2 + 3H_2 \rightleftharpoons 2NH_3$$

Solution

(a) If, for 1 kmol of CH_4 fed to the reactor, α kmol is converted and β kmol of CO_2 is formed, then at a total pressure, P, a material balance is:

	In (kmol)	(kmol)	Out (mole fraction)	(partial pressure)
CH_4	1	$(1-\alpha)$	$(1-\alpha)/(6+2\alpha)$	$(1-\alpha)P/(6+2\alpha)$
H_2O	5	$(5-\alpha-\beta)$	$(5-\alpha-\beta)/(6+2\alpha)$	$(5-\alpha-\beta)P/(6+2\alpha)$
CO	0	$(\alpha-\beta)$	$(\alpha-\beta)/(6+2\alpha)$	$(\alpha-\beta)P/(6+2\alpha)$
H_2	0	$(3\alpha+\beta)$	$(3\alpha+\beta)/(6+2\alpha)$	$(3\alpha+\beta)P/(6+2\alpha)$
CO_2	0	β	$\beta/(6+2\alpha)$	$\beta P/(6+2\alpha)$
Total	6	$6+2\alpha$	1	P

based on the stoichiometry of the reactions:

$$CH_4 + H_2O \rightleftharpoons CO_2 + 3H_2 \qquad \text{(I)}$$

and:

$$CO + H_2O \rightleftharpoons CO_2 + H_2 \qquad \text{(II)}$$

For reaction I:

$$K_{pI} = (P_{CO} \times P_{H_2}^3)/(P_{CH_4} \times P_{H_2O}) = \frac{(\alpha - \beta)(3\alpha + \beta)^3 P^2}{(1 - \alpha)(5 - \alpha - \beta)(6 + 2\alpha)^2}$$

If the units of P are bar and $K_{pI} = 1430$ then:

$$\frac{(\alpha - \beta)(3\alpha + \beta)^3}{(1 - \alpha)(5 - \alpha - \beta)(6 + 2\alpha)^2} = (1430/30^2) = 1.59$$

Substituting $\alpha = 0.950$ and $\beta = 0.44$, then the left-hand side of this equation = 1.61 and hence the equation is satisfied.

For reaction II:

$$K_{pII} = (P_{CO_2} \times P_{H_2})/(P_{CO} \times P_{H_2O}) = \frac{\beta(3\alpha + \beta)}{(\alpha - \beta)(5 - \alpha - \beta)} = 0.784$$

Again, substituting $\alpha = 0.950$ and $\beta = 0.44$, the left-hand side of this equation = 0.786 and hence the equation is satisfied.

(b) The reaction:
$$CH_4 + 2H_2O \rightleftharpoons CO_2 + 4H_2 \qquad \text{(III)}$$

need not be considered because it is *not a further independent* reaction. It can in fact be obtained by the addition of reactions I and II and correspondingly: $K_{pI} \times K_{pII} = K_{pIII}$.

(c) If solid and pure carbon is to be in equilibrium with the product gas mixture:

$$2CO \rightleftharpoons CO_2 + C \qquad \text{(IV)}$$

then the equilibrium constant for this reaction, which can apply only if solid C is present, is: $K_{pIV} = P_{CO_2}/P_{CO}^2$. With P in bar, $K_{pIV} = 0.0276$, and from the material balance:

$$\frac{P_{CO_2}}{P_{CO}^2} = \left(\frac{\beta(6 + 2\alpha)}{(\alpha - \beta)^2}\right)\left(\frac{1}{P}\right) = \left[\frac{0.44(6 + 2 \times 0.950)}{(0.950 - 0.44)^2}\right]\left(\frac{1}{30}\right) = 0.45$$

Thus it may be seen that, in the reaction mixture the partial pressure of CO_2, P_{CO_2}, relative to P_{CO}, greatly exceeds the value that would be required to satisfy K_{pIV} and therefore solid C cannot co-exist at equilibrium. The high value of P_{CO_2} may be thought of as driving reaction IV completely to the left.

(d) It may be noted that reaction I involves an *increase* in the number of moles:

$$CH_4 + H_2O \rightleftharpoons CO + 3H_2$$
$$\text{(2 moles)} \qquad \text{(4 moles)}$$

Qualitatively, Le Chatelier's principle indicates that increasing the pressure will tend to decrease the fractional conversion of CH_4 at equilibrium. At first it may therefore seem surprising that the reformer should be operated at a pressure as high as 30 bar.

The reason for this lies in the relatively high cost of gas compression, which depends on the *ratio* of the inlet and outlet pressures. The methane feed to the reformer will probably be available at a pressure much above 1 bar and similarly with the steam. Le Chatelier's principle indicates also that excess steam, as used in practice, will favour a higher conversion of methane compared with the stoichiometric proportions of the reactants. The same conclusion follows quantitatively from K_{pI} for which the equation involves the total pressure P.

At a pressure of 30 bar and with excess steam the fractional conversion of methane in the reformer is reasonably satisfactory. The high pressure of 30 bar will favour the removal of carbon dioxide, following the shift reaction: $CO + H_2O \rightleftharpoons CO_2 + H_2$, and reduce the cost of compressing the purified hydrogen to a value, typically in the range 50–200 bar, required for ammonia synthesis.

PROBLEM 1.3

An aromatic hydrocarbon feedstock consisting mainly of *m*-xylene is to be isomerised catalytically in a process for the production of *p*-xylene. The product from the reactor consists of a mixture of *p*-xylene, *m*-xylene, *o*-xylene and ethylbenzene. As part of a preliminary assessment of the process, calculate the composition of this mixture if equilibrium were established over the catalyst at 730 K.

Equilibrium constants at 730 K are:

$$m\text{-xylene} \rightleftharpoons p\text{-xylene}, \quad K_p = 0.45$$

$$m\text{-xylene} \rightleftharpoons o\text{-xylene}, \quad K_p = 0.48$$

$$m\text{-xylene} \rightleftharpoons \text{ethylbenzene}, \quad K_p = 0.19$$

Why is it unnecessary to consider reactions such as:

$$o\text{-xylene} \underset{k_r}{\overset{k_f}{\rightleftharpoons}} p\text{-xylene?}$$

Solution

The following equilibria at 730 K and total pressure P may be considered, noting that the positions of the equilibria are not dependent on the value of P.

m-xylene $\rightleftharpoons$ ethylbenzene:

$$K_E = (P_E/P_M) = (y_E P)/(y_M P) = (y_E/y_M) \quad \text{(i)}$$

m-xylene $\rightleftharpoons$ *o*-xylene:

$$K_O = (P_O/P_M) = (y_O P)/(y_M P) = (y_O/y_M) \quad \text{(ii)}$$

m-xylene $\rightleftharpoons$ *p*-xylene:

$$K_P = (P_p/P_M) = (y_P P)/(y_M P) = (y_P/y_M) \quad \text{(iii)}$$

where P_M, P_E, P_O and P_P are the partial pressures of the various components and y_M, y_E, y_O and y_P are the mole fractions.

Thus:
$$(y_E/y_M) = 0.19, \quad (y_O/y_M) = 0.48 \quad \text{and} \quad (y_P/y_M) = 0.45$$

Noting that:
$$y_E + y_O + y_P + y_M = 1$$

then:
$$\underline{\underline{y_M = 0.473, \quad y_E = 0.30, \quad y_O = 0.225 \quad \text{and} \quad y_P = 0.212}}.$$

It is unnecessary to consider other equations for the set of equilibrium such as:

$$o\text{-xylene} \rightleftharpoons p\text{-xylene} \tag{i}$$

$$o\text{-xylene} \rightleftharpoons \text{ethylbenzene} \tag{ii}$$

$$p\text{-xylene} \rightleftharpoons \text{ethylbenzene} \tag{iii}$$

because these are not *independent* equations, since they can be derived from combinations of the equations already considered. For example, subtracting equation (ii) from equation (iii) gives equation (iv) and, correspondingly, dividing the equilibrium constant K_P by K_O gives:

$$K_P/K_O = \left(\frac{P_P}{P_M}\right)\left(\frac{P_M}{P_O}\right) = \frac{P_P}{P_C} = K'_P \text{ the equilibrium constant for equilibrium I.} \tag{iv}$$

Thus the complete set of equilibrium may be depicted as:

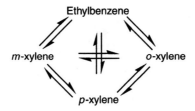

Any three of these equations may be taken as independent relations *provided that* the ones chosen involve *all* the species present.

PROBLEM 1.4

The alkylation of toluene with acetylene in the presence of sulphuric acid is carried out in a batch reactor. 6000 kg of toluene is charged in each batch, together with the required amount of sulphuric acid and the acetylene is fed continuously to the reactor under pressure. Under circumstances of intense agitation, it may be assumed that the liquid is always saturated with acetylene, and that the toluene is consumed in a simple pseudo first-order reaction with a rate constant of 0.0011 s^{-1}.

If the reactor is shut down for a period of 900 s (15 min) between batches, determine the optimum reaction time for the maximum rate of production of alkylate, and calculate this maximum rate in terms of mass P toluene consumed per unit time.

Solution

The toluene is consumed in a simple pseudo first-order reaction with a rate constant of 0.0011 s^{-1}

It may be assumed that the volume of the liquid phase does not change appreciably as the reaction proceeds, although, in practice there will be some departure from this assumption. If the reaction is considered complete at the stage when 1 kmol C_2H_2 (molecular mass = 26 kg/kmol) has been added to 1 kmol C_7H_8 (molecular mass = 92 kg/kmol), then per 1 kmol of toluene, the total mass in the reactor will have increased from 92 kg initially to 118 kg of product having a mass density similar to that of the original toluene.

For a first-order reaction, the integrated form of the rate equation for a constant volume batch reactor, from Table 1.1 in Volume 3, is:

$$t = \frac{1}{k_1} \ln \frac{C_0}{(C_0 - x)}$$

Writing this in terms of the fractional conversion $\alpha = x/C_0$, then:

$$t = \frac{1}{k_1} \ln \frac{1}{(1 - \alpha)}$$

To find the reaction time corresponding to the maximum production rate the method outlined in Section 1.6.3 of Volume 3 is adopted as follows. From the relation:

$$t = \frac{1}{0.0011} \ln \left(\frac{1}{(1 - \alpha)} \right)$$

the following data are obtained:

Fractional conversion, α	0.3	0.4	0.5	0.6	0.65	0.70	0.75	0.80
Time (s)	324	464	630	833	954	1095	1260	1463

These data are plotted in Figure 1a.

A tangent to the curve is then drawn from the point on the time axis at $-(15 \times 60) = -900$ s. The reaction time at the tangent point is 1050 s, at which the fractional conversion is 0.68.

Thus toluene is consumed at the rate of:

$$\frac{(6000 \times 0.68)}{(1050 + 900)} = 2.09 \text{ kg/s}.$$

and the maximum production rate in terms of toluene consumed = 2.09 kg/s (7530 kg/h)

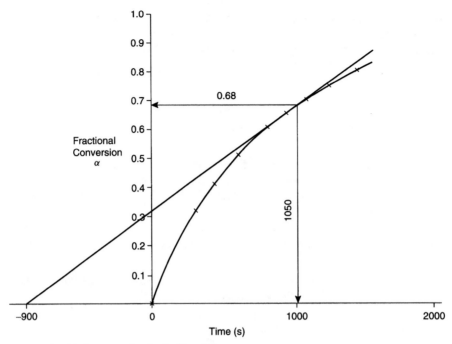

Figure 1a. Graphical construction for Problem 1.4

PROBLEM 1.5

Methyl acetate is hydrolysed by water according to the reaction:

$$CH_3 \cdot COOCH_3 + H_2O \underset{k_r}{\overset{k_f}{\rightleftharpoons}} CH_3 \cdot COOH + CH_3OH$$

A rate equation is required for this reaction taking place in dilute solution. It is expected that reaction will be pseudo first-order in the forward direction and second-order in reverse. The reaction is studied in a laboratory batch reactor starting with a solution of methyl acetate and with no products present. In one test, the initial concentration of methyl acetate was 0.05 kmol/m³ and the fraction hydrolysed at various times subsequently was:

Time (s)	0	1350	3060	5340	7740	∞
Fractional conversion	0	0.21	0.43	0.60	0.73	0.90

(a) Write down the rate equation for the reaction and develop its integrated form applicable to a batch reactor.
(b) Plot the data in the manner suggested by the integrated rate equation, confirm the order of reaction and evaluate the forward and reverse rate constants, k_f and k_r.

Solution

At the stage where χ mole of methyl acetate has been converted per unit volume of reaction mixture, the concentrations of the various species are:

$$CH_3 \cdot COOCH_3 + H_2O \underset{k_r}{\overset{k_f}{\rightleftharpoons}} CH_3 \cdot COOH + CH_3OH$$
$$(C_0 - \chi) \qquad\qquad\qquad \chi \qquad\qquad \chi$$

If the forward reaction is pseudo first order and the reverse reaction second order, then, as discussed in Sections 1.4.4 and 1.4.5 in Volume 3, the rate equation may be written as:

$$\mathscr{R} = \frac{d\chi}{dt} = k_f(C_0 - \chi) - k_r\chi^2 \qquad\qquad (i)$$

At equilibrium the value of χ is χ_e and the net rate of advancement of the reaction is zero, or $\dfrac{d\chi}{dt} = 0$.

Thus:
$$0 = k_f(C_0 - \chi_e) - k_r\chi_e^2$$

or:
$$\frac{\chi_e^2}{(C_0 - \chi_e)} = \frac{k_f}{k_r} = K_c, \text{ the equilibrium constant.} \qquad\qquad (ii)$$

Substituting $\dfrac{k_f}{K_c}$ for k_r in the rate equation, then:

$$\frac{d\chi}{dt} = k_f(C_0 - \chi) - \frac{k_f}{K_c}\chi^2$$

and hence, the integrated form applicable to a batch reactor is given by:

$$t = \frac{1}{k_f}\int_0^\chi \frac{d\chi}{(C_0 - \chi) - \chi^2/K_c} \qquad\qquad (iii)$$

It is convenient at this stage to introduce the numerical values.

The experimental data indicate that, after a long period of time such that equilibrium is established, the fractional conversion is given by:

$$\alpha_e = 0.90 \quad \text{and} \quad \chi_e = \alpha_e C_0 = (0.90 \times 0.05) = 0.045 \text{ kmol/m}^3$$

and, from equation (ii):

$$K_c = \frac{0.045^2}{(0.05 - 0.045)} = \underline{\underline{0.405}}$$

For the experimental data given, the integral to be evaluated is:

$$t = \frac{1}{k_f}\int_0^\chi \frac{d\chi}{(0.05 - \chi) - \dfrac{\chi^2}{0.405}} = \frac{1}{k_f}\int_0^\chi \frac{d\chi}{(0.05 - \chi - 2.47\chi^2)}$$

The factors for the denominator are:
$$t = \frac{1}{k_f} \int_0^\chi \frac{d\chi}{2.47(0.45 + \chi)(0.045 - \chi)}$$

Expressing this in partial fractions, then:
$$t = \frac{1}{1.222 k_f} \int_0^\chi \left[\frac{1}{(0.45 + \chi)} + \frac{1}{(0.045 - \chi)} \right] \cdot d\chi$$

$$t = \frac{1}{1.222 k_f} [\ln(0.45 + \chi) - \ln(0.045 - \chi)]_0^\chi$$

and:
$$t = \frac{1}{1.222 k_f} \ln \frac{(0.45 + \chi)}{(0.045 - \chi)} + \text{constant.} \tag{iv}$$

The experimental data are plotted in the form indicated by equation (iv), noting that $\chi = \alpha C_0 = 0.05\alpha$, or:

$$\ln \frac{(0.45 + \chi)}{(0.045 - \chi)} \text{ is plotted against } t,$$

This should give a straight line of slope $1.222 \, k_f$.

The data are processed as follows and plotted in Figure 1b.

Time (s)	0	1350	3060	5340	7740	∞
Fractional conversion	0	0.21	0.43	0.60	0.73	0.90
χ (kmol/m^3)	0	0.0105	0.0215	0.0300	0.0365	0.0450
$\ln \frac{(0.45 + \chi)}{(0.045 - \chi)}$	2.303	2.591	3.00	3.466	4.047	∞

The slope of the line $= 0.224 \times 10^{-3}$ s^{-1}

Thus: $k_f = (0.224 \times 10^{-3}/1.222) = \underline{0.183 \times 10^{-3}}$ s^{-1}

From equation (ii), $K_c = k_f / k_r$

and: $k_r = k_f / K_c = (0.183 \times 10^{-3}/0.405)$
$= \underline{0.452 \times 10^{-3}}$ m^3/kmol s.

As an alternative to the introduction of the numerical values after equation (iii), it is possible to proceed with the integration of equation (iii) algebraically as follows:

$$t = \frac{1}{k_f} \int_0^\chi \frac{d\chi}{(C_0 - \chi) - \frac{\chi^2}{K_c}}$$

$$= \frac{K_c}{k_f} \int_0^\chi \frac{d\chi}{\left[\frac{K_c + K_c(1 + 4C_0/K_c)^{0.5}}{2} + \chi \right] \left[\frac{-K_c + K_c(1 + 4C_0/K_c)^{0.5}}{2} - \chi \right]}$$

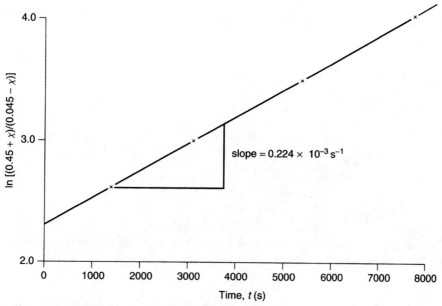

Figure 1b. Graphical construction for Problem 1.5

$$= \frac{1}{k_f(1+4C_0/K_c)^{0.5}} \int_0^\chi \times \left[\frac{1}{\left\{ \frac{K_c + K_c(1+4C_0/K_c)^{0.5}}{2} + \chi \right\}} \right.$$

$$\left. + \frac{1}{\left\{ \frac{-K_c + K_c(1+4C_0/K_c)^{0.5}}{2} - \chi \right\}} \right]$$

$$= \frac{1}{k_f(1+4C_0/K_c)^{0.5}} \left[\ln \frac{\left[\frac{K_c + K_c(1+4C_0/K_c)^{0.5}}{2} + \chi \right]}{\left[\frac{-K_c + K_c(1+4C_0/K_c)^{0.5}}{2} - \chi \right]} \right]_0^\chi$$

$$= \frac{1}{k_f(1+4C_0/K_c)^{0.5}} \left\{ \ln \frac{\left[\frac{K_c + K_c(1+4C_0/K_c)^{0.5}}{2} + \chi \right]}{\left[\frac{-K_c + K_c(1+4C_0/K_c)^{0.5}}{2} - \chi \right]} \right.$$

$$\left. - \ln \frac{[1+(1+4C_0/K_c)^{0.5}]}{[-1+(1+4C_0/K_c)^{0.5}]} \right\}$$

Introducing the numerical values and noting that $(1 + 4C_0/K_c)^{0.5} = \left(1 + \dfrac{4 \times 0.05}{0.405}\right)^{0.5}$
$= 1.222$, then:

$$\dfrac{1}{k_f}\int_0^\chi \dfrac{d\chi}{(0.05 - \chi) - \dfrac{\chi^2}{0.405}}$$

$$= \dfrac{1}{1.222\, k_f}\left\{\ln\dfrac{\left[\dfrac{0.405 + 0.405 \times 1.222}{2} + \chi\right]}{\left[\dfrac{-0.405 + 0.405 \times 1.222}{2} - \chi\right]} - \ln\dfrac{[1 + 1.222]}{[-1 + 1.222]}\right\}$$

$$= \dfrac{1}{1.222\, k_f}\left\{\ln\dfrac{0.45 + \chi}{0.045 - \chi} - \ln 10\right\}$$

This approach is rather more complicated, although it has the advantage that it permits exploration of the effect of the change in numerical values.

PROBLEM 1.6

Styrene is to be produced by the catalytic dehydrogenation of ethylbenzene according to the reaction:

$$C_6H_5 \cdot CH_2 \cdot CH_3 \rightleftharpoons C_6H_5 \cdot CH{:}CH_2 + H_2$$

The rate equation for this reaction takes the form:

$$\mathscr{R} = k\left(P_{Et} - \dfrac{1}{K_p} P_{St} P_H\right)$$

where P_{Et}, P_{St} and P_H are partial pressures of ethylbenzene, styrene and hydrogen respectively.

The reactor will consist of a number of tubes each of 80 mm diameter packed with catalyst with a bulk density of 1440 kg/m^3. The ethylbenzene will be diluted with steam, the feed rates are ethylbenzene 1.6×10^{-3} kmol/m^{2}s and steam 29×10^{-3} kmol/m^{2}s. The reactor will be operated at a mean pressure of 120 kN/m^2 (1.2 bar) and the temperature will be maintained at 833 K (560°C) throughout. If the fractional conversion of ethylbenzene is to be 0.45, estimate the length and number of tubes required to produce 0.231 kg/s (20 tonne styrene/day).

At 833 K, $k = 6.6 \times 10^{-11}$ kmol/(N/m^2) s kg catalyst
$(= 6.6 \times 10^{-6}$ kmol (kg catalyst bar s) and $K_p = 1.0 \times 10^4$ N/m^2.

Solution

The fractional conversion of ethylbenzene is 0.45 or 1 kmol of ethylbenzene feed produces 0.45 kmol styrene. Taking the molecular mass of styrene as 104 kg/kmol, then:

1 kmol ethylbenzene feed produces $(0.45 \times 104) = 46.8$ kg styrene.

Hence, the feed rate of ethylbenzene required $= 1 \times (0.231/46.8)$

$$= \underline{0.00495 \text{ kmol/s}}.$$

The feed rate of ethylbenzene per unit cross-sectional area $= 1.6 \times 10^{-3}$ kmol/m^{2}s.

Thus:

cross-sectional area of all the tubes in the reactor $= (0.00495/0.0016) = 3.09$ m^2

cross-sectional area of one tube $= (\pi \times 0.080^2)/4 = 0.00503$ m^2

and: number of tubes required $= (3.09/0.00503) = \underline{615}$

In estimating the length of the tubes, the mass of catalyst, W, is calculated from the design equation for a tubular reactor as:

$$\frac{W}{F_A} = \int_0^{\alpha_f} \frac{d\alpha}{\mathscr{R}_A^m} \qquad \text{(equation 1.35, Volume 3)}$$

although, as a first stage, the rate of reaction, $\mathscr{R}_A^m$, must be expressed in terms of the fractional conversion, α.

Basis: 1 kmol ethylbenzene entering the reactor at a total pressure of P bar.

	In (kmol)	(kmol)	Out (mole fraction)	(partial pressure)
ethylbenzene	1.0	$1 - \alpha$	$\dfrac{(1-\alpha)}{(19.1+\alpha)}$	$\dfrac{(1-\alpha)P}{(19.1+\alpha)}$
styrene	—	α	$\alpha/(19.1+\alpha)$	$\alpha P/(19.1+\alpha)$
hydrogen	—	α	$\alpha/(19.1+\alpha)$	$\alpha P/(19.1+\alpha)$
steam	18.1	18.1	$18.1/19.1+\alpha)$	$18.1P/(19.1+\alpha)$
TOTAL	19.1	$19.1 + \alpha$	1	P

Thus:
$$\mathscr{R}^m = k\left[\frac{(1-\alpha)P}{(19.1+\alpha)} - \frac{\alpha^2}{(19.1+\alpha)^2}\frac{P^2}{K_P}\right]$$

Noting that $P = 1.2$ bar and $K_p = 0.1$ bar, then:

$$\mathscr{R}^m = (1.2k/(19.1+\alpha)^2)(19.1 - 18.1\alpha - 13\alpha^2)$$

In equation 1.35:

$$\frac{W}{F_A} = \frac{1}{1.2k}\int_0^{\alpha_f} \frac{(19.1+\alpha)^2}{(19.1 - 18.1\alpha - 13\alpha^2)}d\alpha$$

where $\alpha_f = 0.45$.

Thus:
$$\frac{W}{F_A} = \frac{1}{1.2k}\int_0^{\alpha_f}\left[-0.0769 + \frac{28.18 + 2.832\alpha}{1.469 - 1.392\alpha - \alpha^2}\right]\cdot d\alpha$$

$$= \frac{1}{1.2k}\int_0^{\alpha_f}\left[-0.0769 + \frac{10.79}{(0.702-\alpha)} + \frac{7.96}{(2.094+\alpha)}\right]\cdot d\alpha$$

$$= \frac{1}{1.2k}[-0.0769\alpha - 10.79\ln(0.702-\alpha) + 7.96\ln(2.094+\alpha)]_0^{0.45}$$

$$= 12.57/1.2k$$

Noting that $F_A = 0.00495$ kmol/s and $k = 6.6 \times 10^{-6}$ kmol/kg catalyst bar s, then:

$$(W/0.00495) = 12.57/(1.2 \times 6.6 \times 10^{-6})$$

and: $\underline{W = 7856\text{ kg}}$

Total volume of catalyst required $= (7856/1440) = 5.46\text{ m}^3$
Volume of catalyst in each tube $= (5.46/615) = 0.00887\text{ m}^3$
Thus: Length of each tube $= 0.00887/[(\pi \times 0.080^2)/4] = \underline{1.76\text{ m}}$

PROBLEM 1.7

Ethyl formate is to be produced from ethanol and formic acid in a continuous flow tubular reactor operated at a constant temperature of 303 K (30°C). The reactants will be fed to the reactor in the proportions 1 mole HCOOH: 5 moles C_2H_5OH at a combined flowrate of 0.0002 m^3/s (0.72 m^3/h). The reaction will be catalysed by a small amount of sulphuric acid. At the temperature, mole ratio, and catalyst concentration to be used, the rate equation determined from small-scale batch experiments has been found to be:

$$\mathscr{R} = kC_F^2$$

where: $\mathscr{R}$ is formic acid reacting/(kmol/m^{3}s)

C_F is concentration of formic acid kmol/m^3, and

$k = 2.8 \times 10^{-4}$ m^3/kmol s.

The density of the mixture is 820 kg/m^3 and this may be assumed constant throughout. Estimate the volume of the reactor required to convert 70 per cent of the formic acid to the ester.

If the reactor consists of a pipe of 50 mm i.d. what will be the total length required? Determine also whether the flow will be laminar or turbulent and comment on the significance of this in relation to the estimate of reactor volume. The viscosity of the solution is 1.4×10^{-3} N s/m^2.

Solution

Although the ethanol fed to the reactor is partly consumed in the reaction, the rate equation indicates that the rate of reaction depends only on the concentration of the formic acid.

Thus:
$$\mathscr{R} = kC_F^2$$

In this liquid phase reaction, it may be assumed that the mass density of the liquid is unaffected by the reaction, allowing the material balance for the tubular reactor to be applied on a volume basis (Section 1.7.1, Volume 3) with plug flow.

Thus:
$$\frac{V_t}{v} = \int_0^{\chi_f} \frac{d\chi}{\mathscr{R}} \qquad \text{(equation 1.37)}$$

In this case:
$$\mathscr{R} = kC_F^2 = k(C_0 - \chi)^2,$$

where C_0 is the concentration of formic acid in the feed.

Thus:
$$\frac{V_t}{v} = \int_0^{\chi_f} \frac{d\chi}{k(C_0 - \chi)^2} = \frac{1}{k}\left[\frac{1}{(C_0 - \chi_f)} - \frac{1}{C_0}\right]$$

In terms of fractional conversion $\chi_f = \alpha_f C_0$, then:
$$\frac{V_t}{v} = \frac{1}{kC_0}\left[\frac{1}{(1-\alpha_f)} - 1\right] = \frac{1}{kC_0}\left(\frac{\alpha_f}{1-\alpha_f}\right)$$

For HCOOH, the molecular mass = 46 kg/kmol.

For C_2H_5OH, the molecular mass = 46 kg/kmol.

Thus: 1 kmol HCOOH is present in $(1 \times 46) + (5 \times 46)$

$$= 276 \text{ kg feed mixture.}$$

or: $(276/820) = 0.337$ m^3 feed mixture.

Thus: $C_0 = (1/0.337) = 2.97$ kmol/m^3

The volume of the reactor required to convert a fraction of the feed of 0.7 is given by:

$$V_t = (0.72/3600)(1/(2.8 \times 10^{-4} \times 2.97))(0.7/(1-0.7))$$

$$= \underline{0.561 \text{ m}^3}$$

The equivalent length of a 50 mm ID pipe is then:

$$0.561/[(\pi/4)0.050^2] = \underline{286 \text{ m}}$$

say 29 lengths, each of 10 m length, connected by U-bends.

The mean velocity in the tube, $u = (0.72/3600)/(\pi/4)0.050^2 = 0.102$ m/s and, the Reynolds Number is then:

$$Re = u\rho d/\mu = (0.102 \times 820 \times 0.050)/(1.4 \times 10^{-3})$$

$$= \underline{2980}$$

confirming turbulent flow and the validity of the assumed plug flow.

PROBLEM 1.8

Two stirred tanks are available, one 100 m^3 in volume, the other 30 m^3 in volume. It is suggested that these tanks be used as a two-stage CSTR for carrying out a liquid phase reaction A + B → product. The two reactants will be present in the feed stream in equimolar proportions, the concentration of each being 1.5 kmol/m^3. The volumetric flowrate of the feed stream will be 0.3×10^{-3} m^3/s. The reaction is irreversible and is of first order with respect to each of the reactants **A** and **B**, i.e. second order overall, with a rate constant 1.8×10^{-4} m^3/kmol s.

(a) Which tank should be used as the first stage of the reactor system, the aim being to effect as high an overall conversion as possible?
(b) With this configuration, calculate the conversion obtained in the product stream leaving the second tank after steady conditions have been reached.

If there is any doubt regarding which tank should be used as the first stage, the conversions for both configurations should be calculated and compared. Accurate calculations may be required in order to distinguish between the two.

Solution

(a) From the treatment of reactor output given in Volume 3, Section 1.9, it may be seen that placing the smaller tank first would be an advantage. Bringing the feed into the smaller tank will give a relatively high concentration of reactants in this tank. If the overall order of the reaction is greater than 1 (the overall order being 2 in this example), high concentrations at some stage in the system will lead to higher rates of reaction and larger final fractional conversion compared with any alternative configuration.

The case for preferring the smaller tank first is difficult to sustain however by strictly logical argument. The most convincing way is to calculate the fractional conversion for both configurations as the problem suggests.

For a flow of feed of v m^3/s with a concentration of either **A** or **B** — the concentrations of **A** and **B** are equal throughout — of C_0 kmol/m^3, with the flow passing through tanks of volumes V_1 and V_2, then a steady-state mass balances gives:

Tank 1: $\quad vC_0 - v(C_0 - \chi_1) - V_1 k(C_0 - \chi_1)^2 = 0$
$\quad\quad\quad\quad$ (in) $\quad$ (out) $\quad\quad$ (reaction) $\quad\quad$ (accumulation)

or: $\quad v\chi_1 = V_1 k(C_0 - \chi_1)^2$

If $\chi_1 = \alpha_1 C_0$, where α is the fractional conversion, then:

$$v\alpha_1 C_0 = V_1 k C_0^2 (1 - \alpha_1)^2$$

or: $\quad\quad\quad\quad\quad \alpha_1 = (V_1/v) k C_0 (1 - \alpha_1)^2$

Tank 2: $\quad\quad v(C_0 - \chi_1) - v(C_0 - \chi_2) = V_2 k (C_0 - \chi_2)^2 = 0$

or: $\quad\quad\quad\quad\quad v(\chi_2 - \chi_1) = V_2 k (C_0 - \chi_2)^2$

or: $\quad\quad\quad\quad\quad (\alpha_2 - \alpha_1) = (V_2/v) k C_0 (1 - \alpha_2)^2$

In both cases, $v = 0.3 \times 10^{-3}$ m^3/s, $k = 1.8 \times 10^{-4}$ m^3/kmol s and $C_0 = 1.5$ kmol/m^3.

For configuration 1: $V_1 = 30$ m^3, $V_2 = 100$ m^3

Tank 1: $\quad\quad\quad\alpha_1 = (30/0.3 \times 10^{-3})1.8 \times 10^{-4} \times 1.5(1 - \alpha_1)^2$

or: $\quad\quad\quad\quad\quad\alpha_1 = 27(1 - \alpha_1)^2$

Thus: $\quad\quad\quad\quad\alpha_1^2 - 2.03704\alpha_1 + 1 = 0$

Noting that $\alpha_1 < 1$, then:

$$\alpha_1 = 0.8252$$

Tank 2: $\quad\quad(\alpha_2 - \alpha_1) = (100/0.3 \times 10^{-3}) \times 1.8 \times 10^{-4} \times 1.5(1 - \alpha_2)^2$

or: $\quad\quad\quad\alpha_2 - 0.8252 = 90(1 - \alpha_2)^2$

Thus: $\quad\quad\quad\alpha_2^2 - 2.01111\alpha_2 + 1.009169 = 0$

Again, noting that $\alpha_2 < 1$ then:

$$\alpha_2 = \underline{\underline{0.961}}$$

For configuration 2: $V_1 = 100$ m^3, $V_2 = 30$ m^3.

Tank 1: $\quad\quad\quad\alpha_1 = (100/0.3 \times 10^{-3})1.8 \times 10^{-4} \times 1.5(1 - \alpha_1)^2$

and: $\quad\quad\quad\alpha_1^2 - 2.01111\alpha_1 + 1 = 0$

Noting that $\alpha_1 < 1$, then:

$$\alpha_1 = 0.9000$$

Tank 2: $\quad\quad(\alpha_2 - \alpha_1) = (30/0.3 \times 10^{-3})1.8 \times 10^{-4} \times 1.5(1 - \alpha_2)^2$

$$\alpha_2 - 0.9 = 27(1 - \alpha_2)^2$$

and: $\quad\quad\quad\alpha_2^2 - 2.03704\alpha_2 + 1.03333 = 0$

Thus: $\quad\quad\quad\alpha_2 = \underline{\underline{0.955}}$

It may be noted that:

(a) There is seemingly very little difference between the final conversions calculated for the two cases and it may be concluded that, although the final conversion will depend on the total volume of the tanks, it is not very sensitive to how that volume is distributed.
(b) In operating a reactor, however, often it is the fraction which is unreacted which is important for downstream separation.

In this problem, these values are:

$$\text{configuration } 1 = 0.039 \text{ kmol/m}^3$$
$$\text{configuration } 2 = 0.045 \text{ kmol/m}^3.$$

PROBLEM 1.9

The kinetics of a liquid-phase chemical reaction are investigated in a laboratory-scale continuous stirred-tank reactor. The stoichiometric equation for the reaction is **A → 2P** and it is irreversible. The reactor is a single vessel which contains 3.25×10^{-3} m^3 of liquid when it is filled just to the level of the outflow. In operation, the contents of the reactor are well stirred and uniform in composition. The concentration of the reactant **A** in the feed stream is 0.5 kmol/m^3. Results of three steady-state runs are:

Feed rate (m^3/s × 10^5)	Temperature (K)	(°C)	Concentration of **P** in outflow (kmol/m^3)
0.100	298	25	0.880
0.800	298	25	0.698
0.800	333	60	0.905

Determine the constants in the rate equation:

$$\mathscr{R}_A = \mathscr{A}\exp(-E/RT)C_A^p.$$

Solution

The rate equation:

$$\mathscr{R}_A = \mathscr{A}\exp(-E/RT)C_A^p$$

involves three parameters:

p — the order of reaction, $\mathscr{A}$ — the frequency factor and E — the energy of activation.

Substituting the three sets of data to $\mathscr{R}_A$, C_A and T in this equation provides three simultaneous equations in which there are three unknowns, A, E and p, and hence, in principle, a solution may be found. In practice more than three sets of experimental data is desirable in order to validate the experimental technique and the consistency of the measurements.

For a tank volume, V m^3, and a flow of feed of v m^3/s of concentration of **A** of C_0, a mass balance over the tank at steady state for material **A** gives:

$$vC_0 - vC_A - V\mathscr{R}_A = 0$$
$$\text{(in)} \quad \text{(out)} \quad \text{(reaction)}$$

or:
$$\mathscr{R}_A = (v/V)(C_0 - C_A)$$

It may be noted that the experimental data give the concentration of product **P** but not that of reactant **A**, although these may be linked as follows.

For the reaction: **A ⟶ 2P**

then, for 1 m^3 of feed solution, C_P kmol of product is formed from $(C_P/2)$ kmol of **A** leaving $(C_0 - C_P/2)$ kmol of unreacted **A** or:

$$C_A = (C_0 - C_P/2)$$

Using this equation and the mass balance, the following data may be obtained:

Run	Feed rate $(m^3/s \times 10^5)$	Temperature (K)	Concentration, C_A (kmol/m^3)	Rate of reaction, $\mathcal{R}_A$ (kmol/m^{3}s $\times 10^3$)
1	0.100	298	0.06	0.135
2	0.800	298	0.151	0.859
3	0.800	333	0.048	1.114

In the tests, Runs 1 and 2 were carried out at the same temperature and hence:

$$(\mathcal{A}\exp(-E/RT))_1 = (\mathcal{A}\exp(-E/RT))_2$$

and: $(\mathcal{R}_{A2}/\mathcal{R}_{A1}) = (C_{A2}/C_{A1})^p$

Thus: $p = \ln(\mathcal{R}_{A2}/\mathcal{R}_{A1})/\ln(C_{A2}/C_{A1})$

or: $p = \ln((0.859 \times 10^{-3})/(0.135 \times 10^{-3}))/\ln(0.151/0.06)$

$$= 2.005$$

Taking the order of reaction to be exactly 2, although there is no absolute necessity for the order to be an integer, the rate constants for each of the runs, 1 and 2, may now be calculated.

Run 1: $k_1 = \mathcal{R}_{A1}/C_{A1}^2 = (0.135 \times 10^{-3})/0.06^2 = 0.0375$ m^3/kmol s.

Run 2: $k_2 = \mathcal{R}_{A2}/C_{A2}^2 = (0.859 \times 10^{-3})/0.151^2 = 0.0377$ m^3/kmol s.

giving a mean value at 298 K = $\underline{\underline{0.0376 \text{ m}^3/\text{kmol s}}}$.

Similarly, for Run 3 at 333 K with an order of reaction of $p = 2$:

$$k_3 = \mathcal{R}_{A3}/C_{A3}^2 = (1.114 \times 10^{-3})/0.048^2 = \underline{\underline{0.495 \text{ m}^3/\text{kmol s}}}$$

At 298 K:
$$k_{12} = \mathcal{A}\exp(-E/298R) \quad \text{and} \quad \ln k_{12} = \ln \mathcal{A} - E/298R$$

At 333 K:
$$k_3 = \mathcal{A}\exp(-E/333R) \quad \text{and} \quad \ln k_3 = \ln \mathcal{A} - E/333R$$

Subtracting:

$$\ln \frac{k_3}{k_{12}} = \frac{E}{R}\left(\frac{1}{298} - \frac{1}{333}\right) \quad \text{where } R = 8.314 \text{ kJ/kmol K}$$

Thus: $E = 8.314 \ln(0.495/0.0376)/((1/298) - (1/333))$

$$= \underline{\underline{60{,}700 \text{ kJ/kmol}}}$$

Using the data from Run 3:

$$k = \mathscr{A}\exp(-E/RT)$$

or: $\quad 0.495 = \mathscr{A}\exp(-60,700/(8.314 \times 333))$

or: $\quad \underline{\underline{\mathscr{A} = 1.66 \times 10^9 \text{ m}^3/\text{kmol s}}}$

The complete rate equation is therefore:

$$\underline{\underline{\mathscr{R}_A = 1.66 \times 10^9 \exp(-60.7 \times 10^6/RT)C_A^2}}$$

PROBLEM 1.10

A reaction $A + B \rightarrow P$, which is first-order with respect to each of the reactants, with a rate constant of 1.5×10^{-5} m³/kmol s, is carried out in a single continuous flow stirred-tank reactor. This reaction is accompanied by a side reaction $2B \rightarrow Q$, where Q is a waste product, the side reaction being second-order with respect to B, with a rate constant of 11×10^{-5} m³/kmol s.

An excess of A is used for the reaction, the feed rates to the tank being 0.014 kmol/s of A and 0.0014 kmol/s of B. Ultimately reactant A is recycled whereas B is not. Under these circumstances the overflow from the tank is at the rate of 1.1×10^{-3} m³/s, while the capacity of the tank is 10 m³.

Calculate (a) the fraction of B converted into the desired product P, and (b) the fraction of B converted into Q.

If a second tank of equal capacity becomes available, suggest with reasons in what manner it might be incorporated (a) if A but not B is recycled as above, and (b) if both A and B are recycled.

Solution

The reactions are:

$$A + B \longrightarrow P, \text{ Rate, with respect to } A \text{ or } B = k_P C_A C_B$$
$$2B \longrightarrow Q, \text{ Rate with respect to } B \text{ but not } Q = k_Q C_B^2$$

For flows of a_0 kmol/s and b_0 kmol/s of A and B respectively into a vessel of volume V with an outflow of v kmol/s, then a mass balance at steady-state gives:

Component A: $\quad a_0 \;-\; vC_A \;-\; Vk_P C_A C_B \;=\; 0 \quad$ (i)
$\qquad\qquad\qquad$ (in) $\quad$ (out) $\quad$ (reaction) $\qquad$ (accumulation)

Component B: $\quad b_0 \;-\; vC_B \;-\; V(k_P C_A C_B + k_Q C_B^2) = \; 0 \quad$ (ii)

It is necessary to solve equations (i) and (ii) simultaneously in order to obtain C_A and C_B, although, if C_A in terms of C_B from equation (i) is substituted in equation (ii), a cubic equation results. Noting that component A is in excess, the cubic equation may be avoided by, as a first approximation, neglecting the reaction term in equation (i) as compared with the flow term, or:

$$C_A \approx a_0/v = (0.014/1.1 \times 10^{-3}) = 12.73 \text{ kmol/m}^3$$

If this value is substituted into equation (ii), then:

$$0.0014 - 1.1 \times 10^{-3} C_B - 10[(1.5 \times 10^{-5} \times 12.73 C_B) + (11 \times 10^{-5} C_B^2)] = 0$$

and: $\quad C_B^2 + 2.736 C_B - 1.273 = 0$

Disregarding the negative root, then:

$$C_B = 0.405 \text{ kmol/m}^3$$

This value for C_B may now be substituted into equation (i), in order to obtain an 'improved' value of C_A. Thus:

$$0.014 - (1.1 \times 10^{-3} C_A') - (10 \times 1.5 \times 10^{-5} \times 0.405 C_A') = 0$$

or: $\quad C_A' = \underline{12.06 \text{ kmol/m}^3}$

Equation (ii) may now be used to give a more accurate value for C_B. Thus:

$$0.0014 - (1.1 \times 10^{-3} C_B') - 10[(1.5 \times 10^{-5} \times 12.06 C_B') + (11 \times 10^{-5} C_B'^2)] = 0$$

or: $\quad C_B'^2 + 2.645 C_B - 1.273 = 0$

and: $\quad C_B' = \underline{0.416 \text{ kmol/m}^3}$

These values of C_A and C_B are used for the rest of the calculations. It may be noted that the method of successive substitution used here is not the best or the most reliable method of iteration. In this particular example however, the convergence is rapid leading to final values of:

$$C_A = 12.0438 \text{ kmol/m}^3 \quad \text{and} \quad C_B = 0.41613 \text{ kmol/m}^3.$$

If there were no reaction, that is no reaction term in equation (ii), then the concentration of **B** in the outflow would be given by:

$$C_{B_0} = (0.0014/1.1 \times 10^{-3}) = 1.273 \text{ kmol/m}^3$$

Thus the total amount of **B** converted by both reactions is:

$$(1.273 - 0.416) = 0.857 \text{ kmol/m}^3$$

Considering the two reaction terms in equation (ii), then, in unit time:
(material converted to **P**)/(total material converted)

$$= k_P C_A C_B / (k_P C_A C_B + k_Q C_B^2) \qquad \text{(iii)}$$

$$= (1.5 \times 10^{-5} \times 12.06 \times 0.416) / [(1.5 \times 10^{-5} \times 12.06 \times 0.416)$$
$$+ (11 \times 10^{-5} \times 0.416^2)]$$

$$= \underline{0.798}$$

Summarising:
For 1 m^3 solution leaving the reactor, 0.416 kmol **B** remains unreacted.

$$(0.857 \times 0.798) = 0.684 \text{ kmol } \mathbf{B} \text{ reacts to produce } \mathbf{P}.$$

$$(0.857(1 - 0.798)) = 0.173 \text{ kmol } \mathbf{B} \text{ reacts to produce } \mathbf{Q}.$$

and: Total amount of **B** fed to the system $= (0.416 + 0.684 + 0.173)$

$$= 1.273 \text{ kmol}.$$

(a) The fraction of **B** which is converted into the desired product **P**

$$= (0.684/1.273) = \underline{0.537}$$

(b) The fraction of **B** which is converted into the waste product **Q**

$$= (0.173/1.273) = \underline{0.136}$$

By definition, the relative yield is given by equation (iii), where:

$$\text{relative yield} = \underline{0.798}$$

By definition, the operational yield is the fraction of **B** which is converted to the desired product, **P**, or:

$$\text{operational yield} = \underline{0.537}$$

The question of using a second tank and whether it should be connected in series or parallel is now considered.

(a) If **B** is *not recycled*, that is not recovered, then unreacted **B** is lost. The aim should be therefore to achieve the highest fractional conversion of **B** to **P**, that is the highest operational yield, even though a substantial amount of **B** may go to form **Q**. High concentrations of **B** will favour a high degree of conversion even though the reaction to **Q**, being of a higher order with respect to **B** than the reaction to **P**, is favoured by high concentrations of **B**. Such high concentration of **B** will obtained if a second tank is installed in series with the first tank.

(b) If **B** *is recycled*, that is recovered, then unreacted **B** is not wasted. The aim should be, of course, to maximise the relative yield of **B** and, because the unwanted reaction to **Q** is suppressed at low concentrations of **B**, the tanks should be connected in parallel so that the feed stream is further diluted compared to the system which utilises one tank only.

The argument for the case where **B** is not recycled presupposes that, at a high concentration of **B**, the advantage of increasing the rate of reaction to **P** outweighs the disadvantage of also increasing the amount which reacts to form **Q**. This may not be the situation in all circumstances, since it depends on the magnitudes of k_P and k_Q. If the argument put forward is not found to be convincing, then the way forward is to carry out calculations for both configurations as in the solution to Problem 1.8.

PROBLEM 1.11

A substance **A** reacts with a second substance **B** to give a desired product **P**, but **B** also undergoes a simultaneous side reaction to give an unwanted product **Q** as follows:

$$A + B \longrightarrow P; \quad \text{rate} = k_P C_A C_B$$
$$2B \longrightarrow Q; \quad \text{rate} = k_Q C_B^2$$

where C_A and C_B are the concentrations of **A** and **B** respectively.

A single continuous stirred tank reactor is used for these reactions. **A** and **B** are mixed in equimolar proportions such that each has the concentration C_0 in the combined stream fed at a volumetric flowrate v to the reactor. If the rate constants above are $k_P = k_Q = k$ and the total conversion of **B** is 0.95, that is the concentration of **B** in the outflow is $0.05 C_0$, show that the volume of the reactor will be $69 \, v/k C_0$ and that the relative yield of **P** will be 0.82, as for case α in Figure 1.24, Volume 3.

Would a simple tubular reactor give a larger or smaller yield of **P** than the C.S.T.R.? What is the essential requirement for a high yield of **P**? Suggest any alternative modes of contacting the reactants **A** and **B** which would give better yields than either a single C.S.T.R. or a simple tubular reactor.

Solution

A material balance over the tank for material **A** at steady state gives:

$$\underset{\text{(in)}}{vC_{A_0}} - \underset{\text{(out)}}{vC_A} - \underset{\text{(reaction)}}{k_P C_A C_B V} = \underset{\text{(accumulation)}}{0}$$

The fractional conversion of **B**, $\beta = 0.95$. If the fractional conversion of **A** is α, noting that $\alpha < \beta$ since **A** is consumed in the first reaction only, whereas **B** is consumed in both, then:

$$C_A = C_{A_0}(1 - \alpha)$$
$$C_B = C_{B_0}(1 - \beta)$$
$$C_{A_0} = C_{B_0} = C_0$$

and hence the material balance is:

$$vC_0 - vC_0(1 - \alpha) - k_P V C_0^2 (1 - \alpha)(1 - \beta) = 0$$

or:
$$\alpha = (k_P C_0 V/v)(1 - \alpha)(1 - \beta) \qquad (i)$$

Similarly, a material balance for material **B** is:

$$\underset{\text{(in)}}{vC_{B_0}} - \underset{\text{(out)}}{vC_B} - \underset{\text{(reaction)}}{V(k_P C_A C_B + k_Q C_B^2)} = \underset{\text{(accumulation)}}{0}$$

or:
$$vC_0 - vC_0(1 - \beta) - V C_0^2 [k_P (1 - \alpha)(1 - \beta) + k_Q (1 - \beta)^2] = 0$$

Thus:
$$\beta = (C_0 V/v)(1 - \beta)[k_P (1 - \alpha) + k_Q (1 - \beta)] \qquad (ii)$$

259

For the single stirred tank, the relative yield of the desired product, ϕ, is given by the ratio of the rates of reaction, or:

$$\phi = \mathscr{R}_P/(\mathscr{R}_P + \mathscr{R}_Q) = (k_P C_A C_B)/(k_P C_A C_B + k_Q C_B^2)$$
$$= (k_P C_A)/(k_P C_A + k_Q C_B) \tag{iii}$$

When, as in the problem, the reactant proportions are chosen such that $C_{A_0} = C_{B_0} = C_0$, then:

$$\phi = \frac{k_P C_0(1-\alpha)}{k_P C_0(1-\alpha) + k_Q C_0(1-\beta)} = \frac{k_P(1-\alpha)}{k_P(1-\alpha) + k_Q(1-\beta)} \tag{iv}$$

For the special case of $k_P = k_Q = k$, the equations become:

$$\alpha = \frac{kC_0 V}{v}(1-\alpha)(1-\beta) \tag{v}$$

$$\beta = \frac{kC_0 V}{v}(1-\beta)[1-\alpha+1-\beta] \tag{vi}$$

$$\phi = \frac{1-\alpha}{[1-\alpha+1-\beta]} \tag{vii}$$

Since the fractional conversion of **B**, $\beta = 0.95$, then:

from equation (v): $\alpha = \left(\dfrac{kC_0 V}{v}\right)(1-\alpha) \times 0.05$

or: $20\alpha = \left(\dfrac{kC_0 V}{v}\right)(1-\alpha)$ \hfill (viii)

From equation (vi): $0.95 = \left(\dfrac{kC_0 V}{v}\right) \times 0.05[1-\alpha+1-0.95]$

or: $19 = \left(\dfrac{kC_0 V}{v}\right)(1.05-\alpha).$ \hfill (ix)

Dividing to eliminate the term $\left(\dfrac{kC_0 V}{v}\right)$ then:

$$(20\alpha/19) = (1-\alpha)/(1.05-\alpha)$$

or: $\alpha^2 - 2\alpha + 0.95 = 0$

Solving, noting that $\alpha < 0$: $\underline{\alpha = 0.776}$

From equation (ix): $(kC_0 V/v) = (19/(1.05 - 0.776)) = \underline{\underline{69.4}}$

Thus,   Volume of the reactor, $V = (69v/kC_0)$

From equation (vii):

$$\text{Relative yield of } \mathbf{P}, \phi = \frac{(1-0.776)}{(1-0.776+1-0.95)} = \underline{\underline{0.818}}$$

It may be noted that these calculations confirm the values shown in Figure 1.24, in Volume 3.

The *essential requirement* for a high relative yield of **P** can be deduced from the orders of the two competing reactions. With respect to **B** the undesired reaction to **Q** has an order of two whilst the desired reaction to **P** has an order of one and hence the order to **Q** is greater than the order to **P**.

Thus the reaction to **Q** will tend to be *suppressed* at *low concentration* of **B** and the reaction to **P** will be favoured by high concentration of **A**.

In a *simple tubular reactor*, the concentration of both reactants will be high at the inlet and the question is whether the disadvantage of a high concentration of **B** outweighs the *advantage* of a high concentration of **A**. This is nearly always the case because the *relative yield*, $\phi = \dfrac{\mathscr{R}_P}{\mathscr{R}_P + \mathscr{R}_Q} = \dfrac{k_P C_A}{k_P C_A + k_Q C_B}$ which decreases monotonically with increasing C_B irrespective of the particular values of C_A, k_P, k_Q. In the special case where $k_P = k_Q$, a simple tubular reactor gives a poor relative yield of the desired product **P**, as indicated in Figure 1.24 in Volume 3.

Alternative modes of contacting which give better yields are also indicated in Figure 1.24 in Volume 3. The principle is to use multiple injection points for **B** along the reactor so that, although concentrations of **A** are high in the initial stages of the reaction, the concentrations of **B** are comparatively low.

SECTION 3-2

Flow Characteristics of Reactors — Flow Modelling

PROBLEM 2.1

A batch reactor and a single continuous stirred-tank reactor are compared in relation to their performance in carrying out the simple liquid phase reaction $A + B \rightarrow$ products. The reaction is first order with respect to each of the reactants, that is second order overall. If the initial concentrations of the reactants are equal, show that the volume of the continuous reactor must be $1/(1 - \alpha)$ times the volume of the batch reactor for the same rate of production from each, where α is the fractional conversion. Assume that there is no change in density associated with the reaction and neglect the shutdown period between batches for the batch reactor.

In qualitative terms, what is the advantage of using a series of continuous stirred tanks for such a reaction?

Solution

For the batch reactor of volume V_b

The rate equation is:
$$\mathscr{R} = ka^2$$
or:
$$dx/dt = k(a - x)^2$$

After time t, when $x = \alpha a_0$:
$$kt = \int dx/(a_0 - x)^2$$
$$= 1/(a_0 - x) + \text{constant.}$$

When $t = 0, x = 0$ and the constant $= -1/a_0$.

Thus: $kt = 1/(a_0 - x) - 1/a_0 = (1/a_0)((a_0 - a_0 + x)/(a_0 - x))$
$$= (1/a_0)(x/(a_0 - x))$$
$$= \alpha/(a_0(1 - \alpha))$$

Thus: rate of output, $V_b x/t = V_b \alpha a_0 / t$
$$= \underline{\underline{V_b k a_0^2 (1 - \alpha)}}$$

For the continuous stirred-tank reactor of volume V_c

A steady-state balance on reactant **A** gives:

$$va_0 - v(a_0 - x) - V_c k(a_0 - x)^2 = 0$$
$$\text{(in)} \quad \text{(out)} \qquad \text{(reacted)}$$

Thus: $\quad vx - V_c k(a_0 - x)^2 = 0$

The rate of output $= vx = V_c k(a_0 - x)^2 = V_c k a_0^2 (1-\alpha)^2$

Thus, for the same rate of output: $V_b = V_c(1-\alpha)$

or: $\quad \underline{V_c/V_b = 1/(1-\alpha)}$

Qualitatively, the advantage of using a series of equal-sized tanks for the process is that the total volume of the series will be less than the volume of a single tank assuming the same conversion.

This is considered in detail in Section 1.10, Volume 3.

PROBLEM 2.2

Explain carefully the *dispersed plug-flow model* for representing departure from ideal plug flow. What are the requirements and limitations of the tracer response technique for determining *Dispersion Number* from measurements of tracer concentration at only one location in the system? Discuss the advantages of using two locations for tracer concentration measurements.

The residence time characteristics of a reaction vessel are investigated under steady-flow conditions. A pulse of tracer is injected upstream and samples are taken at both the inlet and outlet of the vessel with the following results:

Inlet sample point:

Time (s)	30	40	50	60	70	80	90	100	110	120
Tracer concentration (kmol/m³) × 10³	<0.05	0.1	3.1	10.0	16.5	12.8	3.7	0.6	0.2	<0.05

Outlet sample point:

Time (s)	110	120	130	140	150	160	170	180	190	200	210	220
Tracer concentration (kmol/m³) × 10³	<0.05	0.1	0.9	3.6	7.5	9.9	8.6	5.3	2.5	0.8	0.2	<0.05

Calculate (a) the mean residence time of tracer in the vessel and (b) the dispersion number. If the reaction vessel is 0.8 m in diameter and 12 m long, calculate also the volume flowrate through the vessel and the dispersion coefficient.

Solution

For equally-spaced sampling intervals, the following approximate approach may be used.

$$\text{Mean residence time, } \bar{t} = \Sigma C_i t_i / \Sigma C_i$$

$$\text{Variance, } \sigma^2 = (\Sigma C_i t_i^2 / \Sigma C_i) - \bar{t}^2$$

If, for convenience, all times are divided by 100, then the following data are obtained.

Inlet sampling point

t_i	0.3	0.4	0.5	0.6	0.7	0.8	0.9	1.0	1.1	
C_i	—	0.1	3.1	10.0	16.5	12.8	3.7	0.6	0.2	$\Sigma = 47.0$
$C_i t_i$	—	0.04	1.55	6.0	11.55	10.24	3.33	0.6	0.22	$\Sigma = 33.53$
$C_i t_i^2$	—	0.016	0.775	3.6	8.085	8.192	2.997	0.6	0.242	$\Sigma = 24.507$

$$\bar{t}_1 = (33.53/47.0) = 0.7134 \quad \text{or} \quad 71.34 \text{ s}$$

$$\sigma_1^2 = (24.507/47.0) - 0.7134^2 = 0.01248 \quad \text{or} \quad 124.8 \text{ s}^2$$

Outlet sampling point

t_i	1.2	1.3	1.4	1.5	1.6	1.7	1.8	1.9	2.0	2.1	
C_i	0.1	0.9	3.6	7.5	9.9	8.6	5.3	2.5	0.8	0.2	$\Sigma = 39.4$
$C_i t_i$	0.12	1.17	5.04	11.25	15.84	14.62	9.54	4.75	1.6	0.42	$\Sigma = 64.35$
$C_i t_i^2$	0.144	1.521	7.056	16.875	25.344	24.854	17.172	9.025	3.2	0.882	$\Sigma = 106.073$

$$\bar{t}_2 = (64.35/39.4) = 1.633 \quad \text{or} \quad 163.3 \text{ s}$$

$$\sigma_2^2 = (106.073/39.4) - 1.633^2 = 0.02471 \quad \text{or} \quad 247.1 \text{ s}^2$$

For two sample points: $\quad (\Delta \sigma^2)/(\Delta \bar{t})^2 = 2(D/uL) \quad$ (equation 2.21)

Thus the Dispersion Number, $(D/uL) = (247.1 - 124.8)/[2(163.3 - 71.34)^2]$

$$= \underline{\underline{0.0072}}$$

The mean residence time, $(\bar{t}_2 - \bar{t}_1) = (163.3 - 71.34)$

$$= \underline{\underline{92 \text{ s}}}$$

The volume of the vessel, $V = 12(\pi 0.8^2/4) = 6.03 \text{ m}^3$

and hence: $\quad$ volume flowrate, $v = (6.03/92) = \underline{\underline{0.0656 \text{ m}^3/\text{s}}}$

The velocity in the pipe, $u = (L/t)$

$$= (12/92) = 0.13 \text{ m/s}$$

and hence: $\quad$ Dispersion coefficient, $D = (0.0072 \times 0.13 \times 12)$

$$= \underline{\underline{0.0113 \text{ m}^2/\text{s}}}$$

SECTION 3-3

Gas–Solid Reactions and Reactors

PROBLEM 3.1

An approximate design procedure for packed tubular reactors entails the assumption of plug flow conditions through the reactor. Discuss critically those effects which would:

(a) invalidate plug flow assumptions, and
(b) enhance plug flow.

Solution

The general question of whether or not plug flow can be attained is discussed in Volume 3, Section 1.7. (Tubular Reactors) and the special case of Plug-Flow (Fermenters) is considered in Chapter 5, Section 5.11.3. A more detailed consideration of dispersion in packed bed reactors and those effects which enhance and invalidate plug flow is given in Chapter 3, Section 3.6.1.

PROBLEM 3.2

A first-order chemical reaction occurs isothermally in a reactor packed with spherical catalyst pellets of radius R. If there is a resistance to mass transfer from the main fluid stream to the surface of the particle in addition to a resistance within the particle, show that the effectiveness factor for the pellet is given by:

$$\eta = \frac{3}{\lambda R}\left[\frac{\coth \lambda R - 1/\lambda R}{1 + (2\lambda R/Sh')(\coth \lambda R - 1/\lambda R)}\right]$$

where: $\lambda = (k/D_e)^{1/2}$ and $Sh' = \dfrac{h_D d_p}{D_e}$,

k is the first-order rate constant per unit volume of particle,
D_e is the effective diffusivity, and
h_D is the external mass transfer coefficient.

Discuss the limiting cases pertaining to this effectiveness factor.

Solution

A shell of radii r and $r + \delta r$ within the solid pellet particle of radius R may be considered. Making a material balance for the reactant, then:

Rate of transfer in at radius $r + \delta r$ = rate of transfer out at radius r + reaction rate.

or: $\quad 4\pi(r+\delta r)^2 D_e (dC/dr)_{r+\delta r} = 4\pi r^2 D_e dC/dr + 4\pi r^2 \delta r k C$

where k is the chemical rate constant and C the total molar concentration.

Expanding the left-hand term in this equation and neglecting $(\delta r)^2$ terms and higher then:

$$d^2 C/dr^2 + (2/r)(dC/dr) - (k/D_e)C = 0 \tag{i}$$

There are two boundary conditions:

A: When $r = 0$, $dC/dr = 0$ and C remains finite
B: When $r = R$, $D_e dC/dr = h_D(C_0 - C_i)$

where C_0 is the concentration in the bulk gas and C_i the concentration at the interface, $r = R$. It is convenient to define $\lambda = (k/D_e)^{0.5}$ as in equation 3.13 and the Sherwood Number as $Sh = h_D d_p/D_e = 2h_D R/D_e$.

At $r = R$: $\quad (2R/Sh)dC/dr + C_i = C_0$

Solving equation (i):

$$C = (1/r)(Ae^{\lambda r} + Be^{-\lambda r})$$

From boundary condition A:

$$A = -B$$

and: $\quad C = A/r \sinh \lambda r$

Substituting in boundary condition B:

$$A = RC_0 \bigg/ \left[\frac{2}{Sh}(\lambda R \cosh \lambda R - \sinh \lambda R) + \sinh \lambda R \right]$$

Thus: $\quad C = RC_0 \sinh \lambda R \bigg/ \left[\frac{2r}{Sh}(\lambda R \cosh \lambda R - \sinh \lambda R) + \sinh \lambda R \right] \tag{ii}$

From Example 3.2 in Volume 3, for a sphere, the effectiveness factor is given by:

$$\eta = \frac{4\pi R^2 D_e (dC/dr)_R}{(4/3)\pi R^3 k C_0}$$

$$= (3/\lambda^2 R)(dC/dr)_R/C_0$$

From the relationship between C and r given by equation (ii), then:

$$\underline{\underline{\eta = (3/\lambda R)(\coth \lambda R - 1/\lambda R)/((2\lambda R)/Sh)(\coth \lambda R - 1/\lambda R) + 1)}}$$

PROBLEM 3.3

Two consecutive first-order reactions:

$$A \xrightarrow{k_1} B \xrightarrow{k_2} C$$

occur under isothermal conditions in porous catalyst pellets. Show that the rate of formation of **B** with respect to **A** at the exterior surface of the pellet is:

$$\frac{(k_1/k_2)^{1/2}}{1+(k_1/k_2)^{1/2}} - \left(\frac{k_2}{k_1}\right)^{1/2}\frac{C_B}{C_A}$$

when the pellet size is large, and:

$$1 - \frac{k_2}{k_1}\frac{C_B}{C_A}$$

when the pellet size is small. C_A and C_B represent the concentrations of **A** and **B** respectively at the exterior surface of the pellet, and k_1 and k_2 are the specific rate constants of the two reactions.

Comparing these results, what general conclusions can be deduced concerning the selective formation of **B** on large and small catalyst pellets?

Solution

If the flat plate model for the catalyst pellet as shown in Figure 3.2, Volume 3, is assumed, then a material balance gives:

For component **A**: $D_e(d^2C_A/dx^2) - k_1 C_A = 0$ (equation 3.10)

For component **B**: $D_e(d^2C_A/dx^2) - k_1 C_A - k_2 C_B = 0$

There are two boundary conditions:

A: When $x = L$, $C_A = C_{A\infty}$ and $C_B = C_{B\infty}$

B: When $x = 0$, $dC_A/dx = dC_B/dx = 0$

Solving these two equations:

$$C_A/C_{A\infty} = \cosh\lambda_1 x / \cosh\lambda_1 L, \quad \text{where:} \ \lambda_1 = \sqrt{(k_1/D_e)}$$

and: $$C_B/C_{A\infty} = [k_1/(k_1 - k_2)]\left[\frac{\cosh\lambda_2 x}{\cosh\lambda_2 L} - \frac{\cosh\lambda_1 x}{\cosh\lambda_1 L}\right] + \frac{C_{B\infty}}{C_{A\infty}}\frac{\cosh\lambda_2 x}{\cosh\lambda_2 L}$$

where: $\lambda_2 = \sqrt{(k_2/D_e)}$ (equation 3.46)

If the selectivity is defined as the rate of formation of **B** at the catalyst exterior surface compared to the rate of formation of **A** at the surface, then:

$$\frac{-D_e(dC_B/dx)_{x=L}}{-D_e(dC_A/dx)_{x=L}} = \left(\frac{k_1}{k_1-k_2}\right)\left(1-\frac{\phi_2 \tanh \phi_2}{\phi_1 \tanh \phi_1}\right) - \frac{C_{B\infty}}{C_{A\infty}} \frac{\phi_2 \tanh \phi_2}{\phi_1 \tanh \phi_1}$$

(equation 3.47)

where $\phi_1 = L\lambda_1$ and $\phi_2 = L\lambda_2$.

For large particles, where both ϕ_1 and ϕ_2 are large, the selectivity becomes:

$$\underline{\underline{\frac{\sqrt{(k_1/k_2)}}{1+\sqrt{(k_1/k_2)}} - \frac{C_B}{C_A}\sqrt{(k_2/k_1)}}}$$

which is the limiting form of equation 3.47.

For small particles, the other asymptote in equation 3.47 gives the selectivity as:

$$\underline{\underline{\left(1 - \frac{k_2}{k_1}\frac{C_B}{C_A}\right)}}$$

PROBLEM 3.4

A packed tubular reactor is used to produce a substance **D** at a total pressure of 100 kN/m² (1 bar) utilising the exothermic equilibrium reaction:

$$\mathbf{A} + \mathbf{B} \rightleftharpoons \mathbf{C} + \mathbf{D}$$

0.01 kmol/s (36 kmol/h) of an equimolar mixture of **A** and **B** is fed to the reactor and plug flow conditions within the reactor may be assumed.

What are the optimal isothermal temperature for operation and the corresponding reactor volume for a final fractional conversion z_f of 0.68. Is this the best way of operating the reactor?

The forward and reverse kinetics are second order with the velocity constant $k_1 = 4.4 \times 10^{13} \exp(-105 \times 10^6/RT)$ and $k_2 = 7.4 \times 10^{14} \exp(-125 \times 10^6/RT)$ respectively. k_1 and k_2 are expressed in m³/kmol s and **R** in J/kmol K.

Solution

For the reaction:

$$\mathbf{A} + \mathbf{B} \rightleftharpoons \mathbf{C} + \mathbf{D}$$

The mole fractions at any cross-section, starting with an equimolar mixture of **A** and **B** are:

$(1-z)/2$ moles **A**, $(1-z)/2$ moles **B**, $(z/2)$ moles **C**   and $(z/2)$ moles **D**.

where z is the fractional conversion and the concentrations are:

$$C_A = C_B = \frac{(1-z)}{2}\frac{P}{RT}$$

$$C_C = C_D = (z/2)\frac{P}{RT}.$$

The reaction rate is then:

$$\mathscr{R} = \left(\frac{P}{2RT}\right)^2 (k_1(1-z)^2 - (k_2 z^2))$$

For plug flow, the reactor volume, $V = F \int dz/\mathscr{R}$

where F is the rate of feed to the reactor.

Substituting for $\mathscr{R}$ and integrating with respect to z at constant temperature then:

$$V = F\left(\frac{2RT}{P}\right)^2 \int_0^{z_f} \frac{dz}{k_1(1-z)^2 - k_2 z^2}$$

$$= F\left(\frac{2RT}{P}\right)^2 \frac{K}{k_1} \int_0^{z_f} \frac{dz}{K(1-z)^2 - z^2}$$

$$\frac{1}{K(1-z)^2 - z^2} = \frac{1}{(\sqrt{K}(1-z)+z)(\sqrt{K}(1-z)-z)}$$

$$= \frac{1}{(\sqrt{K} - (\sqrt{K}+1)z)(\sqrt{K} - (\sqrt{K}-1)z)}$$

If $\alpha = \sqrt{K}$, $\beta = \sqrt{K}+1$ and $\gamma = \sqrt{K}-1$, then:

$$\frac{1}{K(1-z)^2 - z^2} = \frac{1}{(\alpha - \beta z)(\alpha - \gamma z)} = \frac{\gamma}{\alpha(\gamma-\beta)(\alpha-\gamma z)} - \frac{\beta}{\alpha(\gamma-\beta)(\alpha-\beta z)}$$

The integral then becomes:

$$\frac{1}{\alpha(\gamma-\beta)} \left[\int_0^{z_f} \frac{\gamma\, dz}{(\alpha-\gamma z)} - \int_0^{z_f} \frac{\beta\, dz}{(\alpha-\beta z)} \right]$$

Noting that $(\gamma - \beta) = (\sqrt{K}-1) - (\sqrt{K}+1) = -2$, the integral is:

$$\frac{1}{\alpha(\gamma-\beta)} \ln\left[\frac{\alpha - \beta z_f}{\alpha - \gamma z_f}\right]$$

Thus:
$$= \frac{1}{2\alpha} \ln\left[\frac{\alpha - \beta z_f}{\alpha - \gamma z_f}\right]$$

Reactor volume, $V = (F/2)(2RT/P)^2 (\alpha/k_1)[\ln(\alpha - \gamma z_f)/(\alpha - \beta z_f)]$

Substituting $\mathbf{R} = 8314$ J/kmol K, $P = 1 \times 10^5$ N/m^2 and $F = 0.010$ kmol/s then:

$$V = 0.000138 T^2 (\alpha/k_1)[\ln(\alpha - \gamma z_f)/(\alpha - \beta z_f)]$$

$$k_1 = 4.4 \times 10^{13} e^{-(105 \times 10^6 / 8314 T)}$$

and:
$$K = (4.4/7.4) \times 10^{-1} e^{(20 \times 10^6 / 8314 T)}.$$

Making calculations for $T = 540$, 545 and 550 K, the following data are obtained:

T (K)	540	545	550
$k_1(\text{s}^{-1})$	3064	3798	4688
K	5.116	4.911	4.718
α	2.262	2.216	2.172
β	3.262	3.216	3.172
γ	1.262	1.216	1.172
$\ln(\alpha - \gamma z_f)/(\alpha - \beta z_f)$	3.466	3.346	4.516
$V\,(\text{m}^3)$	0.103	0.080	0.087

from which the optimum isothermal temperature = 545 K

and the reactor volume = 0.080 m³

SECTION 3-4

Gas–Liquid and Gas–Liquid–Solid Reactors

PROBLEM 4.1

What is the significance of the parameter $\beta = (k_2 C_{BL} D_A)^{0.5}/k_L$ in the choice and the mechanism of operation of a reactor for carrying out a second-order reaction, rate constant k_2, between a gas **A** and a second reactant **B** of concentration C_{BL} in a liquid? In this expression, D_A is the diffusivity of **A** in the liquid and k_L is the liquid-film mass transfer coefficient. What is the 'reaction factor' and how is it related to β?

Carbon dioxide is to be removed from an air stream by reaction with a solution containing 0.4 kmol/m³ of NaOH at 258 K (25°C) at a total pressure of 110 kN/m² (1.1 bar). A column packed with 25 mm Raschig rings is available for this purpose. The column is 0.8 m internal diameter and the height of the packing is 4 m. Air will enter the column at a rate of 0.015 kmol/s (total) and will contain 0.008 mole fraction CO_2. If the NaOH solution is supplied to the column at such a rate that its concentration is not substantially changed in passing through the column, calculate the mole fraction of CO_2 in the air leaving the column. Is a packed column the most suitable reactor for this operation?

Data: Effective interfacial area for 25 mm packing = 280 m²/m³
Mass transfer film coefficients:
  Liquid, $k_L = 1.3 \times 10^{-4}$ m/s
  Gas, $k_G a = 0.052$ kmol/m³ s bar
For a 0.4 kmol/m³ concentration of NaOH at 298 K:
  Solubility of CO_2: $P_A = \mathcal{H} C_A$ where P_A is partial pressure of CO_2, C_A is the equilibrium liquid-phase concentration and $\mathcal{H} = 32$ bar m³/kmol.
  Diffusivity of CO_2, $D = 0.19 \times 10^{-8}$ m²/s

The second-order rate constant for the reaction $CO_2 + OH^- = HCO_3^-$, $k_2 = 1.35 \times 10^4$ m³/kmol s. Under the conditions stated, the reaction may be assumed pseudo first-order with respect to CO_2.

Solution

The reaction factor is such that:
$$N_A = k_L C_{Ai} f_i$$

For a fast, first-order reaction in the film:

$$f_i = \beta/\tanh\beta \quad \text{(equation 4.13)}$$

where:
$$\beta = (k_2 C_{BL} D_A)^{0.5}/k_L$$

In the present case:

$$\beta = (1.35 \times 10^4 \times 0.4 \times 0.19 \times 10^{-8})^{0.5}/(1.3 \times 10^{-4}) = 24.6$$

This value of β confirms the regime of a fast reaction occurring mainly in the film for which a packed column is a suitable reactor.

For large values of β, $\tanh\beta = 1$ and:

$$N_A = k_L C_{Ai} \beta = k'_L C_{Ai}$$

where k'_L is the effective liquid film mass transfer coefficient enhanced by the chemical reaction; that is:

$$k'_L = k_L \beta$$

In the present case:

$$k'_L = (1.3 \times 10^{-4} \times 24.6) = 32 \times 10^{-4} \text{ m/s}$$

Combining the liquid- and gas-film resistances and replacing k_L by k'_L since the mass transfer is enhanced by the reaction, then:

$$1/K_G a = 1/k_G a + H/k'_L a$$

or:
$$1/K_G a = (1/0.052) + (32/(32 + 10^{-4} \times 280))$$

and:
$$K_G a = \underline{0.0182 \text{ kmol/m}^3\text{s bar}}$$

If G kmol/m²s is the carrier gas flowrate per unit area and y is the mole fraction of CO_2, equal approximately to the mole ratio, then a balance across an element of the column δh high, gives:

$$-G\delta y = K_G a(P - P^*)\delta h$$

where P^* is the partial pressure that would be in equilibrium with the bulk liquid. Noting that, in this case, $P^* = 0$ and $P = yP_T$, where P_T is the total pressure, then:

$$-G\delta y = K_G a y P_T \delta h$$

Thus:
$$h = (G/K_G a P_T) \int_{y_{out}}^{y_{in}} \frac{dy}{y}$$

$$= (G/K_G a P_T) \ln(y_{in}/y_{out})$$

In this case:

$$G = 0.015/(\pi 0.8^2/4) = 0.0298 \text{ kmol/m}^2\text{s}$$

$$G/(K_G a P_T) = 0.0298/(0.0182 \times 1.1) = 1.49 \text{ m}$$

Thus:
$$\ln(y_{in}/y_{out}) = (4/1.49) = 2.68$$

$$(y_{in}/y_{out}) = 14.6$$

and:
$$y_{out} = (0.008/14.6) = \underline{0.00055.}$$

PROBLEM 4.2

A pilot-scale reactor for the oxidation of o-xylene by air according to the following reaction has been constructed and its performance is being tested.

$$\text{o-xylene} + 1.5\,O_2 = \text{2-methylbenzoic acid} + H_2O$$

The reactor, an agitated tank, operates under a pressure of 1.5 kN/m² (15 bar) and at 433 K (160°C). It is charged with a batch of 0.06 m³ of o-xylene and air introduced at the rate of 0.0015 m³/s (5.4 m³/h) measured at reactor conditions. The air is dispersed into small bubbles whose mean diameter is estimated from a photograph to be 0.8 mm, and from level sensors in the reactor, the volume of the dispersion produced is found to be 0.088 m³. Soon after the start of the reaction (before any appreciable conversion of the o-xylene) the gas leaving the reactor is analysed (after removal of condensibles) and found to consist of 0.045 mole fraction O_2, 0.955 mole fraction N_2.

Assuming that under these conditions, the rate of the above reaction is virtually independent of the o-xylene concentration and is thus pseudo first-order with respect to the concentration of dissolved O_2 in the liquid, calculate the value of the pseudo first-rate constant.

Data: Estimated liquid-phase mass transfer coefficient, $k_L = 4.0 \times 10^{-4}$ m/s.
Equilibrium data: $P_A = \mathscr{H} C_A$ where P_A is the partial pressure and C_A the equilibrium concentration in liquid, $\mathscr{H} = 127$ m³ bar/kmol
Diffusivity of O_2 in liquid o-xylene, $D = 1.4 \times 10^{-9}$ m²/s.
Gas constant, $\mathbf{R} = 8314$ J/kmol K
Composition of air (molar): $O_2 = 20.9$ per cent, $N_2 = 79.1$ per cent

State clearly any further assumptions made and discuss their validity. Is an agitated tank the most suitable type of reactor for this process.

Solution

Noting that, for an ideal gas, $n = (PV/\mathbf{R}T)$, then:

$$\text{feed rate of air to the reactor} = (15 \times 10^5 \times 5.4)/(8314 \times 433 \times 3600)$$

$$= 0.625 \times 10^{-3} \text{ kmol/s}$$

Air is 20.9 per cent oxygen and hence:

$$\text{oxygen in the feed} = (0.625 \times 10^{-3} \times 20.9/100)$$

$$= 0.1306 \times 10^{-3} \text{ kmol/s}$$

nitrogen in the feed = nitrogen in the exit gas = 0.791 kmol/kmol feed

Thus: oxygen leaving the reactor = $(0.791 \times 0.045/0.955) = 0.0372$ kmol/kmol feed and:

$$\text{Flow of oxygen from the reactor} = 0.625 \times 10^{-3}(0.791 \times 0.045/0.955)$$
$$= 0.0233 \times 10^{-3} \text{ kmol/s}$$

Thus:
$$\text{oxygen reacting} = (0.1306 \times 10^{-3}) - (0.0233 \times 10^{-3})$$
$$= 0.1073 \times 10^{-3} \text{ kmol/s}$$

From which:

$$\text{fraction of oxygen fed which reacts} = (0.1073 \times 10^{-3})/(0.1306 \times 10^{-3}) = 0.822$$

and:

$$\text{rate of oxygen per unit volume dispersed} = (0.1073 \times 10^{-3})/(88 \times 10^{-3})$$
$$= \underline{0.00123 \text{ kmol/m}^3\text{s}.}$$

The volume fraction of gas in dispersion, $\varepsilon_G = (88 - 60)/88 = 0.318$

The mean bubble diameter is:

$$d_b = 6\varepsilon_G/a \qquad \text{(equation 4.30)}$$

Thus: $a = 6\varepsilon_G/d_b = (6 \times 0.318)/(0.8 \times 10^{-3}) = 2390 \text{ m}^2/\text{m}^3$

For a pseudo first-order reaction, the rate with respect to oxygen $= kC$ (per unit volume of liquid). If C_i is the concentration at the bubble interface and C_b the concentration in the bulk, then assuming that:

(a) C_b is significant, that is not equal to zero, and
(b) a pseudo steady-state exists:
Rate of mass transfer of oxygen from the interface to the bulk $\equiv$ Rate of reaction in the bulk or, per unit volume of dispersion,

or: $\qquad k_L a(C_i - C_b) = kC_b(1 - \varepsilon_G)$

that is: $\qquad (C_i - C_b) = (k(1 - \varepsilon_G)/k_L a)C_b$

or: $\qquad C_i = C_b(1 + (k(1 - \varepsilon_G)/k_L a))$

If it is further assumed that:
(c) the gas phase in the bubbles in dispersion is well-mixed and has the same composition as the outlet gas, and
(d) the partial pressure of the o-xylene in the gas phase may be neglected in that the total pressure of 1.5 kN/m² (1.5 bar) is relatively high even at 433 K (160°C), then, using Henry's law:

$$C_i = y_{O_2} P/\mathcal{H} = (0.045 \times 15)/127 = 0.00532 \text{ kmol/m}^3.$$
$$C_b = C_i/[1 + k(1 - \varepsilon_G)/k_L a]$$

and the rate reaction per unit volume of dispersion $\mathscr{R}_d$ is given by:

$$\mathscr{R}_d = kC_i(1-\varepsilon_G)/[1+k(1-\varepsilon_G)/k_La]$$
$$= C_i/[1/k(1-\varepsilon_G) + 1/k_La]$$

Thus: $\quad (C_i/\mathscr{R}_d) = 1/[k(1-\varepsilon_G)] + (1/k_La)$

or: $\quad (0.00532/0.00123) = 1/[k(1-\varepsilon_G)] + 1/(40 \times 10^{-4} \times 2390)$

Thus: $\quad 1/[k(1-\varepsilon_G)] = 3.28$

and: $\quad k = 1/[3.28(1-0.318)] = \underline{\underline{0.447 \text{ s}^{-1}}}$

It may be noted that:

$$C_b = C_i/[1 + (0.447 \times 0.682)/(4 \times 10^{-4} \times 2390)]$$
$$= 0.76 \, C_i$$

which is in line with assumption (a).

PROBLEM 4.3

It is proposed to manufacture oxamide by reacting cyanogen with water using a strong solution of hydrogen chloride which acts as a catalyst according to the reaction:

$$(CN)_2 + 2H_2O \longrightarrow (CONH_2)_2$$

The reaction is pseudo first-order with respect to dissolved cyanogen, the rate constant at the operating temperature of 300 K being 0.19×10^{-3} s^{-1}. An agitated tank will be used containing 15 m^3 of liquid with a continuous flow of a cyanogen–air mixture at 300 kN/m^2 (3 bar) total pressure, composition 0.20 mole fraction cyanogen, and a continuous feed and outflow of the hydrogen chloride solution; the gas feed flowrate will be 0.01 m^3/s total and the liquid flowrate 0.0018 m^3/s. At the chosen conditions of agitation the following estimates have been made:

Liquid-phase mass transfer coefficient	$k_L = 1.9 \times 10^{-5}$ m/s
Gas–liquid interfacial area per unit volume of dispersion	$a = 47$ m^2/m^3
Gas volume fraction in dispersion	$\varepsilon_g = 0.031$
Diffusivity of cyanogen in solution	$D = 0.6 \times 10^{-9}$ m^2/s
Henry law coefficient where P_A is partial pressure and C_A is concentration in liquid at equilibrium	$P_A/C_A = \mathscr{H} = 1.3$ bar m^3/kmol
Gas constant	$R = 8314$ J/kmol K

Assuming ideal mixing for both gas and liquid phases, calculate:

(a) the concentration of oxamide in the liquid outflow (the inflow contains no dissolved oxamide);
(b) the concentration of dissolved but unreacted cyanogen in the liquid outflow, and

(c) the fraction of cyanogen removed from the gas stream. What further treatment would you suggest for the liquid leaving the tank before separation of the oxamide?

Calculate the value of β for this system and suggest another type of reactor that might be considered for this process.

Solution

(a) A material balance on cyanogen transferred from gas to liquid gives:

$$v_{G_0}y_0 P/RT - v_G y P/RT - V_d k_L a(C_i - C_L) = 0$$
$$\text{(in)} \qquad \text{(out)} \qquad \text{(transferred)}$$

or:
$$(P/RT)(v_{G_0}y_0 - v_G y) - V_d k_L a[(yP/\mathcal{H}) - C_L] = 0 \qquad \text{(i)}$$

(b) A material balance on the cyanogen in the liquid gives:

$$0 - v_L C_L + V_d k_L a(C_i - C_L) - kVC_L = 0$$
$$\text{(in)} \quad \text{(out)} \qquad \text{(transferred)} \qquad \text{(reaction)}$$

or:
$$V_d k_L a[(yP/\mathcal{H}) - C_L] - (v_L + kV)C_L = 0 \qquad \text{(ii)}$$

It may be noted that V_d is the volume of dispersion and:

$$V = (1 - \varepsilon_G)V_d$$

or:
$$V_d = V/(1 - \varepsilon_G) = V/0.969$$

(c) A material balance for the oxamide gives:

$$0 - v_L C_M + kVC_L = 0$$
$$\text{(in)} \quad \text{(out)} \quad \text{(reaction)} \qquad \text{(iii)}$$

(d) A material balance for the air, which is required because the volume flow out, v_G, is less than the flowrate in, v_{G_0}, gives:

$$v_{G_0}(1 - y_0) = v_G(1 - y) \qquad \text{(iv)}$$

or:
$$v_G = v_{G_0}(1 - y_0)/(1 - y)$$
$$= 0.01(1 - 0.20)/(1 - y) = 0.008/(1 - y).$$

From equation (i):

$$[3 \times 10^5/(8314 \times 300)]\{(0.01 \times 0.2) - [0.008y/(1 - y)]\}$$
$$-[(15 \times 1.9 \times 10^{-5} \times 47)/0.969][(3y/1.3) - C_L] = 0$$

or:
$$0.12[0.002 - 0.008y/(1 - y)] - 0.0138(2.31y - C_L) = 0 \qquad \text{(v)}$$

From equation (ii):

$$0.0138(2.31y - C_L) - [0.0018 + (0.19 \times 10^{-3} \times 15)]C_L = 0$$

or:
$$C_L = (0.0319y/0.0185) = 1.72y \qquad \text{(vi)}$$

Substituting from equation (vi) into equation (v):

$$0.002 - [(0.008y)/(1 - y)] - (0.0138/0.12)(2.31y - 1.72y) = 0$$

or: $\quad 0.002 - 0.078y + 0.068y^2 = 0$

Solving the quadratic equation and noting that $y < 1$, then:

$$y = 0.0265$$

and: $\quad C_L = (1.72 \times 0.0265) = \underline{0.046 \text{ kmol/m}^3}$

From equation (iv):

cyanogen absorbed from the gas stream, $v_G = 0.01(1 - 0.20)/(1 - 0.0265)$

$$= 0.0082 \text{ m}^3/\text{s}.$$

The fraction absorbed $= (v_{G_0}y_0 - v_G y)/v_{G_0}y_0 = 1 - (v_G y/v_{G_0}y_0)$

$$= 1 - (0.0082 \times 0.0265)/(0.01 \times 0.20)$$

$$= \underline{0.89}$$

From equation (iii):

oxamide in the outflow $C_M = (kV/v_L)C_L$

$$= (0.19 \times 10^{-3} \times 15) \times 0.046/0.0018 = \underline{0.073 \text{ kmol/m}^3}$$

$$\beta = (kD)^{0.5}/k_L = (0.19 \times 10^{-3} \times 0.6 \times 10^{-9})^{0.5}/(1.9 \times 10^{-5}) = \underline{0.018}$$

and hence a bubble column is the preferred design.

PROBLEM 4.4

(a) Consider a gas-liquid-solid hydrogenation such as that described in (b) in which the reaction takes place within a porous catalyst particle in a trickle bed reactor. Assume that the liquid containing the compound to be hydrogenated reaches a steady state with respect to dissolved hydrogen immediately it enters the reactor and that the liquid is involatile. Show that the rate of reaction per unit volume of reactor space $\mathscr{R}$ [(kmol H_2 converted)/m^{3}s] for a reaction which is pseudo first-order with respect to hydrogen is given by:

$$\mathscr{R} = \frac{P_A}{\mathscr{H}} \left[\frac{1}{k_L a} + \frac{V_p}{k_s S_x(1-e)} + \frac{1}{k\eta(1-e)} \right]^{-1}$$

where: P_A = Pressure of hydrogen (bar)
$\mathscr{H}$ = Henry Law coefficient (bar m^3/kmol)
$k_L a$ = Gas–liquid volumetric mass transfer coefficient (s^{-1})
k_s = Liquid–solid mass transfer coefficient (m/s)
V_p = Volume of single particle (m^3)
S_x = External surface area of a single particle (m^2)
e = Voidage of the bed (−)

k = First-order rate constant based on volume of catalyst
   [m³/(m³ catalyst) s = s⁻¹]
η = Effectiveness factor (−)

(b) Crotonaldehyde is to be selectively hydrogenated to *n*-butyraldehyde in a process using a palladium catalyst deposited on a porous alumina support in a trickle bed reactor. The particles will be spheres of 5 mm diameter packed into the reactor with a voidage e of 0.4. Estimated values of the parameters listed in (a) are as follows:

$$k_L a = 0.02 \text{ s}^{-1}, k_s = 2.1 \times 10^{-4} \text{ m/s}$$

$$k = 2.8 \text{ m}^3/(\text{m}^3 \text{ cat})\text{s}, \mathcal{H} = 357 \text{ bar m}^3/\text{kmol}$$

Also for spheres the effectiveness factor is given by:

$$\eta = \frac{1}{\phi}\left(\coth 3\phi - \frac{1}{3\phi}\right) \quad \text{where the Thiele modulus,} \quad \phi = \frac{V_p}{S_x}\left(\frac{k}{D_e}\right)^{1/2} \quad \text{(equation 3.19)}$$

For the catalyst, the effective diffusivity, $D_e = 1.9 \times 10^{-9}$ m²/s.

If the pressure of hydrogen in the reactor is 1 bar, calculate $\mathcal{R}$, the rate of reaction per unit volume of reactor, and comment on the relative values of the transfer/reaction resistances involved in the process.

(c) Discuss whether the trickle bed reactor and the conditions described in (b) are the best choices for this process. What alternatives might be considered?

Solution

(a) *Hydrogenation* takes place in a sequence of steps:

1. mass transfer takes place from the gas to the liquid,
2. mass transfer takes place from the liquid to the external surface of the particles,
and 3. diffusion and reaction take place within the particles.

At steady-state, the rates of all these are the same and equal to the overall rate of reaction, $\mathcal{R}$. On the basis of unit volume of reactor space that is the volume of gas, liquid and solid, each of these steps is considered in turn.

1. If $k_L a$ is the volume mass transfer coefficient, that is referring to the whole reactor space, then:

$$\mathcal{R} = k_L a(C_i - C_L).$$

$$C_i = P/\mathcal{H}$$

and hence:
$$\mathcal{R} = k_L a[(P/\mathcal{H}) - C_L] \quad \text{(i)}$$

2. The rate of mass transfer from the liquid to the solid for one particle

$$= k_s S_x (C_L - C_s)$$

Number of particles per unit volume of reactor space $= (1-e)/V_p$

Thus: Rate of mass transfer per unit volume of reactor space,

$$\mathcal{R} = (k_s S_x / V_p)(1-e)(C_L - C_s) \quad \text{(ii)}$$

3. The rate of diffusion and reaction per unit volume of particles $= kC_s\eta$. Since the particles occupy only a fraction $(1-e)$ of the reactor space, then:

Rate of diffusion and reaction per unit volume of reactor space,

$$\mathscr{R} = kC_s\eta(1-e) \qquad \text{(iii)}$$

From equations (i), (ii) and (iii):

$$\mathscr{R}/k_L a = (P/\mathscr{H}) - C_L$$

$$\mathscr{R}/[(k_s S_x/V_p)(1-e)] = C_L - C_s$$

and: $\mathscr{R}/[k\eta(1-e)] = C_s$

Adding: $\mathscr{R}[(1/k_L a) + (V_p/(k_s S_x(1-e))) + 1/(k\eta(1-e))] = P/\mathscr{H}$

and: $$\underline{\mathscr{R} = \left(\frac{P}{\mathscr{H}}\right)\left\{\left(\frac{1}{k_L a}\right) + \frac{V_p}{k_s S_x(1-e)} + \frac{1}{k\eta(1-e)}\right\}^{-1}}$$

(b) *For the crotonaldehyde hydrogenation:*

$$\mathscr{R} = (1/357)[(1/0.02) + V_p/(2.1 \times 10^{-4} S_x(1-0.4)) + 1/(2.8\eta(1-0.4))]^{-1}$$
$$= (1/357)[(1/0.02) + (V_p/S_x)/1.26 \times 10^{-4} + 1/(1.68\eta)]^{-1} \qquad \text{(iv)}$$

For spherical particles:

$$(V_p/S_x) = (\pi d_p^3/6)/(\pi d_p^2) = (d_p/6) = (0.005/6) = 0.000833$$

The effectiveness factor, $\eta = (1/\phi)(\coth 3\phi - 1/3\phi)$

where: $\phi = (V_p/S_x)(k/D_e)^{0.5} = 0.000833(2.8/(1.9 \times 10^{-9}))^{0.5}$

$= 32$, which is large.

Thus: $\eta = (1/32)[\coth(3 \times 32) - 1/(3 \times 32)] = 0.031$, which is low.

Substituting these values in equation (iv), then:

$$\mathscr{R} = (1/357) \quad [(1/0.02) \quad + \quad (1/0.151) \quad + \quad (1/0.052)]^{-1}$$
$$= 0.0028 \qquad (50 \qquad + \qquad 6.6 \qquad + \qquad 19.2)^{-1}$$
$$\qquad\qquad (\text{gas–liquid}) \quad (\text{liquid–solid}) \quad (\text{diffusion + reaction})$$
$$\underline{= 3.7 \times 10^{-5} \text{ kmol/m}^3\text{s.}}$$

PROBLEM 4.5

Aniline present as an impurity in a hydrocarbon stream is to be hydrogenated to cyclohexylamine in a trickle bed catalytic reactor operating at 403 K (130°C).

$$C_6H_5 \cdot NH_2 + 3H_2 \longrightarrow C_6H_{11} \cdot NH_2$$

The reactor, in which the gas phase will be virtually pure hydrogen, will operate under a pressure of 2 MN/m² (20 bar). The catalyst will consist of porous spherical particles 3 mm in diameter, and the voidage, that is the fraction of bed occupied by gas plus liquid, will be 0.4. The diameter of the bed will be such that the superficial liquid velocity will be 0.002 m/s. The concentration of the aniline in the liquid feed will be 0.055 kmol/m³.

(a) From the following data, calculate what fraction of the aniline will be hydrogenated in a bed of depth 2 m. Assume that a steady state between the rates of mass transfer and reaction is established immediately the feed enters the reactor.
(b) Describe, stating the basic equations of a more complete model, how the validity of the steady-state assumption would be examined further.
(c) Is a trickle bed the most suitable type of reactor for this process? If not, suggest with reasons a possibly better alternative.

Data:
The rate of the reaction has been found to be first-order with respect to hydrogen but independent of the concentration of aniline. The first-order rate constant k of the reaction on a basis of kmol hydrogen reacting per m³ of catalyst particles at 403 K (130°C) is 90 s⁻¹.

Effective diffusivity of hydrogen in the catalyst particles with liquid-filled pores, $D_e = 0.84 \times 10^{-9}$ m²/s.
Effectiveness factor η for spherical particles of diameter d_p:

$$\eta = \frac{1}{\phi}\left(\coth 3\phi - \frac{1}{3\phi}\right) \quad \text{where:} \quad \phi = \frac{d_p}{6}\left(\frac{k}{D_e}\right)^{1/2}$$

External surface area of particles per unit volume of reactor = 1200 m²/m³.
Mass transfer coefficient, liquid to particles = 0.10×10^{-3} m/s
Volume mass transfer coefficient, gas to liquid (basis unit volume of reactor), $(k_L a) = 0.02$ s⁻¹
Henry's law coefficient $\mathcal{H} = P_A/C_A$ for hydrogen dissolved in feed liquid = 2240 bar/(kmol/m³).

Solution

(a) The effectiveness factor for the catalyst is given by:

$$\eta = (1/\phi)(\coth 3\phi - 1/3\phi)$$

where:
$$\phi = (d_p/6)(k/D_e)^{0.5}$$

In this case: $\phi = (0.003/6)(90/(0.84 \times 10^{-9}))^{0.5} = 164$
and: $\eta = 1/\phi = (1/164) = 0.0061$ (which is low).

Noting that the rate of reaction is independent of the aniline concentration, then assuming that steady-state is established immediately on entry:

Rate of transfer of H_2 from gas to liquid step (1) = rate of transfer of H_2 from liquid to the catalyst surface step (2) = rate of reaction within the catalyst step (3).

Converting the rate of reaction per unit volume of particles to rate of reaction per unit volume of reactor, then for step (3):

$$\text{Rate of reaction per unit volume of reactor, } \mathcal{R} = kC_s\eta(1-e)$$

Equating steps (1) and (2):

$$k_L a(C_i - C_L) = k_s a_s(C_L - C_s) \tag{i}$$

Equating steps (2) and (3):

$$k_s a_s(C_L - C_s) = kC_s\eta(1-e) \tag{ii}$$

$$C_i = P/\mathcal{H} = (20/2240) = 0.0089$$

and hence, in equation (i):

$$0.02(0.0089 - C_L) + (0.1 \times 10^{-3} \times 1200)(C_L - C_s) \tag{iii}$$

In equation (ii):

$$(0.1 \times 10^{-3} \times 1200)(C_L - C_s) = (90 \times C_s \times 0.0061)(1 - 0.4) \tag{iv}$$

From equation (iii):

$$(C_L - C_s) = 2.75 C_s \quad \text{and} \quad C_L = 3.75 C_s.$$

Substituting in equation (iv) gives:

$$0.02(0.0089 - C_L) = 0.12(C_L - C_L/3.75)$$

and: $$C_L = 1.65 \times 10^{-3} \text{ kmol/m}^3.$$

and: $$C_s = (0.00165/3.75) = 0.44 \times 10^{-3} \text{ kmol/m}^3$$

From step (3):

$$\text{Rate of reaction, } \mathcal{R} = (90 \times 0.44 \times 10^{-3} \times 0.0061)(1 - 0.4)$$

$$= 0.145 \times 10^{-3} \text{ kmol } H_2 \text{ reacting/m}^3 \text{ bed s.}$$

From the stoichiometry:

Rate of reaction of aniline = $(0.145 \times 10^{-3})/3 = 0.048 \times 10^{-3}$ kmol/m^3 bed s.

The superficial velocity of the liquid in the bed = 0.002 m/s or 0.002 m^3/m^{2}s

Thus: Aniline feed rate = $(0.002 \times 0.055) = 0.11 \times 10^{-3}$ kmol/m^{2}s.

Noting that the reaction rate of aniline is independent of the position in the bed, for bed, 2 m deep, 1 m^2 in area, then:

$$\text{aniline reacting} = (0.048 \times 10^{-3} \times 2) = 0.096 \times 10^{-3} \text{ kmol/m}^2\text{s}$$

and: fraction of aniline reacted = $(0.096 \times 10^{-3}/0.11 \times 10^{-3}) = \underline{0.87}$

(b) For a more complete model, a section of bed of depth δz and unit cross-sectional are considered.

A material balance for hydrogen in the liquid phase gives:

$$u_L C_L - u_L(C_L + (dC_L/dz)\delta z) + k_L a(C_i - C_L)\delta z - k_s a_s(C_L - C_s)\delta z = 0$$
$$\text{(in)} \qquad \text{(out)} \qquad \text{(gas to liquid)} \qquad \text{(liquid to solid)}$$

or: $\quad -u_L(dC_L/dz) + k_L a(C_i - C_L) - k_s a_s(C_L - C_s) = 0$

Since the rate of hydrogen transfer from the liquid to the solid is equal to the rate of reaction in the solid, then:

$$k_s a_s(C_L - C_s)\delta z = k C_s \eta (1 - \varepsilon)\delta z$$

Finally, a material balance on the aniline in the liquid gives:

$$u_L C_{LA} - u_L(C_{LA} + (dC_{LA}/dz)\delta z) - k C_s \eta (1 - e)\delta z/3 = 0$$

or: $\quad -u_L(dC_{LA}/dz) = k C_s \eta (1 - e)/3.$

(c) The trickle bed reactor is probably not the most suitable because of the very low value of the effectiveness factor and a suspended-bed catalyst system with a smaller particle size would be a much better option.

PROBLEM 4.6

Describe the various mass transfer and reaction steps involved in a three-phase gas–liquid–solid reactor. Derive an expression for the overall rate of a catalytic hydrogenation process where the reaction is pseudo first-order with respect to the hydrogen with a rate constant k (based on unit volume of catalyst particles).

Aniline is to be hydrogenated to cyclohexylamine in a suspended-particle agitated-tank reactor at 403 K (130°C) at which temperature the value of k is 90 s^{-1}. The diameter d_p of the supported nickel catalyst particles will be 0.1 mm and the effective diffusivity D_e for hydrogen when the pores of the particle are filled with aniline is 1.9×10^{-9} m^2/s.

For spherical particles the effectiveness factor is given by:

$$\eta = \frac{1}{\phi}\left(\coth 3\phi - \frac{1}{3\phi}\right) \quad \text{where} \quad \phi = \frac{d_p}{6}\left(\frac{k}{D_e}\right)^{1/2}$$

The proposed catalyst loading, that is the ratio by volume of catalyst to aniline, is to be 0.03. Under the conditions of agitation to be used, it is estimated that the gas volume fraction in the three-phase system will be 0.15 and that the volumetric gas–liquid mass transfer coefficient (also with respect to unit volume of the whole three-phase system) $k_L a$, 0.20 s^{-1}. The liquid–solid mass transfer coefficient is estimated to be 2.2×10^{-3} m/s and the Henry's law coefficient $\mathscr{H} = P_A/C_A$ for hydrogen in aniline at 403 K (130°C) = 2240 bar m^3/kmol where P_A is the partial pressure in the gas phase and C_A is the equilibrium concentration in the liquid.

(a) If the reactor is operated with a partial pressure of hydrogen equal to 1 MN/m^2 (10 bar) calculate the rate at which the hydrogenation will proceed per unit volume of the three-phase system.

(b) Consider this overall rate in relation to the operating conditions and the individual transfer resistances. Discuss the question of whether any improvements might be made to the conditions specified for the reactor.

Solution

For a first-order reaction, the reaction rate, $\mathscr{R} = kC_s\eta$ where C_s is the concentration at the particle surface. On the basis of unit volume of the three-phase dispersion, the reaction rate becomes $\mathscr{R}_t$ kmol/m^{3}s.

For mass transfer from the gas to the liquid:

$$\mathscr{R}_t = k_L a(C_i - C_L) \tag{i}$$

where C_i is the concentration of the gas–liquid interface and C_L in the bulk liquid.

For mass transfer from the liquid to the particle surface:

$$\mathscr{R}_t = k_s a_p \varepsilon_p (C_L - C_s) \tag{ii}$$

where a_p is the external surface per unit volume of particles and ε_p the volume fraction of the solid particles.

For diffusion and reaction within the particles:

$$\mathscr{R}_t = kC_s \eta \varepsilon_p \tag{iii}$$

For spherical particles:

$$a_p = \pi d_p^2/(\pi d_p^3/6) = 6/d_p$$

Rearranging and adding equations (i), (ii) and (iii) gives:

$$\mathscr{R}_t \left(\underbrace{\frac{1}{k_L a}}_{\text{(gas–liquid)}} + \underbrace{\frac{1}{k_s(6/d_p)\varepsilon_p}}_{\text{(liquid–solid)}} + \underbrace{\frac{1}{k\eta\varepsilon_p}}_{\text{(diffusion + reaction)}} \right) = C_i$$

Writing $\mathscr{R}_t = K_v C_i$, then:

$$(1/K_v) = (1/k_L a) + 1/(k_s \varepsilon_p(6/d_p)) + 1/k\eta\varepsilon_p$$

For transfer from the gas to the liquid:

$$1/k_L a = (1/0.20) = \underline{\underline{5.0 \text{ s}}}$$

The ratio: volume of solid/volume of liquid = 0.03

or: $\varepsilon_p/\varepsilon_L = 0.03$

$\varepsilon_G = 0.15$

and hence: $(\varepsilon_p + \varepsilon_L) = (1 - 0.15) = 0.85$

$\varepsilon_p(1 + 1/0.03) = 0.85$

and: $\varepsilon_p = 0.0248$

Thus: $1/(k_s(6/d_p)\varepsilon_p) = 1/[2.2 \times 10^{-3}(6/0.1 \times 10^{-3})0.0248] = \underline{0.305 \text{ s}}$

The Thiele Modulus, $\phi = (d_p/6)(k/D_e)^{0.5}$
$$= (0.1 \times 10^{-3}/6)(90/1.9 \times 10^{-9})^{0.5} = 3.63$$

and hence: $\eta = (1/3.63)(\coth(3 \times 3.63) - 1/(3 \times 3.63)) = 0.251$

Thus: $1/k\eta\varepsilon_p = 1/(90 \times 0.251 \times 0.0248) = \underline{1.78 \text{ s}}$

and: $1/K_v = (5.0 + 0.305 + 1.78) = 7.09 \text{ s}$
$$C_i = P/\mathscr{H} = (10/2240) = 4.46 \times 10^{-3} \text{ kmol/m}^3$$

and hence:

Rate of hydrogenation, $\mathscr{R}_t = K_v C_i = (4.46 \times 10^{-3}/7.09) = \underline{0.63 \times 10^{-3} \text{ kmol/m}^3\text{s}}$

Considering the three resistances, that of the gas to liquid transfer is the greatest. $k_L a$ might be improved, although, since $k_L a$ and ε_G are already fairly high, there is probably little scope for this. Solids loading might be increased and the overall rate of reaction could be increased by increasing the overall pressure.

SECTION 3-5

Biochemical Reaction Engineering

PROBLEM 5.1

The residence time, based on fresh feed, in an activated-sludge waste-water treatment unit is 21.2 Ms (5.9 h). The fresh feed has a BOD of 275 mg/l and the settler produces a recycle stream containing 6000 mg/l. Using a sludge of age 6 days, calculate the recycle ratio and the final effluent BOD, assuming that it contains no biomass, given that the yield coefficient Y is 0.54 and that the specific growth rate of the sludge is given by:

$$\mu = \frac{\mu_m S}{K_s + S} - k_d \qquad \text{(equation 5.70)}$$

where S is the substrate concentration (BOD), $\mu_m = 0.47$ h^{-1}, $K_s = 89$ mg (BOD)/l and the endogenous respiration coefficient $k_d = 0.009$ h^{-1}.

Solution

The flow diagram is given in Figure 5a.

Sludge age, θ_c = Biomass in reactor/net rate of biomass generation.

$$= XV/[(\mu_m SXV/(K_s + S)) - k_d XV]$$
$$= (K_s + S)/(\mu_m S - k_d(K_s + S))$$

From which:

$$S = K_s(1 + k_d\theta_c)/[\theta_c(\mu_m - k_d) - 1]$$

Thus: Final concentration, $S = \dfrac{89(1 + 0.009 \times 6 \times 24)}{6 \times 24(0.47 - 0.009) - 1}$

$$= \underline{3.12 \text{ mg (BOD)}/l}$$

From equation 5.151, Volume 3 a material balance for the substrate across the aeration tank gives:

$$F_0 S_0 + F_R S - (F_0 + F_R)S - (1/Y)\mu_m SXV/(K_s + S) = V \, dS/dt$$

At steady state, $dS/dt = 0$.

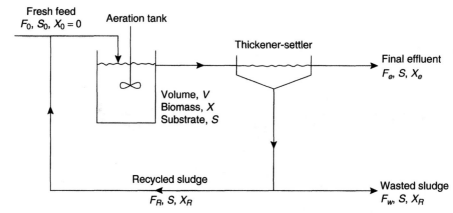

Figure 5a. Flow diagram for Problem 5.1

Dividing throughout by F_0, the recycle ratio, $R = F_R/F_0$ and the hydraulic residence time, $\theta = V/F_0$, then:

$$S_0 + RS - S - RS - [\mu_m SX\theta/Y(K_s + S)] = 0$$

and:

$$X = [(S_0 - S)Y(K_s + S)]/\mu_m S\theta$$

Hence:

$$X = (275 - 3.12)0.54(89 + 3.12)/(0.47 \times 3.12 \times 5.9)$$
$$= 1569 \text{ mg/l}$$

Thus, the concentration factor β for the thickener–settler is:

$$\beta = X_R/X$$
$$= (6000/1569) = \underline{\underline{3.82}}$$

From Equation 5.152, a material balance for biomass over the aeration tank gives:

$$F_0 X_0 + F_R X_R - (F_0 + F_R)X + (\mu_m SXV/(K_s + S)) - k_d XV = V(dX/dt)$$

Since X_0 is very much smaller than X_R, it may be assumed that $X_0 = 0$.

At steady state, $dX/dt = 0$ and hence:

$$R\beta - 1 - R + (\mu_m S/(K_s + S))\theta - k_d \theta = 0$$

Thus:

$$R = [1 - \theta((\mu_m S/(K_s + S)) - k_d)]/(\beta - 1)$$

Hence:

$$\text{Recycle ratio, } R = \left[1 - 5.9\left(\frac{0.47 \times 3.12}{89 + 3.12} - 0.009\right)\right] \Big/ (3.82 - 1)$$
$$= \underline{\underline{0.34}}$$

PROBLEM 5.2

A continuous fermenter is operated at a series of dilution rates though at constant, sterile, feed concentration, pH, aeration rate and temperature. The following data were obtained when the limiting substrate concentration was 1200 mg/l and the working volume of the fermenter was 9.8 l. Estimate the kinetic constants K_m, μ_m and k_d as used in the modified Monod equation:

$$\mu = \frac{\mu_m S}{K_s + S} - k_d$$

and also the growth yield coefficient Y.

Feed flowrate (l/h)	Exit substrate concentration (mg/l)	Dry weight cell density (mg/l)
0.79	36.9	487
1.03	49.1	490
1.31	64.4	489
1.78	93.4	482
2.39	138.8	466
2.68	164.2	465

Solution

The flow diagram is as Figure 5.56 in Volume 3, where the inlet and outlet streams are defined as F_0, X_0, S_0 and F_0, X, S respectively.

$$\text{The accumulation} = \text{input} - \text{output} + \text{rate of formation,}$$

which for the biomass gives:

$$V(dX/dt) = FX_0 - FX + V(\mu_m SX)/(K_s + S) - k_d XV \qquad \text{(equation 5.126)}$$

and for the substrate:

$$V(dS/dt) = FS_0 - FS - V\mu_m SX/Y(K_s + S) \qquad \text{(equation 5.127)}$$

At steady state, $dS/dt = 0$.

Taking the dilution rate, $D = F/V$, then the balance for the substrate becomes:

$$S_0 - S - \{\mu_m SX/[DY(K_s + S)]\} = 0$$

or:
$$X/D(S_0 - S) = (K_s Y/\mu_m)/S + Y/\mu_m \qquad \text{(i)}$$

Similarly, for the biomass:

$$X_0 - X + \{\mu_m SX/[D(K_s + S)]\} - k_d X/D = 0$$

Since the feed is sterile, $X_0 = 0$ and:

$$DX = [\mu_m SX/(K_s + S)] - k_d X.$$

From the material balance for substrate:

$$(\mu_m S X)/(K_s + S) = DY(S_0 - S)$$

and substitution gives:

$$DX = DY(S_0 - S) - k_d X$$

or:
$$(S_0 - S)/X = k_d/DY - 1/Y \qquad \text{(ii)}$$

From equation (i), it is seen that a plot of $X/D(S_0 - S)$ and $1/S$ will produce a straight line of slope $K_s Y/\mu_m$ and intercept Y/μ_m.

From equation (ii), it is seen that a plot of $(S_0 - S)/X$ against $1/D$ will produce a straight line of slope k_d/Y and intercept $1/Y$.

The data are calculated as follows:

Feed flowrate (F l/h)	Exit substrate (S mg/l)	$1/S$ (l/mg)	$X/D(S_0 - S)$ (h)	$1/D$ (h)	$(S_0 - S)/X$ (–)
0.79	36.9	0.0271	5.19	12.41	2.388
1.03	49.1	0.0204	4.05	9.51	2.349
1.31	64.4	0.0155	3.22	7.48	2.322
1.78	93.4	0.0107	2.40	5.51	2.296
2.39	138.8	0.0072	1.80	4.10	2.277
2.68	164.2	0.0061	1.64	3.66	2.228

The data are then plotted in Figure 5b from which:

$$K_s Y/\mu_m = 170, \quad Y/\mu_m = 0.59, \quad k_d/Y = 0.0133 \quad \text{and} \quad 1/Y = 2.222.$$

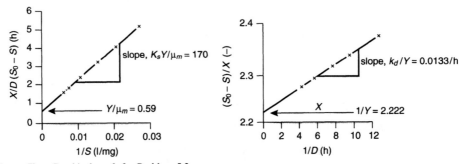

Figure 5b. Graphical work for Problem 5.2

Thus:

yield coefficient, $Y = (1/2.222) = 0.45$

endogenous respiration coefficient, $k_d = (0.45 \times 0.0133) = \underline{0.006 \text{ h}^{-1}}$

$$\mu_m = (0.45/0.59) = \underline{0.76 \text{ h}^{-1}}$$

and:
$$K_s = (170 \times 0.76/0.45) = \underline{300 \text{ mg/l}}$$

PROBLEM 5.3

When a pilot-scale fermenter is run in continuous mode with a fresh feed flowrate of 65 l/h, the effluent from the fermenter contains 12 mg/l of the original substrate. The same fermenter is then connected to a settler–thickener which has the ability to concentrate the biomass in the effluent from the tank by a factor of 3.2, and from this a recycle stream of concentrated biomass is set up. The flowrate of this stream is 40 l/h and the fresh feed flowrate is at the same time increased to 100 l/h. Assuming that the microbial system follows Monod kinetics, calculate the concentration of the final clarified liquid effluent from the system. $\mu_m = 0.15$ h^{-1} and $K_s = 95$ mg/l.

Solution

A flow diagram is given in Figure 5c.

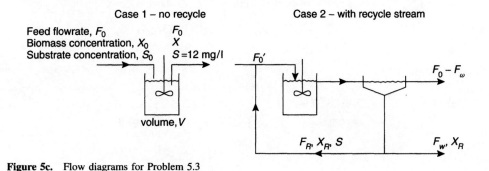

Figure 5c. Flow diagrams for Problem 5.3

Case 1: No recycle
A material balance for biomass over the aeration tank gives:

biomass in feed − biomass in effluent + biomass formed by growth

= accumulation of biomass

or:
$$F_0 X_0 - F_0 X + \mu X V = V(dX/dt) \qquad \text{(equation 5.126)}$$

Taking the dilution rate, $D = F_0/V$, noting that $X_0 = 0$ and that at steady state, $dX/dt = 0$, then:

$$D = \mu$$

and assuming Monod kinetics apply:

$$D = \mu_m S/(K_s + S) \qquad \text{(equation 5.132)}$$

Thus: $D = (0.15 \times 12)/(95 + 12) = 0.0168$ h^{-1}

The volume of the aeration tank, $V = F_0/D$

or: $V = (6.5/0.0168) = \underline{\underline{3864 \text{ l}}}$

Case 2: With recycle
The material balance for biomass over the aeration tank becomes:

biomass in feed + biomass in recycle − biomass leaving the settler − biomass formed
= accumulation of biomass

or:
$$F_0 X_0 + F_R X_R - (F_0 + F_R)X + \mu X V = V(dX/dt) \quad \text{(equation 5.152)}$$

For sterile feed, $X_0 = 0$ and, at steady state:
$$\frac{F_R}{F_0}\frac{X_R}{X} - \left(1 - \frac{F_R}{F_0}\right) + \frac{\mu V}{F_0} = 0 \quad \text{(equation 5.153)}$$

If β, the concentrating effect of the thickener–settler is (X_R/X), then:

Specific growth rate, $\mu = (F_0/V)(1 + (F_R/F_0)(1 - \beta))$

or:
$$\mu = [1 + (40/100)(1 - 3.2)](100/3864)$$
$$= \underline{0.0031 \text{ h}^{-1}}$$

From equation 5.133:
$$S = K_s \mu/(\mu_m - \mu)$$
$$= (95 \times 0.0031)/(0.15 - 0.0031) = \underline{2.0 \text{ mg/l}}$$

PROBLEM 5.4

When a continuous culture is fed with substrate of concentration 1.00 g/l, the critical dilution rate for washout is 0.2857 h^{-1}. This changes to 0.0983 h^{-1} if the same organism is used but the feed concentration is 3.00 g/l. Calculate the effluent substrate concentration when, in each case, the fermenter is operated at its maximum productivity.

Solution

At incipient washout, the critical dilution rate, D_{crit}, is related to the Monod constants by:
$$D_{\text{crit}} = \mu_m S_0/(K_s + S_0) \quad \text{(equation 5.148)}$$

where S_0 is the concentration of substrate in the feed.
Rearranging:
$$\mu_m = D_{\text{crit}}(K_s + S_0)/S_0$$

and for the initial conditions:
$$\mu_m = 0.2857(K_s + 1.0)/1.0 - 0.2857 K_s + 0.2857 \quad \text{(i)}$$

For the increased feed rate:

$$\mu_m = 0.09833 K_s + 0.295 \qquad \text{(ii)}$$

From equations (i) and (ii):

$$K_s = 0.0496 \text{ g/l} \quad \text{and} \quad \mu_m = 0.30 \text{ h}^{-1}.$$

The maximum cell productivity occurs at an optimum dilution rate, D_{opt}, given by:

$$D_{opt} = \mu_m \left[1 - \sqrt{\left(\frac{K_s}{K_s + S_0}\right)} \right] \qquad \text{(equation 5.140)} \quad \text{(iii)}$$

and the substrate concentration for any dilution rate below the critical value is given by:

$$S = DK_s/(\mu_m - D) \qquad \text{(iv)}$$

Thus, for the initial conditions:

$$D_{opt} = 0.30 \left\{ 1 - \sqrt{[0.0496/(0.0496 + 1.00)]} \right\} = 0.235 \text{ h}^{-1}$$

and:

$$S_1 = (0.235 \times 0.0496)/(0.30 - 0.235) = 0.18 \text{ g/l}$$

For the increased flowrate:

$$D_{opt} = 0.30 \left\{ 1 - \sqrt{[0.0496/(0.0496 + 3.00)]} \right\} = \underline{0.262 \text{ h}^{-1}}$$

and:

$$S_2 = (0.262 \times 0.0496)/(0.30 - 0.262) = \underline{\underline{0.34 \text{ g/l}}}$$

PROBLEM 5.5

Two continuous stirred-tank fermenters are arranged in series such that the effluent of one forms the feed stream of the other. The first fermenter has a working volume of 100 l and the other has a working volume of 50 l. The volumetric flowrate through the fermenters is 18 h^{-1} and the substrate concentration in the fresh feed is 5 g/l. If the microbial growth follows Monod kinetics with $\mu_m = 0.25 \text{ h}^{-1}$, $K_s = 0.12$ g/l, and the yield coefficient is 0.42, calculate the substrate and biomass concentrations in the effluent from the second vessel. What would happen if the flow were from the 50 l fermenter to the 100 l fermenter?

Solution

The flow diagram for this operation is shown in Figure 5.62. Together with the relevant nomenclature.

A material balance for biomass over the first fermenter, as discussed in section 5.11.3, leads to the equation:

$$D_1 = \mu_1 \qquad \text{(equation 5.131)}$$

where D_1 is the dilution rate in the first vessel and μ_1 is the specific growth rate for that vessel.

Assuming Monod kinetics to apply:

$$D_1 = \mu_m S_1/(K_s + S_1) \qquad \text{(equation 5.132)}$$

where S_1 is the steady-state concentration of substrate in the first vessel, where:

$$S_1 = D_1 K_s/(\mu_m - D_1) \qquad \text{(equation 5.133)}$$

If $D_1 = F/V_1 = (18/100) = 0.18 \text{ h}^{-1}$, then:

$$S_1 = (0.18 \times 0.12)/(0.25 - 0.18) = 0.309 \text{ g/l}$$

Since the feed to this fermenter is sterile, $X_0 = 0$ and from equation 5.134, Volume 3 the steady-state concentration of biomass in the first vessel is given by:

$$X_1 = Y(S_0 - S_1) \qquad \text{(equation 5.134)}$$
$$= 0.42(5 - 0.309) = \underline{1.97 \text{ g/l}}$$

In a similar way, a mass-balance over the second vessel gives:

$$D_2 = \mu_2 X_2/(X_2 - X_1) \qquad \text{(equation 5.167)}$$

where D_2 is the dilution rate in the second vessel, μ_2 is the specific growth rate in that vessel and X_2 the steady-state concentration of biomass.

The yield coefficient to the second vessel is then:

$$Y = (X_2 - X_1)/(S_1 - S_2)$$

where S_2 is the steady-state concentration of substrate in that vessel.

Thus:
$$X_2 = X_1 + Y(S_1 - S_2)$$
$$= 1.97 + 0.42(0.309 - S_2)$$
$$= 2.1 + 0.42 S_2$$

Substituting this equation for X_2 into equation 5.167, together with values for D_2, μ_2 and X_1 leads to a quadratic equation in S_2:

$$0.128 S_2^2 + 1.379 S_2 - 0.01555 = 0$$

from which:

$$S_2 = \underline{0.0113 \text{ g/l}}$$

and:
$$X_2 = 2.1 + (0.42 \times 0.0113) = \underline{2.1 \text{ g/l}}$$

When the tanks are reversed, that is with fresh feed entering the 50 litre vessel, then the dilution rate for this vessel will be as before, 0.36 h^{-1}, although the critical dilution rate will now be:

$$D_{\text{crit}} = \mu_m S_0/(K_s + S_0) \quad \text{(equation 5.148)}$$
$$= (0.25 \times 5)/(0.12 + 5) = 0.244 \text{ h}^{-1}$$

This is lower than the dilution rate imposed and washout of the smaller vessel would take place. The concentrations of substrate and biomass in the final effluent would eventually be those attained if only the 100 litre vessel existed, that is:

$$\text{concentration of biomass} = \underline{\underline{1.97 \text{ g/l}}}$$

$$\text{concentration of substrate} = \underline{\underline{0.309 \text{ g/l}}}.$$

SECTION 3-7

Process Control

PROBLEM 7.1

After being in use for some time, a pneumatic three-term controller as shown in Figure 7.118, Volume 3, develops a significant leak in the partition between the integral bellows and the proportional bellows. It is known that the rate of change of pressure in the integral bellows due to the leak is half that due to air flow through the integral restrictor. Show that the leak does not affect the form of the output response of the controller and that the ratio of the gain of the controller with the leak to that of the same controller before the leak developed is given by:

$$\frac{3\tau_2 + \tau_1}{2\tau_2 + \tau_1}$$

where τ_1 and τ_2 are the time constants of the integral and derivative restrictors respectively.

Solution

The relevant diagram is included as Figure 7a.

The change in separation of the flapper from the nozzle at X is due to the net movement of B and C and the relative lengths of l_1 and l_2.

$$\text{Movement of } B = k_2(p_1 - p_2)$$
$$\text{Movement of } C = -k_1\varepsilon$$

where k_1 and k_2 are constants and ε, the error, is the difference in movement between E, the set-point, and F, the measured value.

Hence:

$$\text{net movement of flapper at } X = -k_1\varepsilon\left(\frac{l_2}{l_1 + l_2}\right) + k_2(p_1 - p_2)\left(\frac{l_1}{l_1 + l_2}\right)$$

The change in output pressure is proportional to this, or:

$$\Delta P = C\left[-k_1\varepsilon\left(\frac{l_2}{l_1 + l_2}\right) + k_2(p_1 - p_2)\left(\frac{l_1}{l_1 + l_2}\right)\right] \quad \text{(i)}$$

where C is the amplification factor.

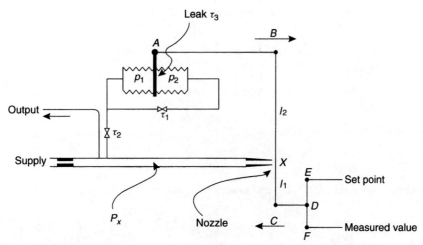

Figure 7a. Diagram for problem 7.1

If C is large, then from equation (i):

$$p_1 - p_2 = \left(\frac{k_1}{k_2}\right)\left(\frac{l_2}{l_1}\right)\varepsilon = K\varepsilon \qquad \text{(ii)}$$

where K is a constant for a given controller mechanism.

If $p_1 > p_2$, then air will flow from the proportional bellows through the integral restrictor and through the leak. As the rate of change of pressure in the bellows is proportional to this difference is pressure,

then:
$$\frac{dp_2}{dt} = \frac{1}{\tau_1}(p_1 - p_2) + \frac{1}{\tau_3}(p_1 - p_2) = \frac{1}{\tau_A}(p_1 - p_2) \qquad \text{(iii)}$$

where τ_1 and τ_3 are the time constants for the integral restrictor and the leak respectively.

For the derivative restrictor and proportional bellows:

$$\frac{dp_1}{dt} = \frac{1}{\tau_2}(p - p_1) - \frac{1}{\tau_1}(p_1 - p_2) - \frac{1}{\tau_3}(p_1 - p_2)$$

$$= \frac{1}{\tau_2}(p - p_1) - \frac{1}{\tau_A}(p_1 - p_2) \qquad \text{(iv)}$$

where P is the output pressure for $t > 0$.

From equations (ii) and (iii):

$$\frac{dp_2}{dt} = \frac{K}{\tau_A}\varepsilon \qquad \text{(v)}$$

Thus:
$$p_2 - P_x = \frac{K}{\tau_A}\int_0^t \varepsilon\, dt \quad (p_2 = P_x \text{ at } t = 0) \qquad \text{(vi)}$$

295

Substituting equations (v) and (vi) into equation (iii):

$$\frac{K}{\tau_A}\varepsilon = \frac{1}{\tau_A}\left(p_1 - P_x - \frac{K}{\tau_A}\int_0^t \varepsilon\, dt\right)$$

Thus:
$$p_1 = P_x + K\varepsilon + \frac{K}{\tau_A}\int_0^t \varepsilon\, dt \tag{vii}$$

and, with constant P_x:

$$\frac{dp_1}{dt} = K\frac{d\varepsilon}{dt} + \frac{K}{\tau_A}\varepsilon \tag{viii}$$

Substituting equations (iii), (vii), and (viii) into equation (iv):

$$K\frac{d\varepsilon}{dt} + \frac{K}{\tau_A}\varepsilon = \frac{1}{\tau_2}\left(p - P_x - K\varepsilon - \frac{K}{\tau_A}\int_0^t \varepsilon\, dt\right) - \frac{K}{\tau_A}\varepsilon$$

$$p - P_x = K\left[\left(\frac{2\tau_2}{\tau_A} + 1\right)\varepsilon + \frac{1}{\tau_A}\int_0^t \varepsilon\, dt + \tau_2\frac{d\varepsilon}{dt}\right]$$

This is the change in the controller output at time t and is in standard **PID** form. Hence the gain of controller with the leak is:

$$(\text{Gain})_1 = K\left(\frac{2\tau_2 + \tau_A}{\tau_A}\right)$$

Following the same procedure without the leak (that is $\tau_3 = \infty$) gives $\tau_A = \tau_1$, from equation (iii). Thus, for no leak:

$$(\text{Gain})_2 = K\left(\frac{2\tau_2 + \tau_1}{\tau_1}\right)$$

Hence:
$$\frac{(\text{Gain})_1}{(\text{Gain})_2} = \frac{\tau_1(2\tau_2 + \tau_A)}{\tau_A(2\tau_2 + \tau_1)} \tag{ix}$$

The rate of change of pressure in the integral bellows due to the leak is, however, half that due to flow through the integral restrictor, or:

$$\tau_3 = 2\tau_1$$

and:
$$\frac{1}{\tau_A} = \frac{1}{\tau_1} + \frac{1}{2\tau_1} = \frac{3}{2\tau_1} \quad \text{and} \quad \tau_A = \frac{2}{3}\tau_1$$

From equation (ix):

$$\frac{(\text{Gain})_1}{(\text{Gain})_2} = \frac{\tau_1(2\tau_2 + \frac{2}{3}\tau_1)}{\frac{2}{3}\tau_1(2\tau_2 + \tau_1)}$$

$$= \frac{(3\tau_2 + \tau_1)}{(2\tau_2 + \tau_1)}$$

PROBLEM 7.2

A mercury thermometer having first-order dynamics with a time constant of 60 s is placed in a bath at 308 K (35°C). After the thermometer reaches a steady state it is suddenly placed in a bath at 313 K (40°C) at $t = 0$ and left there for 60 s, after which it is immediately returned to the bath at 308 K (35°C).

(a) Draw a sketch showing the variation of the thermometer reading with time.
(b) Calculate the thermometer reading at $t = 30$ s and at $t = 120$ s.
(c) What would be the reading at $t = 6$ s if the thermometer had only been immersed in the 313 K bath for less than 1 s before being returned to the 308 K bath?

Solution

(a) At $t = 0$ the thermometer is subjected to a step change of 5 deg K.

$$\frac{\overline{\vartheta}_{t_1}}{\overline{\vartheta}_b} = \frac{1}{1 + 60\,s}$$

$$\overline{\vartheta}_b = \frac{5}{s} \quad \text{at} \quad t = 0$$

Thus:
$$\overline{\vartheta}_{t_1} = \left(\frac{5}{s}\right)\left(\frac{1}{1 + 60\,s}\right)$$

$$= \frac{A}{s} + \frac{B}{1 + 60\,s}$$

Thus: $A = 5, \quad B = -300$

$$\overline{\vartheta}_{t_1} = \frac{5}{s} - \frac{300}{1 + 60\,s} = \frac{5}{s} - \frac{5}{1/60 + s}$$

$$\overline{\vartheta}_{t_1} = 5 - 5e^{-t/60}$$
$$= 5(1 - e^{-t/60})$$

After 60 s, $\overline{\vartheta}_{t_1} = 5(1 - e^{-1}) = 3.16$ deg K

Hence after 60 s, the thermometer will read $(308 + 3.16) = \underline{\underline{311.16 \text{ K}}}$.

At $t = 60$ s, a further negative step change is imposed of 3.16 deg K. The thermometer will respond immediately to this as it has only first order dynamics.

Thus:
$$\vartheta_{t_2} = 3.16(1 - e^{-t/60})$$

(b) At $t = 30$ s, the thermometer reading is:

$$\vartheta_{t_1} = 5(1 - e^{-30/60}) = 1.97 \text{ deg K}$$

and: thermometer reading $= \underline{\underline{309.97 \text{ K}}}$

At $t = 120$ s, since a step decrease is applied at $t = 60$ s, then:

$$\vartheta_{t_2} = 3.16(1 - e^{-60/60}) = 2 \text{ deg K}$$

Thus: Thermometer reading at 120 s = $(311.16 - 2)$

$$= \underline{309.16 \text{ K}}$$

(c) At $t = 0$, the thermometer is immersed for less than 1 s in the 313 K bath, hence assuming that an impulse applied is:

$$\frac{\overline{\vartheta_{t_1}}}{\overline{\vartheta_b}} = \frac{1}{1 + 60 \text{ s}}$$

From Section 7.8.5:

$$F(t)_{\text{Impulse}} = \frac{d}{dt}\{F(t)\}_{\text{step}}$$

Thus:

$$= \frac{d}{dt}\{5(1 - e^{-t/60})\} = \frac{1}{12}e^{-t/60}$$

Thus at $t = 6$ s

$$\vartheta_{t_1} = \frac{1}{12}e^{-0.1} = 0.075 \text{ deg K}$$

and the thermometer will read $\underline{308.075 \text{ K}}$

The variation of temperature reading with time is shown in Figure 7b.

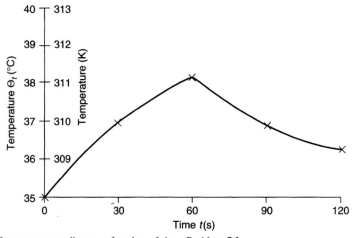

Figure 7b. Thermometer reading as a function of time, Problem 7.2

PROBLEM 7.3

A tank having a cross-sectional area of 0.2 m² is operating at a steady state with an inlet flowrate of 10^{-3} m³/s. Between the liquid heads of 0.3 m and 0.09 m the flow-head characteristics are given by:

$$Q_2 = 0.002Z + 0.0006$$

where Q_2 is the outlet flowrate and Z is the liquid level. Determine the transfer functions relating (a) inflow and liquid level, (b) inflow and outflow.

If the inflow increases from 10^{-3} to 1.1×10^{-3} m^3/s according to a step change, calculate the liquid level 200 s after the change has occurred.

Solution

(a) A mass balance gives:

$$Q_1\rho - Q_2\rho = \frac{d}{dt}(\rho V) = \frac{d}{dt}(\rho A Z)$$

where Q_1 is the inlet flow, ρ the fluid density, V the volume of fluid in the tank and A the area of the base (ρ and A are constants).

Thus:
$$Q_1 - Q_2 = A\frac{dZ}{dt} \quad \text{(i)}$$

At steady-state:
$$Q_1' - Q_2' = 0 \quad \text{(ii)}$$

Subtracting equation (ii) from equation (i), then:

$$(Q_1 - Q_1') - (Q_2 - Q_2') = A\frac{dZ}{dt} = A\frac{d\mathscr{Z}}{dt}$$

$$\mathscr{Q}_1 - \mathscr{Q}_2 = A\frac{d\mathscr{Z}}{dt} \quad \text{(iii)}$$

where: $\quad \mathscr{Q}_1 = Q_1 - Q_1', \quad \mathscr{Q}_2 = Q_2 - Q_2', \quad \mathscr{Z} = Z - Z'$

But: $\quad Q_2 = 0.002Z + 0.0006$

Thus: $\quad \mathscr{Q}_2 = 0.002\mathscr{Z} \quad \text{(iv)}$

From equations (iii) and (iv): $\quad \mathscr{Q}_1 - 0.002\mathscr{Z} = A\frac{d\mathscr{Z}}{dt}$

Transforming: $\quad \overline{\mathscr{Q}}_1 - 0.002\overline{\mathscr{Z}} = A\frac{d\overline{\mathscr{Z}}}{dt}$

Hence, the transfer function, $\quad \mathbf{G}_A = \overline{\mathscr{Z}}/\overline{\mathscr{Q}}_1 = \dfrac{500}{1 + 100\,s} \quad \text{(v)}$

(b) From equations (iv) and (v): $\quad \mathbf{G}_B = \overline{\mathscr{Q}}_2/\overline{\mathscr{Q}}_1 = \dfrac{1}{1 + 100\,s} \quad \text{(vi)}$

For a step change in Q_1 of 10^{-4} m^3/s:

$$\overline{\mathscr{Q}}_1 = \frac{10^{-4}}{s}$$

Thus, from equation (v):
$$\bar{\mathscr{Z}} = \frac{10^{-4}}{s} \cdot \frac{500}{1 + 100s} = \frac{5 \times 10^{-4}}{s(0.01 + s)}$$

$$= 0.05\left(\frac{1}{s} - \frac{1}{s + 0.01}\right)$$

Thus: $\mathscr{Z} = 0.05(1 - e^{-0.01t})$

At $t = 200$ s: $\mathscr{Z} = 0.05(1 - e^{-2}) = 0.043$ m.

At the original steady-state: $Q'_2 = 0.002Z' + 0.006 = Q'_1$

Thus: $0.002Z' + 0.0006 = 0.001$

and: original steady-state level $Z'_1 = 0.2$ m

Hence, level at $t = 200$ s is $(0.2 + 0.043) = \underline{\underline{0.243 \text{ m}}}$

PROBLEM 7.4

A continuous stirred tank reactor is fed at a constant rate F m³/s. The reaction is:

$$A \longrightarrow B$$

which proceeds at a rate of:
$$\mathscr{R} = kC_0$$

where: $\mathscr{R}$ = kmoles **A** reacting/m³ mixture in tank (s),
k = first order reaction velocity constant (s⁻¹), and
C_0 = concentration of **A** in reactor (kmol/m³).

If the density ρ and volume V of the reaction mixture in the tank are assumed to remain constant, derive the transfer function relating the concentration of **A** in the reactor at any instant to that in the feed stream C_i. Sketch the response of C_0 to an impulse in C_i.

Solution

If the volume V is constant then outlet flow = inlet flow = F m³/s. Assuming the fluid is well-mixed, then outlet concentration = C_0 kmol/m³.

A mass balance on component A gives:

$$FC_i - FC_0 - \mathscr{R}V = \frac{d}{dt}(VC_0)$$

$$FC_i - (F + kV)C_0 = V\frac{dC_0}{dt} \quad \text{(i)}$$

At steady-state: $FC'_i - (F + kV)C'_0 = 0$ (ii)

Subtracting equation (ii) from equation (i): $F\mathscr{C}_i - (F + kV)\mathscr{C}_0 = V\frac{d\mathscr{C}_0}{dt}$

where C_i, C_0 are deviation variables.

Transforming: $$F\overline{\mathscr{C}}_i - (F+kV)\overline{\mathscr{C}}_0 = Vs\overline{\mathscr{C}}_0$$

The transfer function $= \mathbf{G}(s) = \overline{\mathscr{C}}_0/\overline{\mathscr{C}}_i = \dfrac{F}{(F+kV)+Vs}$

$$= \underline{\underline{\dfrac{F/(F+kV)}{1+[V/(F+kV)]s}}}$$

For impulse of area (magnitude) A_I:

$$\overline{\mathscr{C}}_i = A_I \qquad \text{(equation 7.78)}$$

Thus: $\overline{\mathscr{C}}_0 = \dfrac{A_I K}{1+\tau s} \qquad$ where $\quad A = F/(F+kV)$

and $\quad \tau = V/(F+kV)$

Thus: $\mathscr{C}_0 = \dfrac{A_I K}{\tau} e^{-t/\tau}$

The response of C_0 will be as in Figure 7.30, Volume 3, with $\tau C_0/A_I K$ as the abscissa and t/τ as the ordinate.

PROBLEM 7.5

Liquid flows into a tank at the rate of Q m^3/s. The tank has three vertical walls and one sloping outwards at an angle β to the vertical. The base of the tank is a square with sides of length x m and the average operating level of liquid in the tank is Z_0 m. If the relationship between liquid level and flow out of the tank at any instant is linear, develop an expression for the time constant of the system.

Solution

For constant density throughout, a mass balance gives:

$$Q - Q_{\text{out}} = dV/dt \qquad \text{(i)}$$

where: $\qquad V = x^2 Z + \tfrac{1}{2}(xZ^2 \tan \beta)$

and Z is the liquid level at any instant.

This is a non-linear relationship. Linearising using Taylor's series, as given in equation 7.24, Volume 3:

$$V = V_0 + \left[\dfrac{dV}{dZ}\right]_{Z_0}(Z - Z_0) + \cdots = k_1 + k_2 \mathscr{Z} \qquad \text{(ii)}$$

where: $\qquad V_0 = x^2 Z_0 + \dfrac{1}{2}xZ_0^2 \tan \beta = k_1$

$$\left[\dfrac{dV}{dZ}\right]_{Z_0} = x^2 + xZ_0 \tan \beta = k_2 \quad \text{and} \quad \mathscr{Z} = Z - Z_0$$

301

From equations (i) and (ii): $\quad Q - Q_{out} = \dfrac{d}{dt}(k_1 + k_2 \mathscr{Z}) = k_2 \dfrac{d\mathscr{Z}}{dt}$ \hfill (iii)

At steady-state: $\quad Q' - Q'_{out} = 0$ \hfill (iv)

Subtracting equation (iv) from equation (iii): $\quad \overline{Q} - \overline{Q}_{out} = k_2 \dfrac{d\overline{\mathscr{Z}}}{dt}$ \hfill (v)

There is a linear relation between $\overline{Q}_{out}$ and $\overline{\mathscr{Z}}$ and $\overline{Q}_{out} = k_3 \overline{\mathscr{Z}}$ \hfill (vi)

From equations (v) and (vi): $\quad \overline{Q} - k_3 \overline{\mathscr{Z}} = k_2 \dfrac{d\overline{\mathscr{Z}}}{dt}$

Transforming: $\quad \overline{\overline{Q}} - k_3 \overline{\overline{\mathscr{Z}}} = k_2 s \overline{\overline{\mathscr{Z}}}$

Hence, the transfer function $\mathbf{G}(s) = \overline{\overline{\mathscr{Z}}}/\overline{\overline{Q}} = \dfrac{1}{k_3 + k_2 s}$

$$= \dfrac{1/k_3}{1 + (k_2/k_3)s}$$

and the process time constant is $\underline{\underline{k_2/k_3}}$.

PROBLEM 7.6

Write the transfer function for a mercury manometer consisting of a glass U-tube 0.012 m internal diameter, with a total mercury-column length of 0.54 m, assuming that the actual frictional damping forces are four times greater than would be estimated from Poiseuille's equation. Sketch the response of this instrument when it is subjected to a step change in an air pressure differential of 14,000 N/m² if the original steady differential was 5000 N/m². Draw the frequency-response characteristics of this system on a Bode diagram.

Solution

From Section 7.5.4, Volume 3, for a manometer in which fractional damping is four times that predicted by Poiseuille's equation:

$$G(s) = \dfrac{K_{MT}}{\tau^2 s^2 + 2\zeta \tau s + 1} \quad \text{(equation 7.52)}$$

where: $\quad \tau = \sqrt{\left(\dfrac{l}{2g}\right)} = \sqrt{\left(\dfrac{0.54}{2 \times 9.81}\right)} \approx 0.17 \text{ s}$

and: $\quad \zeta = 4 \times \dfrac{8\mu}{d^2 \rho}\sqrt{\left(\dfrac{2l}{g}\right)}$

At 298 K: $\quad \mu_{Hg} = 1.6 \times 10^{-3}$ Ns/m² and $\rho_{Hg} = 13{,}530$ kg/m³

Thus:
$$\zeta = \left(\frac{32 \times 1.6 \times 10^{-3}}{0.012^2 \times 13{,}530}\right)\sqrt{\left(\frac{2 \times 0.54}{9.81}\right)} = 8.7 \times 10^{-3}$$

$$K_{MT} = \frac{1}{2\rho g} = \frac{1}{(2 \times 13{,}530 \times 9.81)} = 3.8 \times 10^{-6} \text{ m}^4/\text{kg s}^2$$

Clearly $\zeta < 1$ and the step response may be calculated from equation 7.82, Volume 3, where $M = 14{,}000$ N/m². $\zeta \approx 0$ however and thus the step response will approach a continuous oscillation with constant amplitude as shown in Figure 7.28, Volume 3.

The frequency response may be determined from equations 7.94 and 7.95, Volume 3.

PROBLEM 7.7

The response of an underdamped second-order system to a unit step change may be shown to be:

$$Y(t) = 1 - \frac{1}{\sqrt{(1-\zeta^2)}} \exp(-\zeta t/\tau) \left\{ \zeta \sin\left[\sqrt{(1-\zeta^2)}\frac{t}{\tau}\right] \right.$$
$$\left. + \sqrt{(1-\zeta^2)} \cos\left[\sqrt{(1-\zeta^2)}\frac{t}{\tau}\right] \right\}$$

Prove that the overshoot for such a response is given by:

$$\exp\{-\pi\zeta/\sqrt{(1-\zeta^2)}\}$$

and that the decay ratio is equal to the (overshoot)².

A forcing function, whose transform is a constant K is applied to an under-damped second-order system having a time constant of 0.5 min and a damping coefficient of 0.5. Show that the decay ratio for the resulting response is the same as that due to the application of a unit step function to the same system.

Solution

The overshoot is represented by the maximum of the first peak of the system response, as shown in Figure 7.57. This maximum is given by:

$$\frac{dY(t)}{dt} = 0$$

or:
$$-\left[\frac{1}{\sqrt{(1-\zeta^2)}} \exp\left(-\frac{\zeta t}{\tau}\right)\right]\left[\frac{\zeta\sqrt{(1-\zeta^2)}}{\tau} \cos\left(\sqrt{(1-\zeta^2)}\frac{t}{\tau}\right)\right.$$
$$\left. -\frac{(1-\zeta^2)}{\tau} \sin\left(\sqrt{(1-\zeta^2)}\frac{t}{\tau}\right)\right] + \left[\zeta \sin\left(\sqrt{(1-\zeta^2)}\frac{t}{\tau}\right)\right.$$
$$\left. + \sqrt{(1-\zeta^2)} \cos\left(\sqrt{(1-\zeta^2)}\frac{t}{\tau}\right)\right]\left[\frac{\zeta}{\tau} \cdot \frac{1}{\sqrt{(1-\zeta^2)}} \exp\left(-\frac{\zeta t}{\tau}\right)\right] = 0$$

For finite t and $\alpha = \dfrac{\sqrt{(1-\zeta^2)}}{\tau}$ then:

$$(1 - \zeta^2 + \zeta^2) \sin \alpha t + [-\zeta\sqrt{(1-\zeta^2)} + \zeta\sqrt{(1-\zeta^2)}] \cos \alpha t = 0$$

Thus:
$$\sin \alpha t = 0$$
$$t = \dfrac{n\pi}{\alpha} \qquad n = 0, 1, 2 \ldots$$

For the first peak, $n = 1$ and $t = \pi/\alpha$

Hence:
$$Y(t)_{\max|n=1} = 1 - \dfrac{1}{\sqrt{(1-\zeta^2)}} \exp\left[-\dfrac{\pi\zeta}{\sqrt{(1-\zeta^2)}}\right] [\zeta \sin \pi + \sqrt{(1-\zeta^2)} \cos \pi]$$

$$= 1 + \exp\left[-\dfrac{\pi\zeta}{\sqrt{(1-\zeta^2)}}\right] \qquad \text{(i)}$$

The final value of $Y(t) = Y(\infty) = 1$

Thus, the overshoot of first peak from equation 1 is $\exp\left[-\dfrac{\pi\zeta}{\sqrt{(1-\zeta^2)}}\right]$ (ii)

For the second peak, $n = 3$ and $t = \dfrac{3\pi}{\alpha}$

Thus:
$$Y(t)_{\max|n=3} = 1 - \dfrac{1}{\sqrt{(1-\zeta^2)}} \exp\left[-\dfrac{3\pi\zeta}{\sqrt{(1-\zeta^2)}}\right] [\zeta \sin 3\pi + \sqrt{(1-\zeta^2)} \cos 3\pi]$$

$$= 1 + \exp\left[-\dfrac{3\pi\zeta}{\sqrt{(1-\zeta^2)}}\right]$$

$$\text{Decay ratio} = \dfrac{\text{overshoot peak 2}}{\text{overshoot peak 1}} = \exp\left[-\dfrac{3\pi\zeta}{\sqrt{(1-\zeta^2)}}\right] \bigg/ \exp\left[-\dfrac{\pi\zeta}{\sqrt{(1-\zeta^2)}}\right]$$

$$= \left\{\exp\left[-\dfrac{\pi\zeta}{\sqrt{(1-\zeta^2)}}\right]\right\}^2$$

$$= \{\text{overshoot (peak 1)}\}^2 \qquad \text{(iii)}$$

$\tau = 0.5$ min, $\quad \zeta = 0.5 \quad \sqrt{(1-\zeta^2)} = 0.866 \quad \alpha = 1.73$

Thus:
$$Y(t) = 1 - 1.15 e^{-t}(0.5 \sin 1.73t + 0.866 \cos 1.73t)$$

The forcing function of transform K is an impulse of magnitude K. The response to an impulse given by equation 7.99, Volume 3 is: $\dfrac{dY(t)}{dt} \times K = X(t)$

$$= -1.15 K e^{-t}(0.866 \cos 1.73t - 1.5 \sin 1.73t)$$
$$+ (0.5 \sin 1.73t + 0.866 \cos 1.73t)(1.15 K e^{-t})$$

$$\dfrac{X(t)}{1.15K} = e^{-t}(2 \sin 1.73t) \qquad \text{(iv)}$$

The overshoot for an impulse response is given by: $\dfrac{d}{dt}\left(\dfrac{X(t)}{1.15\,K}\right) = 0$

That is: $\qquad -2\sin 1.73t + 3.46\cos 1.73t = 0$

$\qquad\qquad \tan 1.73t = 1.73$

Thus: $\qquad 1.73t = \pi/3$ radians for peak 1

and $(2\pi + \pi/3)$ radians for peak 2

Thus: $\qquad t_{\text{peak 1}} = 0.193\pi$ min.

$\qquad t_{\text{peak 2}} = 1.35\pi$ min.

The decay ratio for the impulse response from equation (iv) $= \dfrac{e^{-1.35\pi}}{e^{-0.193\pi}} = \underline{\underline{(e^{-0.193\pi})^6}}$

Decay ratio for equivalent step response $= \left[\exp\left(-\dfrac{\pi\zeta}{\sqrt{(1-\zeta^2)}}\right)\right]^2 = (e^{-0.577\pi})^2$

$\qquad\qquad\qquad = \underline{\underline{(e^{-0.193\pi})^6}}$

and thus the two decay ratios are the same.

PROBLEM 7.8

Air containing ammonia is contacted with fresh water in a two-stage countercurrent bubble-plate absorber. L_n and V_n are the molar flowrates of liquid and gas respectively leaving the nth plate. x_n and y_n are the mole fractions of NH_3 in liquid and gas respectively leaving the nth plate. H_n is the molar holdup of liquid on the nth plate. Plates are numbered up the column.

A. Assuming (a) temperature and total pressure throughout the column to be constant, (b) no change in molar flowrates due to gas absorption, (c) plate efficiencies to be 100 per cent, (d) the equilibrium relation to be given by $y_n = mx_n^* + b$, (e) the holdup of liquid on each plate to be constant and equal to H, and (f) the holdup of gas between plates to be negligible, show that the variations of the liquid compositions on each plate are given by:

$$\dfrac{dx_1}{dt} = \dfrac{1}{H}(L_2 x_2 - L_1 x_1) + \dfrac{mV}{H}(x_0 - x_1)$$

$$\dfrac{dx_2}{dt} = \dfrac{mV}{H}(x_1 - x_2) - \dfrac{1}{H}L_2 x_2$$

where $V = V_1 = V_2$.

B. If the inlet liquid flowrate remains constant, prove that the open-loop transfer function for the response of y_2 to a change in inlet gas composition is given by:

$$\dfrac{\bar{y}_2}{\bar{y}_0} = \dfrac{c^2/(a^2 - bc)}{\{1/(a^2 - bd)\}s^2 + \{2a/(a^2 - bc)\}s + 1}$$

where $\bar{y}_2, \bar{y}_0$ are the transforms of the appropriate deviation variables and:

$$L = L_1 = L_2, \quad a = \frac{L}{H} + \frac{mV}{H}, \quad b = \frac{L}{H}, \quad c = \frac{mV}{H}$$

Discuss the problems involved in determining the relationship between $\bar{y}$ and changes in inlet liquid flowrate.

Solution

The definitions of the various symbols are shown in Figure 7c.

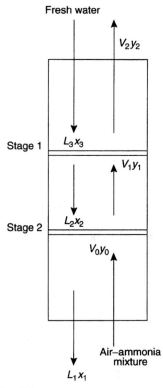

Figure 7c. Nomenclature for Problem 7.8

A. From assumptions (b) and (f): $V_1 = V_2 = V$.

An unsteady-state mass balance for ammonia around stage 1 gives:

$$L_2 x_2 + V_0 y_0 - L_1 x_1 - V_1 y_1 = d(Hx_1)/dt$$
$$= H(dx_1/dt) \tag{i}$$

From assumption (e), H, the total moles on each plate, is constant.

A mass balance for ammonia around stage 2 gives:

$$L_3 x_3 + V_1 y_1 - L_2 x_2 - V_2 y_2 = H \frac{dx_2}{dt} \tag{ii}$$

where: $x_3 = 0$.

from assumptions (d) and (c):

$$y_0 = mx_0^* + c = mx_0 + c$$
$$y_1 = mx_1^* + c = mx_1 + c$$
$$y_2 = mx_2^* + c = mx_2 + c$$

Substituting in equation (i) for y and putting $V = V_1 = V_2$, then:

$$\frac{1}{H}(L_2 x_2 - L_1 x_1) + \frac{1}{H}(Vmx_0 - Vmx_1) = \frac{dx_1}{dt}$$

Hence:

$$\frac{dx_1}{dt} = \frac{1}{H}(L_2 x_2 - L_1 x_1) + \frac{mV}{H}(x_0 - x_1) \tag{iii}$$

Substituting in equation (ii) for y and putting $V = V_1 = V_2$, then:

$$\frac{dx_2}{dt} = \frac{mV}{H}(x_1 - x_2) - \frac{1}{H} L_2 x_2 \tag{iv}$$

B. Substituting $L_1 = L_2 = L$ in equations (iii) and (iv) gives:

from equation (iii):
$$\frac{dx_1}{dt} = \frac{L}{H}(x_2 - x_1) + \frac{mV}{H}(x_0 - x_1)$$
$$= \frac{L}{H} x_2 - \left(L/H + \frac{mV}{H}\right) x_1 + \frac{mV}{H} x_0$$
$$= \beta x_2 - \alpha x_1 + \gamma x_0 \tag{v}$$

From equation (iv):
$$\frac{dx_2}{dt} = -\alpha x_2 + \gamma x_1 \tag{vi}$$

where: $\alpha = L/H + mV/H$, $\beta = L/H$ and $\gamma = \frac{mV}{H}$.

At steady-state, from equations (v) and (vi):

$$0 = \beta x_2' - \alpha x_1' + \gamma x_0'$$

and:
$$0 = -\alpha x_2' + \gamma x_1'$$

where x_0', x_1' and x_2' are steady-state concentrations. Subtracting steady-state relationships from equations (v) and (vi) respectively:

$$\frac{d\mathscr{X}_1}{dt} = \beta \mathscr{X}_2 - \alpha \mathscr{X}_1 + \gamma \mathscr{X}_0 \tag{vii}$$

and:
$$\frac{d\mathscr{X}_2}{dt} = -\alpha \mathscr{X}_2 + \gamma \mathscr{X}_1 \tag{viii}$$

where: $\quad x_0 = X_0 - X_0', \quad x_1 = X_1 - X_1' \quad \text{and} \quad x_2 = X_2 - X_2'$

and:
$$\frac{dx_1}{dt} = \frac{d}{dt}(X_1 - X_1') = \frac{dX_1}{dt}$$
$$\frac{dx_2}{dt} = \frac{d}{dt}(X_2 - X_2') = \frac{dX_2}{dt}$$

Transforming, using equations (vii) and (viii), gives:

$$s x_1(s) = \beta x_2(s) - \alpha x_1(s) + \gamma x_0(s) \qquad \text{(ix)}$$
$$s x_2(s) = -\alpha x_2(s) + \gamma x_1(s). \qquad \text{(x)}$$

From equation (x):
$$x_1(s) = \frac{s+\alpha}{\gamma} x_2(s)$$

Substituting in equation (ix):

$$\frac{s(s+\alpha)}{\gamma} x_2(s) = \beta x_2(c) - \frac{\alpha(s+\alpha)}{\gamma} x_2(s) + \gamma x_0(s)$$

Thus:
$$\frac{x_2(s)}{x_0(s)} = \frac{\gamma^2}{s(s+\alpha) - \beta\gamma + \alpha(s+\alpha)}$$

$$= \frac{\gamma^2}{s^2 + 2\alpha s + (\alpha^2 - \beta\gamma)}$$

But: $\quad y_n = m x_n^* + c = m x_n + c$

Thus: $\quad y_n - y_n' = m(x_n - x_n') \quad \text{or} \quad y_n = m x_n \quad \text{and} \quad y_n(s)^2 m x$

and:
$$\frac{x_2(s)}{x_0(s)} = \frac{m y_2(s)}{m y_0(s)} = \frac{\gamma^2/(\alpha^2 - \beta\gamma)}{[1/(\alpha^2 - \beta\gamma)]s^2 + [2\alpha/(\alpha^2 - \beta\gamma)]s + 1}$$

PROBLEM 7.9

A proportional controller is used to control a process which may be represented as two non-interacting first-order lags each having a time constant of 600 s (10 min). The only other lag in the closed loop is the measuring unit which can be approximated by a distance/velocity lag equal to 60 s (1 min). Show that, when the gain of a proportional controller is set such that the loop is on the limit of stability, the frequency of the oscillation is given by:

$$\tan \omega = \frac{-20\omega}{1 - 100\omega^2}$$

Solution

A block diagram is given in Figure 7d.

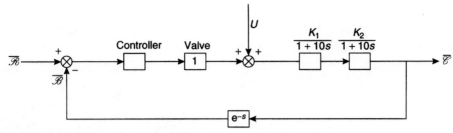

Figure 7d. Block Diagram for Problem 7.9

As discussed in Section 7.10.4, Volume 3, the Bode stability criterion states that the total open loop phase shift is $-\pi$ radians at the limit of stability of the closed loop system.

Phase shift of each first-order lag $= \tan^{-1}(-10\omega)$ radians   (equation 7.92)

Phase shift of DV lag $= -\omega$ radians   (equation 7.97)

Thus: Total phase shift $= \tan^{-1}(-10\omega) + \tan^{-1}(-10\omega) - \omega = -\pi$

$$\tan^{-1}(-10\omega) = 1/2(\omega - \pi)$$

$$-10\omega = \tan[1/2(\omega - \pi)]$$

$$= \pm\sqrt{\left[\frac{1 - \cos(\omega - \pi)}{1 + \cos(\omega - \pi)}\right]}$$

Thus: $$100\omega^2 = \frac{1 + \cos\omega}{1 - \cos\omega}$$

$$\cos\omega = \frac{100\omega^2 - 1}{100\omega^2 + 1}$$

and: $$\tan\omega = \frac{20\omega}{100\omega^2 - 1}$$

$$= -\frac{20\omega}{1 - 100\omega^2}$$

PROBLEM 7.10

A control loop consists of a proportional controller, a first-order control valve of time constant τ_v and gain K_v and a first-order process of time constant τ_1 and gain K_1. Show that, when the system is critically damped, the controller gain is given by:

$$K_c = \frac{(E-1)^2}{4EK_vK_1} \text{ where } E = \frac{\tau_v}{\tau_1}$$

If the desired value is suddenly changed by, an amount ΔR when the controller is set to give critical damping, show that the error ϵ will be given by:

$$\frac{\varepsilon}{\Delta R} = \frac{4E}{(1+E)^2} + \left\{\left[\frac{(1-E)^2}{2E(1+E)}\right]\frac{t}{\tau_1} + \frac{(1-E)^2}{(1+E)^2}\right\}\exp\left[-\left(\frac{1+E}{2E}\right)\frac{t}{\tau_1}\right]$$

Solution

A block diagram for this problem is shown as Figure 7e.

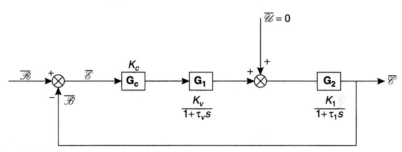

Figure 7e. Block diagram for Problem 7.10

The closed loop transfer function is:

$$\frac{\overline{\mathscr{C}}}{\overline{\mathscr{R}}} = \frac{(K_c)\left(\dfrac{K_v}{1+\tau_v s}\right)\left(\dfrac{K_1}{1+\tau_1 s}\right)}{(1+K_c)\left(\dfrac{K_v}{1+\tau_v s}\right)\left(\dfrac{K_1}{1+\tau_1 s}\right)} \quad \text{(i)}$$

The characteristic equation is:

$$K_c K_v K_1 + (1+\tau_v s)(1+\tau_1 s) = 0 \quad \text{(equation 7.119)}$$

$$\tau_r \tau_1 s^2 + (\tau_v + \tau_1)s + (1 + K_c K_v K_1) = 0$$

For critical damping, the roots of the characteristic equation are equal, hence:

$$(\tau_v + \tau_1)^2 = 4\tau_v \tau_1 (1 + K_c K_v K_1) \quad \text{(ii)}$$

$$\left(\frac{\tau_v}{\tau_1}\right)^2 - 2\left(\frac{\tau_v}{\tau_1}\right) + 1 = 4\frac{\tau_v}{\tau_1} K_c K_v K_1$$

If: $\quad E = \tau_v/\tau_1$

then: $\quad E^2 - 2E + 1 = 4E K_c K_v K_1$

and: $\quad K_c = \dfrac{(E-1)^2}{4E K_v K_1} \quad \text{(iii)}$

From the block diagrams: $\quad \overline{\varepsilon} = \overline{\mathscr{R}} - \overline{\mathscr{B}} = \overline{\mathscr{R}} - \overline{\mathscr{C}}$

and: $\quad \dfrac{\overline{\varepsilon}}{\overline{\mathscr{R}}} = 1 - \dfrac{\overline{\mathscr{C}}}{\overline{\mathscr{R}}} \quad \text{(iv)}$

From equation (iv): $\quad \dfrac{\overline{\mathscr{C}}}{\overline{\mathscr{R}}} = \dfrac{K}{K + (1+\tau_v s)(1+\tau_1 s)},$

where, from equation (iii): $\quad K = K_c K_v K_1 = (E-1)^2/4E$

For a step of magnitude ΔR in $\overline{\mathscr{R}}$, then:

$$\overline{\mathscr{R}} = \Delta R/s \text{ and } \overline{\mathscr{C}} = \left(\frac{\Delta R}{s}\right)\left(\frac{K}{\tau_v\tau_1 s^2 + (\tau_v + \tau_1)s + (1+K)}\right)$$

Thus:
$$\frac{\overline{\mathscr{C}}}{\Delta R} = \left(\frac{1}{s}\right)\left[\frac{(K/\tau_v\tau_1)}{s^2 + \left(\frac{\tau_v + \tau_1}{\tau_v\tau_1}\right)s + \left(\frac{1+K}{\tau_v\tau_1}\right)}\right]$$

As the system is critically damped, the roots of the denominator must be equal, that is it factorises to give $(s+a)^2$,

where:
$$a = \frac{(\tau_v + \tau_1)}{2\tau_v\tau_1}$$

Thus:
$$\frac{\overline{\mathscr{C}}}{\Delta R} = \frac{K}{\tau_v\tau_1}\left[\left(\frac{1}{s}\right)\left(\frac{1}{(s+a)^2}\right)\right]$$

$$= \frac{K}{\tau_v\tau_1 a^2}\left[\frac{1}{s} - \frac{a}{(s+a)^2} - \frac{1}{s+a}\right]$$

Inverting:
$$\frac{\mathscr{C}}{\Delta R}(t) = \frac{K}{\tau_v\tau_1 a^2}[1 - (1+at)e^{-at}]$$

From equation (iii):
$$K = K_c K_v K_1 = \frac{(E-1)^2}{4E}$$

From equation (ii):
$$\tau_v\tau_1 a^2 = \frac{(\tau_v + \tau_1)^2}{4\tau_v\tau_1} = (1+K)$$

Hence:
$$\frac{\mathscr{C}}{\Delta R}(t) = \frac{K}{1+K}[1 - (1+at)e^{-at}]$$

where:
$$\frac{K}{1+K} = \frac{(E-1)^2/4E}{1 + (E-1)^2/4E} = \frac{(E-1)^2}{(E+1)^2}$$

and:
$$at = \left(\frac{1 + 1/E}{2}\right)\left(\frac{t}{\tau_1}\right) = \left(\frac{E+1}{2E}\right)\left(\frac{t}{\tau_1}\right)$$

Hence, from equation (iv):

$$\frac{\varepsilon}{\Delta R} = 1 - \left(\frac{\mathscr{C}}{\Delta R}\right) = \left(\frac{1}{1+K}\right) + \left(\frac{K}{1+K}\right)(1+at)e^{-at}$$

$$= \frac{4E}{(E+1)^2} + \frac{(E-1)^2}{(E+1)^2}\left[\left(1 + \left(\frac{E+1}{2E}\right)\left(\frac{t}{\tau_1}\right)\right)\right]\exp\left[-\left(\frac{E+1}{2E}\right)\left(\frac{t}{\tau_1}\right)\right]$$

PROBLEM 7.11

A temperature-controlled polymerisation process is estimated to have a transfer function of:

$$G(s) = \frac{K}{(s-40)(s+80)(s+100)}$$

Show by means of the Routh–Hurwitz criterion that two conditions of controller parameters define upper and lower bounds on the stability of the feedback system incorporating this process.

Solution

A block diagram for this problem is given in Figure 7f.

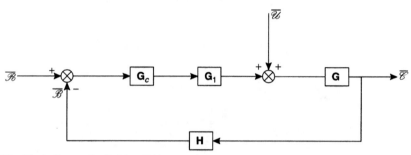

Figure 7f. Block diagram for Problem 7.11

Assuming that G_1 and H are constant and written as K_1 and K_2 respectively, that is the time constants of the final control element and measuring element are negligible in comparison with those of the process, and that the proportional controller has a gain K_C, then:

$$G_c = K_c$$

and:

$$\frac{\mathscr{C}}{\mathscr{R}} = \frac{G_c G_1 G}{1 + G_c G_1 GH}$$

The characteristic equation is: $1 + G_c G_1 GH = 0$

or: $1 + \dfrac{K_c K_1 K_2 K}{(s-40)(s+80)(s+100)} = 0$

Thus: $(s-40)(s+80)(s+100) + K_x = 0$ where $K_x = K_c K_1 K_2 K$.

and: $s^3 + 140s^2 + 800s + (K_x - 320{,}000) = 0$

From Section 7.10.2 in Volume 3, for a stable system $\underline{\underline{K_x \geq 320{,}000}}$

This is confirmed by the Routh array:

$$\begin{array}{ccc} 1 & 800 & 0 \\ 140 & (K_x - 320{,}000) & 0 \\ \left(\dfrac{432{,}000 - K_x}{140}\right) & 0 & 0 \\ (K_x - 320{,}000) & 0 & 0 \end{array}$$

Also, from the Routh array: $\underline{K_x \le 432{,}000}$

PROBLEM 7.12

A process is controlled by an industrial **PI** controller having the transfer function:

$$G_c = \frac{\tau_I s^2 + (K_c + 2\tau_I)s + 2K_c}{2\tau_I^2 s}$$

The measuring and final control elements in the control loop are described by transfer functions which can be approximated by constants of unit gain, and the process has the transfer function:

$$G_2 = \frac{1/\tau_2}{\tau_1 \tau_2 s^2 - (2\tau_1 - \tau_2)s - 2}$$

If $\tau_1 = \tfrac{1}{2}$, show that the characteristic equation of the system is given by:

$$(s + 2)\{\tau_2^2 s^2 + (1/\tau_I - 2\tau_2)s + K_c/\tau_I^2\} = 0$$

Show also that the condition under which this control loop will be stable:

(a) when:
$$K_c \ge \left(\frac{1}{2\tau_2} - \tau_I\right)^2$$

(b) when:
$$K_c < \left(\frac{1}{2\tau_2} - \tau_I\right)^2$$

is that $1/\tau_I > 2\tau_2$, provided that K_c and $\tau_2 > 0$.

Solution

A block diagram for this problem is given in Figure 7g.

$$G_1 = H = 1$$

Hence, the characteristic equation is:

$$1 + G_c G_2 = 0$$

$$1 + \left[\frac{\tau_I s^2 + (K_c + 2\tau_I)s + 2K_c}{2\tau_I^2 s}\right] \cdot \left[\frac{1/\tau_2}{\tau_1 \tau_2 s^2 - (2\tau_1 - \tau_2)s - 2}\right] = 0$$

313

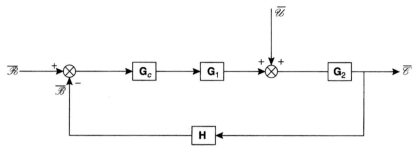

Figure 7g. Block Diagram for Problem 7.12

Putting $\tau_1 = 1/2$, then:

$$\tau_2^2\tau_I^2 s^3 - 2\tau_I^2\tau_2(1-\tau_2)s^2 + \tau_I s^2 + (K_c + 2\tau_I - 4\tau_I^2\tau_2)s + 2K_c = 0 \quad (i)$$

$$(s+2)[\tau_2^2 s^2 + (1/\tau_I - 2\tau_2)s + K_c/\tau_I^2] = 0$$

and: $\quad \tau_2^2 s^3 + 2\tau_2^2 s^2 + 1/\tau_I s^2 - 2\tau_2 s^2 + \left(\dfrac{2}{\tau_I} - 4\tau_2 + \dfrac{K_c}{\tau_I^2}\right)s + \dfrac{2K_c}{\tau_I^2} = 0 \quad$ (ii)

Equations (i) and (ii) are the identical. Hence, the roots of the characteristic equation are:

$$s = -2 \quad \text{or} \quad s = \dfrac{-\left(\dfrac{1}{\tau_I} - 2\tau_2\right) \pm \sqrt{\left[\left(\dfrac{1}{\tau_I} - 2\tau_2\right)^2 - 4\tau_2^2 K_c/\tau_I^2\right]}}{2\tau_2^2}$$

For roots to have negative real parts, that is for the system to be stable, then:

$$\underline{\dfrac{1}{\tau_I} > 2\tau_2}$$

Hence, for a stable system:

$$K_x \leq 432{,}000 \quad \text{and} \quad K_x \geq 320{,}000$$

or: $\quad \underline{\left(\dfrac{320{,}000}{K_1 K_2 K}\right) \leq K_c \leq \left(\dfrac{432{,}000}{K_1 K_2 K}\right)}$

Thus there are two value of K_c defining the upper and lower bounds on the stability of the feedback system.

PROBLEM 7.13

The following data are known for a control loop of the type shown in Figure 7.3a, Volume 3:

(a) The transfer function of the process is given by:

$$G_2 = \dfrac{1}{(0.1s^2 + 0.3s + 0.2)}$$

(b) The steady-state gains of valve and measuring elements are 0.4 and 0.6 units respectively;
(c) The time constants of both valves and measuring element may be considered negligible.

It is proposed to use one of two types of controller in this control loop, either (a) a PD controller whose action approximates to:

$$J = J_0 + K_c \varepsilon + K_D \frac{d\varepsilon}{dt}$$

where J is the output at time t. J_0 is the output $t = 0$, ε is the error, and K_C (proportional gain) = 2 units and K_D = 4 units;

or (b) an *inverse* rate controller which has the action:

$$J = J_0 + K_c \varepsilon - K_D \frac{d\varepsilon}{dt}$$

where J, J_0, ε, K_c and K_D have the same meaning and values as given previously.

Based only on the amount of offset obtained when a step change in load is made, which of these two controllers would be recommended? Would this change if the system stability was taken into account?

Having regard to the form of the equation describing the control action in each case, under what general circumstances would inverse rate control be better than normal PD control?

Solution

A block diagram for this problem is included as Figure 7h.

$$G_1 = 0.4, \quad G_2 = \frac{1}{(0.1s^2 + 0.3s + 0.2)} = \frac{10}{(s+2)(s+1)}$$

$$H = 0.6$$

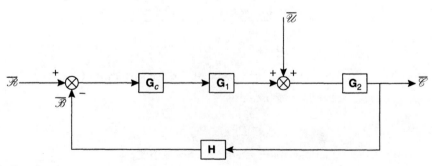

Figure 7h. Block diagram for Problem 7.13

(a) $G_c = K_c\{1 + \tau_D s\} = K_c + K_D s = 2 + 4s = 2(1 + 2s)$

Closed-loop transfer function for load change,

$$\frac{\overline{\mathscr{C}}}{\overline{\mathscr{U}}} = \frac{G_2}{1 + G_c G_1 G_2 H} = \frac{\dfrac{10}{(s+2)(s+1)}}{1 + 2(1+2s) \times 0.4 \left(\dfrac{10}{(s+2)(s+1)}\right) \times 0.6}$$

$$= \frac{10}{(s^2 + 3s + 2 + 4.8 + 9.6s)} = \frac{10}{(s^2 + 12.6s + 6.8)}$$

For unit step in load, $\overline{\mathscr{U}} = 1/s$

$$\overline{\mathscr{C}} = \frac{10}{s(s^2 + 12.6s + 6.8)}$$

Offset $= \mathscr{R}(\infty) - \mathscr{C}(\infty)$ \hfill (equation 7.113)

$\mathscr{R}(\infty) = 0$, that is no change in set point.

$\mathscr{C}(\infty) = \lim_{t \to \infty} \mathscr{C}(t) = \lim_{s \to 0} \left[s\left(\dfrac{1}{s}\right)\left(\dfrac{10}{s^2 + 12.6s + 6.8}\right)\right]$ (final value theorem, given in Section 7.8.)

$$= (10/6.8) \approx 1.5$$

Thus: Offset $= (0 - 1.5) = \underline{\underline{-1.5}}$

(b) $G_c = 2(1 - 2s)$

Thus: $\dfrac{\overline{\mathscr{C}}}{\overline{\mathscr{U}}} = \dfrac{10}{(s^2 + 3s + 2 + 4.8 - 9.6s)} = \dfrac{10}{(s^2 - 6.6s + 6.8)}$

Clearly the offset will be the same as for (a) and no recommendation can be made.

In case (a), the characteristic equation is:

$$s^2 + 12.6s + 6.8 = 0$$

Thus: $s = \dfrac{-12.6 \pm \sqrt{131.6}}{2} = \dfrac{-12.6 \pm 11.5}{2}$

$$= -1.1 \quad \text{or} \quad -24.1$$

Both roots have negative real parts (as expected) and the system is stable.

In case (b), the characteristic equation is:

$$s^2 - 6.6s + 6.8 = 0$$

Which clearly indicates an unstable system. (Section 7.10.2).

Inverse rate control is used in extremely fast processes to introduce lag so that the output is more compatible in terms of time with the final control element.

PROBLEM 7.14

A proportional plus integral controller is used to control the level in the reflux accumulator of a distillation column by regulating the top product flowrate. At time $t = 0$, the desired value of the flow controller which is controlling the reflux is increased by 3×10^{-4} m^3/s. If the integral action time of the level controller is half the value which would give a critically damped response and the proportional band is 50 per cent, obtain an expression for the resulting change in level.

The range of head covered by the level controller is 0.3 m, the range of top product flowrate is 10^{-3} m^3/s and the cross-sectional area of the accumulator is 0.4 m^2. It may be assumed that the response of the flow controller is instantaneous and that all other conditions remain the same.

If there had been no integral action, what would have been the offset in the level in the accumulator?

Solution

Flow and block diagrams for this problem are shown in Figure 7i.

The proportional band (= 50 per cent) = 100 × (error required to move the valve from fully open to fully shut)/(full range of level)

Thus: Error required = $(50 \times 0.3)/100 = 0.15$ m

Variation in controller output corresponding to this error = $0.15 K_c$.

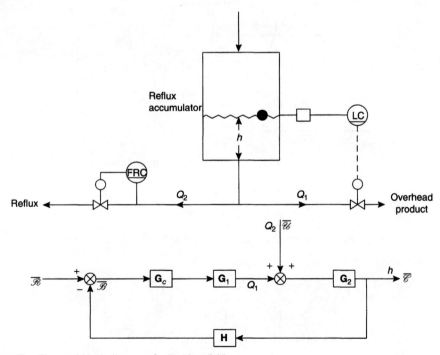

Figure 7i. Flow and block diagrams for Problem 7.14

If the valve is included, then the total variation or range of overhead product flowrate = $0.15K_c$, which is equal to the range of the reflux flowrate, or:

$$10^{-3} = 0.15K_c$$

and:
$$\underline{K_c = 6.67 \times 10^{-3}}$$

For a PI Controller: $\mathbf{G}_c = K_c\left(1 + \dfrac{1}{\tau_I s}\right)$

For the valve, assuming negligible dynamics: $\mathbf{G}_1 = K_v$

For the measuring element, again assuming negligible dynamics: $\mathbf{H} = K_m$

Assuming a linear relationship between h and Q_1, then, from Section 7.5.2:

$$\mathbf{G}_1 = \frac{K}{1 + \tau s},$$

where:
$$K = \frac{\text{steady-state change in level}}{\text{steady-state change in flowrate}}$$
$$= 0.3/10^{-3}$$
$$= 300 \text{ m s/m}^3$$

Thus:
$$\tau = AK = (0.4 \times 300) = 120 \text{ s}.$$

Assuming $K_v = K_m = 1$, then the closed loop transfer function for a load change is given by:

$$\frac{\overline{\mathscr{C}}}{\overline{\mathscr{U}}} = \frac{300/(1 + 120s)}{1 + K_c\left(1 + \dfrac{1}{\tau_I s}\right) K_v K_m K/(1 + \tau s)}$$

$$= \frac{300\tau_I s}{\tau_I \tau s^2 + \tau_I(1 + K_A)s + K_A}, \quad \text{where } K_A = KK_v K_m K_c = 2$$

$$= \frac{300\tau_I s}{120\tau_I s^2 + 3\tau_I s + 2}$$

For a critically-damped system, the roots of characteristic equation are equal, or:

$$9\tau_I^2 = 4 \times 2 \times 120\tau_I$$

and:
$$\tau_I = 106.7 \text{ s}$$

Thus: the integral time of the level controller is $(106.7/2) = \underline{\underline{53 \text{ s}}}$

For a step change in load of 3×10^{-4} m^3/s, then:

$$\overline{\mathscr{U}} = 3 \times 10^{-4} \text{ s}^{-1}$$

Thus:
$$\overline{\mathscr{C}} = (3 \times 10^{-4} \text{ s}^{-1})\left(\frac{300\tau_I s}{120\tau_I s^2 + 3\tau_I s + 2}\right)$$

$$= \frac{2.39s}{s(3180s^2 + 79.5s + 1)}$$

If there is no integral action, then:

$$\frac{\overline{\mathscr{E}}}{\overline{\mathscr{U}}} = \frac{K}{\tau s + (K_A + 1)}$$

From Section 7.9.3:

$$\text{the offset} = \lim_{t \to \infty} \mathscr{R}(t) - \lim_{t \to \infty} \mathscr{E}(t)$$

$$= 0 - \lim_{s \to 0}\left[s \left(\frac{3 \times 10^{-4}}{s}\right) \frac{K}{\tau s + (K_A + 1)}\right] = \underline{\underline{-0.03}}$$

PROBLEM 7.15

Draw the Bode diagrams of the following transfer functions:

(a) $G(s) = \dfrac{1}{s(1 + 6s)}$

(b) $G(s) = \dfrac{(1 + 3s) \exp(-2s)}{s(1 + 2s)(1 + 6s)}$

(c) $G(s) = \dfrac{5(1 + 3s)}{s(s^2 + 0.4s + 1)}$

Comment on the stability of the closed-loop systems having these transfer functions.

Solution

(a) $\qquad G(s) = \dfrac{1}{s(1 + 6s)}$

$$\text{Amplitude Ratio (AR)} = AR_1 \times AR_2 \qquad \text{(equation 7.104)}$$

where: $\quad AR_1$ is the AR of $G_1(s) = \dfrac{1}{s}$

and: $\quad AR_2$ is the AR of $G_2(s) = \dfrac{1}{1 + 6s}$

$$\text{Phase shift} = \psi_1 + \psi_2 \qquad \text{(equation 7.105)}$$

where: $\quad \psi_1 = $ phase shift of $G_1(s) = \dfrac{1}{s}$

and: $\quad \psi_2 = $ phase shift of $G_2(s) = \dfrac{1}{1 + 6s}$

AR_1 (and ψ_1)   On the Bode AR diagram,

In $G_1(s) = \dfrac{1}{s}$ put $s = i\omega$ (Substitution Rule — Section 7.8.4)

Thus: $$G_1(i\omega) = \dfrac{1}{i\omega} = 0 - i\left(\dfrac{1}{\omega}\right)$$

and: $$AR_1 = \dfrac{1}{\omega} \quad \text{and} \quad \psi_1 = \tan^{-1}(-\infty) = -90°$$

$\log AR_1 = -\log \omega$ which is a straight line of slope -1 on the log $AR/\log \omega$ diagram passing through $(1,1)$

ψ is $-90°$ for all values of ω

AR_2 (and ψ_2) $\quad G_2 = \dfrac{1}{1+6s}$ which is a first order transfer function

Thus: $$AR_2 = \dfrac{1}{\sqrt{1+36\omega^2}} \qquad \text{(equation 7.91)}$$

and: $$\psi_2 = \tan^{-1}(-6\omega) \qquad \text{(equation 7.92)}$$

The Bode diagram for a first order system is given in Figure 7.45.

The Bode diagram (Figure 7j) shows plots for $G_1(s)$, $G_2(s)$ and $G(s)$ as amplitude ratio against frequency. Only the asymptotes (Section 7.10.4, Volume 3) are plotted.

For the phase shift plot ψ against ω for G_2 is required.

ω	$\psi° = \tan^{-1}(-6\omega)$
0.02	-7
0.05	-17
0.1	-31
0.167	-45
0.2	-50
0.3	-61
0.6	-74
1.0	-81

As the phase shift does not cut the $-180°$ line, then the equivalent closed-loop system is stable as discussed in Section 7.10.4.

(b) $$G(s) = \dfrac{(1+3s)\exp(-2s)}{s(1+2s)(1+6s)}$$

$$AR = \dfrac{1}{\omega}\sqrt{\left[\dfrac{1+9\omega^2}{(1+4\omega^2)(1+36\omega^2)}\right]}$$

ψ (in radians) $= -2\omega + \tan^{-1}(3\omega) + \tan^{-1}(-2\omega) + \tan^{-1}(-6\omega) - \dfrac{\pi}{2}$

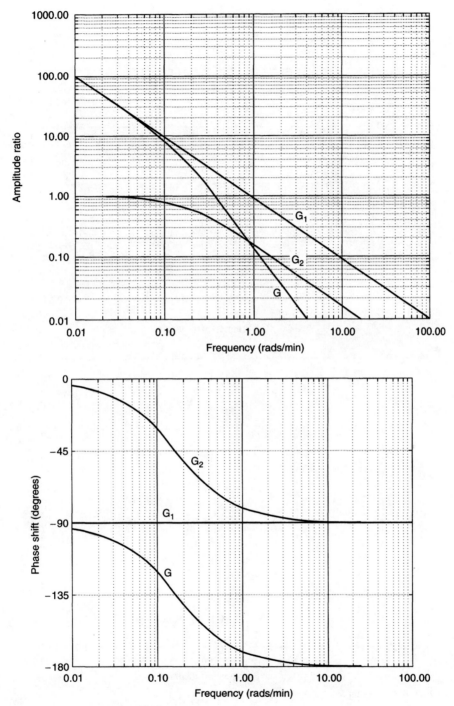

Figure 7j. Bode diagram for Problem 7.15(a)

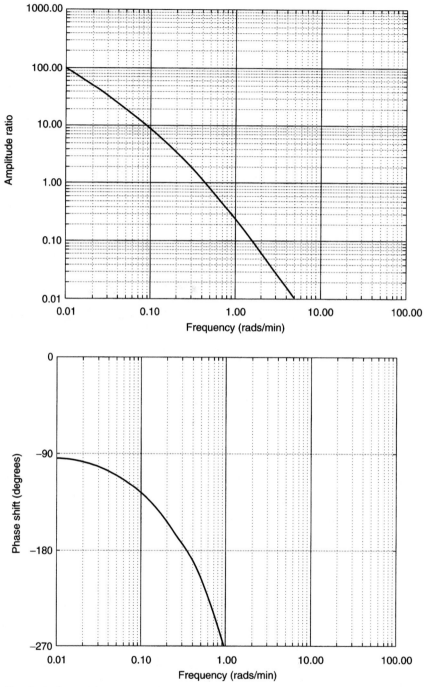

Figure 7k. Bode diagram for Problem 7.15(b)

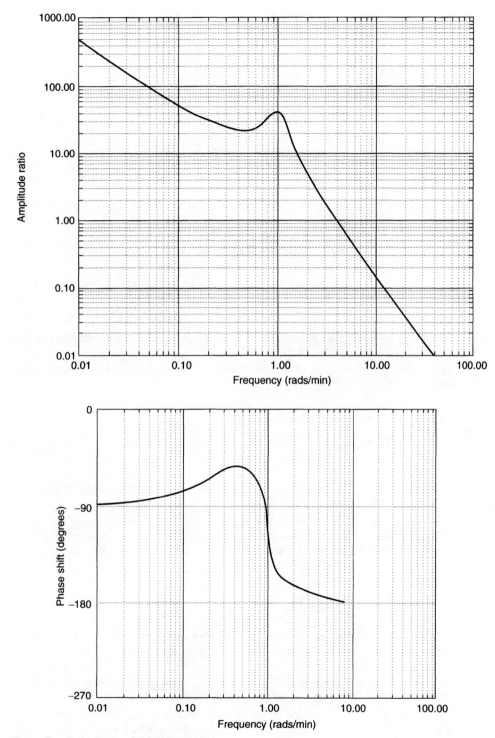

Figure 7l. Bode diagram for Problem 7.15(c)

This is plotted as in Section 7.10.4, Volume 3, AR > 1 at frequency at which $\psi = -180°$.
Thus the system is unstable by Bode criterion.

(c) $$G(s) = \frac{s(1+3s)}{s(s^2 + 0.4s + 1)}$$

$$AR = \frac{5}{\omega}\sqrt{\left[\frac{1+9\omega^2}{(1-\omega^2)^2 + (0.4\omega)^2}\right]}$$

$$\frac{\psi}{(\text{radians})} = \tan^{-1}(3\omega) + \tan^{-1}\left(\frac{-0.4\omega}{1-\omega^2}\right) - \frac{\pi}{2}$$

This is plotted as in Section 7.10.4.

The phase shift plot does not cut the $-180°$ line, and hence the system is stable.

PROBLEM 7.16

The transfer function of a process and measuring element connected in series is given by:

$$(2s+1)^{-2} \exp(-0.4s)$$

(a) Sketch the open-loop Bode diagram of a control loop involving this process and measurement lag (but without the controller).
(b) Specify the maximum gain of a proportional controller to be used in this control system without instability occurring.

Solution

$$G(s) = 0.9e^{-0.4s}/(2s+1)^2$$

This may be split into a DV lag, $e^{-0.4s} = G_1$, and two first-order systems, each $G_2 = 1/(1+2s)$.

Amplitude Ratio Plot
G_2: For the plot of AR and ω for the first-order systems, the LFA is AR = 1.
    The HFA is a straight line of slope -1 passing through AR = 1, $\omega_c = 1/\tau$ where τ is the time constant. $\tau = 2$ and hence $\omega_c = 0.5$.

$$1/(1+2s)^2 = [1/(1+2s)][1/(1+2s)]$$

constitutes two first-order systems in series and hence an overall AR plot may be produced for both together which will be the same as for G_2 other than that the HFA will have a gradient of -2.
    G_1: The AR plot for a DV lag is AR = 1, which does not affect the overall AR plot. Thus, the plot for $e^{-0.4s}/(2s+1)^2$ is the same as the AR plot for $1/(2s+1)^2$.
    The steady-state gain of 0.9 can be included in two ways: either start the overall AR plot at AR = 0.9 instead of AR = 1, or plot AR/0.9 on the vertical scale and start at AR = 1. In the latter case, the gain of the proportional controller calculated for the given

gain margin must be divided by the steady-state gain. The latter procedure will be used here as shown in Figure 7m.

Phase-Shift Plot

G_2: On the basis of two identical first order systems in series, the following is obtained:

ω	ψ for $1/(2s+1)^2$
LFA	$0°$
0.1	$-23°$
0.3	$-62°$
0.5	$-90°$
1.0	$-127°$
5.0	$-169°$
HFA	$-180°$

as $\psi = 2\tan^{-1}(-\omega\tau) = 2\tan^{-1}(-2\omega)$.

G_1: DV lag: $\psi = -\omega\tau$ radians

ω	ψ
0.1	$-2°$
0.5	$-11°$
1.0	$-22°$
2.0	$-44°$

For ω_{co}, ω is found at $\psi = -180°$

Thus: $\omega_{co} = 1.6$ radians/unit time.

From the AR diagram at ω_{co}:

$$\frac{AR}{0.9} = 0.1$$

Hence the maximum allowable gain $= (1/0.1) = 10$ which is for $(AR/0.9)$. Thus the maximum gain for the proportional controller if the system is to remain stable $= (10/0.9) = 11.1$.

Thus, the value of K_c to give a gain margin of $1.8 = (11.1/1.8) = \underline{\underline{6.2}}$

For a phase margin of $35°$, the allowable ψ on the phase-shift diagram $= -180° - (-35°) = -145°$

For this value of ψ, the ω_{co} is 0.93 radians/unit time.

From the AR plot: $\qquad \dfrac{AR}{0.9} = 0.3$

Thus: $\qquad K_c$ for $35°$ phase margin $= \dfrac{1}{(0.3 \times 0.9)}$

$$= \underline{\underline{3.7}}$$

In practice the lower value of $K_c = 3.7$ would be used for safety.

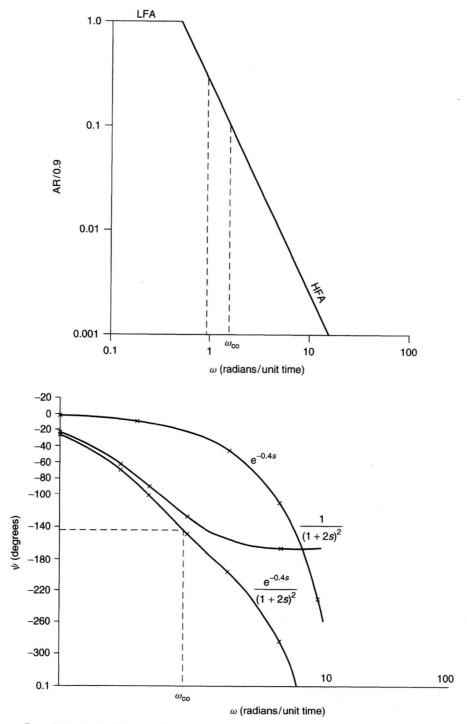

Figure 7m. Bode plot for Problem 7.16

PROBLEM 7.17

A control system consists of a process having a transfer function G_p, a measuring element H and a controller G_C.

If $G_p = (3s + 1)^{-1} \exp(-0.5s)$ and $H = 4.8(1.5s + 1)^{-1}$, determine, using the method of Ziegler and Nichols, the controller settings for P, PI, PID controllers.

Solution

The relevant block-diagram is shown in Figure 7n.

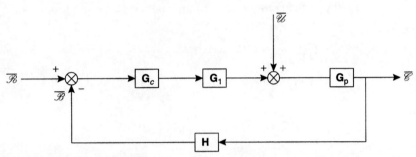

Figure 7n. Block diagram for Problem 7.17

To use the Ziegler-Nichols rules, it is necessary to plot the open-loop Bode diagram without the controller. All other transfer functions are assumed to be unity.

G_p: This consists of a DV lag of $e^{-0.5s}$ and a first order system, $1/(3s + 1)$

With the AR plot, only the first-order system contributes:

LFA: AR = 1

HFA: The slope $= -1$, passing through AR $= 1$, $\omega_c = 1/\tau = 0.33$ radians/minute.

With the ψ plot, for the first order part:

$$\psi = \tan^{-1}(-\omega\tau) = \tan^{-1}(-3\omega)$$

which gives:

ω	ψ
0	0°
0.1	−17°
0.2	−31°
0.33	−48°
0.8	−67°
2.0	−81°

With the DV lag plot:

$$\psi_{DV} = -\omega\tau_{DV} = -0.5\omega \text{ radians.}$$

which gives:

ω	ψ
0.5	$-14°$
1.0	$-29°$
2.0	$-57°$
3.0	$-86°$
5.0	$-143°$
8.0	$-229°$

H *(measuring element):*

This is first order with a time constant of 1.5 min. and a steady-state gain of 4.8.

With the AR plot,

LFA: AR = 1

HFA: The slope is -1 and $\omega_c = (1/1.5) = 0.67$

With the ψ plot,

$$\psi = \tan^{-1}(-\omega\tau) = \tan^{-1}(-1.5\omega)$$

which gives:

ω	ψ
0	$0°$
0.2	$-17°$
0.5	$-37°$
0.67	$-45°$
1.0	$-56°$
2.0	$-72°$
∞	$-90°$

The Bode diagrams are plotted for these in Figure 7o. The asymptotes on the AR plots are summed and the sums on the ψ plots are obtained by linear measurement.

From the overall ψ plots, at $\psi = -180°$, $\omega_{co} = 1.3$ radians/min and the corresponding value of AR = 0.123. By the Ziegler–Nichols procedure, this means that:

$$K_u = \frac{1}{(0.123)} = 8.13$$

Taking into account the steady-state gain of the measuring elements, K_u will be reduced by a factor of 4.8 as both together will give the total gain of the loop.

Hence:
$$K_u = \left(\frac{8.13}{4.8}\right) \approx \underline{\underline{1.7}}$$

and:
$$T_u = \left(\frac{2\pi}{\omega_{co}}\right) = \frac{2\pi}{1.3} \approx \underline{\underline{4.8}} \text{ min.}$$

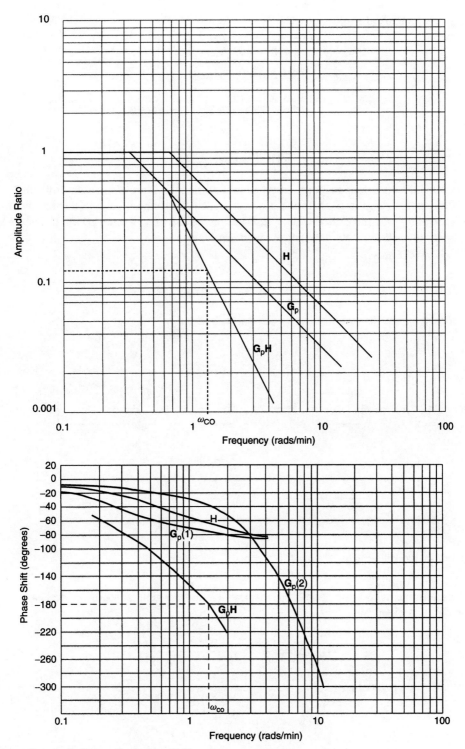

Figure 7o. Bode diagram for Problem 7.17

From the Ziegler–Nichols settings:

$$\text{for P controller } K_c = 0.5K_u = 0.85$$

$$\text{PI } K_c = 0.77 \quad \text{and} \quad \tau_I = 4 \text{ min.}$$

$$\underline{\underline{\text{PID } K_c = 1.0, \quad \tau_I = 2.4 \text{ min}, \quad \tau_s = 0.6 \text{ min}}}$$

PROBLEM 7.18

Determine the open-loop response of the output of the measuring element in Problem 7.17 to a unit step change in input to the process. Hence determine the controller settings for the control loop by the Cohen-Coon and ITAE methods for P, PI and PID control actions. Compare the settings obtained with those in Problem 7.17.

Solution

A block diagram for this problem is shown in Figure 7p.

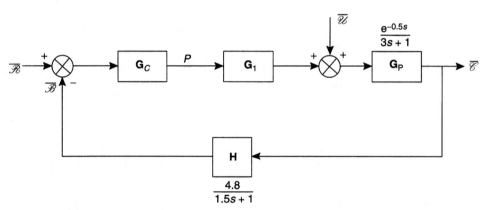

Figure 7p. Block diagram for Problem 7.18

From Problem 7.17, the open-loop transfer function for generation of the process reaction curve is given by:

$$G_{OL}(S) = \frac{\overline{\mathscr{B}}}{\overline{\mathscr{P}}} = \frac{4.8e^{-0.5s}}{(1.5s+1)(3s+1)}$$

For unit step change in P, then:

$$\overline{\mathscr{B}} = \frac{1}{s} \cdot \frac{4.8e^{-0.5s}}{(1.5s+1)(3s+1)}$$

In order to determine $\overline{\mathscr{B}}$ without the dead-time, then:

$$\overline{\mathscr{B}} = \frac{4.8}{s(1.5s+1)(3s+1)} = \frac{1.07}{s(s+0.67)(s+0.33)}$$

Thus:
$$\frac{\overline{\mathscr{B}}}{4.9} = \frac{1}{s} + \frac{1}{s+0.67} - \frac{2}{s+0.33}$$

and:
$$\mathscr{B} = 4.9\{1 + e^{-0.67t} - 2e^{-0.33t}\} \tag{i}$$

Including the dead-time, $\mathscr{B}$ will start to respond according to equation 1, 0.5 minutes after the unit step change in $\mathscr{P}$.

(a) Using the Cohen–Coon method, then from the process reaction curve shown in Figure 7q:

$$\tau_{ad} = 1.13 \text{ min}, \quad \tau_a = 6.0 \text{ min}, \quad K_r = 4.9$$

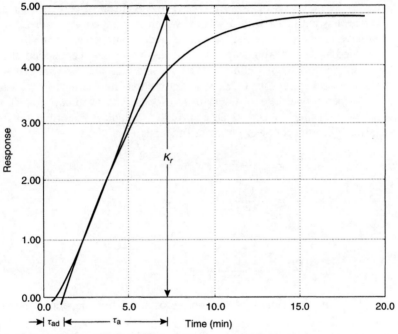

Figure 7q. Process reaction curve for Problem 7.18

From Table 7.4 in Volume 3:

For P action:
$$K_c = \left(\frac{1}{K_r}\right)\frac{\tau_a}{\tau_{ad}}\left(1 + \frac{\tau_{ad}}{3\tau_a}\right) = 1.2$$

For PI action:
$$K_c = 0.99 \quad \text{and} \quad \tau_I = 2.7 \text{ min.}$$

and for PID action, the results are obtained in the same way.

(b) Using the Integral Criteria (ITAE) method, then from Table 7.5 and equations 7.139–7.142 in Volume 3:

For P action:

$$\Upsilon = \sigma_1 \left(\frac{\tau_{ad}}{\tau_a}\right)^{\sigma_2} = 0.49 \left(\frac{1.13}{6}\right)^{-1.084} = 2.99 = K_r K_c$$

Thus:
$$\underline{\underline{K_c = 0.61}}$$

For PI action:
$$\underline{\underline{K_c = 0.90}} \quad \text{and} \quad \underline{\underline{\tau_I = 2.9 \text{ min } (174 \text{ s})}}$$

and for PID action, the results are obtained in the same way.

PROBLEM 7.19

A continuous process consists of two sections A and B. Feed of composition X_1 enters section A where it is extracted with a solvent which is pumped at a rate L_1 to A. The raffinate is removed from A at a rate L_2 and the extract is pumped to a cracking section B. Hydrogen is added at the cracking stage at a rate L_3 whilst heat is supplied at a rate Q. Two products are formed having compositions X_3 and X_4. The feed rate to A and L_1 and L_2 can easily be kept constant, but it is known that fluctuations in X_1 can occur. Consequently a feed-forward control system is proposed to keep X_3 and X_4 constant for variations in X_1 using L_3 and Q as controlling variables. Experimental frequency response analysis gave the following transfer functions:

$$\frac{\bar{x}_2}{\bar{x}_1} = \frac{1}{s+1}, \quad \frac{\bar{x}_3}{\bar{l}_3} = \frac{2}{s+1}$$

$$\frac{\bar{x}_3}{\bar{q}} = \frac{1}{2s+1}, \quad \frac{\bar{x}_4}{\bar{l}_3} = \frac{s+2}{s^2+2s+1}$$

$$\frac{\bar{x}_4}{\bar{q}} = \frac{1}{s+2}, \quad \frac{\bar{x}_3}{\bar{x}_2} = \frac{2s+1}{s^2+2s+1}$$

$$\frac{\bar{x}_4}{\bar{x}_2} = \frac{1}{s^2+2s+1}$$

where $\bar{x}_1, \bar{x}_2, \bar{l}_3, \bar{q}$, etc., represent the transforms of small time-dependent perturbations in X_1, X_2, L_3, Q, etc.

Determine the transfer functions of the feed-forward control scheme assuming linear operation and negligible distance–velocity lag throughout the process. Comment on the stability of the feed-forward controllers you design.

Solution

A diagram of this process is shown in Figure 7r.

$$G_{11} = \bar{x}_3/\bar{x}_1, \quad G_{12} = \bar{x}_3/\bar{S}_2, \quad G_{13} = \bar{x}_3/\bar{Q},$$

$$G_{21} = \bar{x}_4/\bar{x}_1, \quad G_{22} = \bar{x}_4/\bar{S}_2, \quad \text{and} \quad G_{23} = \bar{x}_4/\bar{Q}.$$

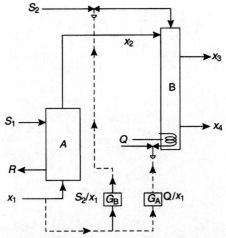

Figure 7r. Process diagram for Problem 7.19

If x_1, s_2 and Q all vary at the same time, then by the principle of superposition:

$$\bar{x}_3 = \mathbf{G}_{11}\bar{x}_1 + \mathbf{G}_{12}\bar{S}_2 + \mathbf{G}_{13}\bar{Q}$$

and:
$$\bar{x}_4 = \mathbf{G}_{21}\bar{x}_1 + \mathbf{G}_{22}\bar{S}_2 + \mathbf{G}_{23}\bar{Q}.$$

The control criterion is that x_3 and x_4 do not vary. That is $\bar{x}_3 = \bar{x}_4 = 0$ if these are deviation variables.

Thus:
$$\mathbf{G}_{11}\bar{x}_1 + \mathbf{G}_{12}\bar{S}_2 + \mathbf{G}_{13}\bar{Q} = 0$$

and:
$$\mathbf{G}_{21}\bar{x}_1 + \mathbf{G}_{22}\bar{S}_2 + \mathbf{G}_{23}\bar{Q} = 0$$

Hence, eliminating $\bar{S}_2$:

$$\mathbf{G}_A = \bar{Q}/\bar{x}_1 = \frac{\mathbf{G}_{22}\mathbf{G}_{11} - \mathbf{G}_{21}\mathbf{G}_{12}}{\mathbf{G}_{12}\mathbf{G}_{23} - \mathbf{G}_{22}\mathbf{G}_{13}}$$

From the given data:

$$\mathbf{G}_{22} = \bar{x}_4/\bar{S}_2 = \frac{s+2}{s^2+2s+1} = \frac{s+2}{(s+1)^2}$$

$$\mathbf{G}_{11} = \bar{x}_3/\bar{x}_1 = \bar{x}_3/\bar{x}_2 \cdot \bar{x}_2/\bar{x}_1 = \frac{2s+1}{(s+1)^3}$$

$$\mathbf{G}_{21} = \bar{x}_4/\bar{x}_1 = \bar{x}_4/\bar{x}_2 \cdot \bar{x}_2/\bar{x}_1 = \frac{1}{(1+s)^3}$$

$$\mathbf{G}_{12} = \bar{x}_3/\bar{S}_2 = \frac{2}{1+s}$$

$$\mathbf{G}_{23} = \bar{x}_4/\bar{Q} = \frac{1}{2+s}$$

$$\mathbf{G}_{13} = \bar{x}_3/\bar{Q} = \frac{1}{1+2s}$$

Thus: $G_A = \overline{Q}/\overline{x}_1 = \dfrac{\dfrac{s+2}{(s+1)^2} \cdot \dfrac{2s+1}{(s+1)^3} - \left[\dfrac{1}{(1+s)^3}\right]\left(\dfrac{2}{1+s}\right)}{\left(\dfrac{2}{1+s}\right)\left(\dfrac{1}{2+s}\right) - \left[\dfrac{s+2}{(s+1)^2}\right]\left(\dfrac{1}{1+2s}\right)}$ and so on.

Similarly, eliminating $\overline{Q}$:

$$G_B = \overline{S}_2/\overline{x}_1 = \dfrac{G_{23}G_{11} - G_2G_{13}}{G_{13}G_{22} - G_{23}G_{12}}$$ and so on.

PROBLEM 7.20

The temperature of a gas leaving an electric furnace is measured at X by means of a thermocouple. The output of the thermocouple is sent, via a transmitter, to a two-level solenoid switch which controls the power input to the furnace. When the outlet temperature of the gas falls below 673 K (400°C) the solenoid switch closes and the power input to the furnace is raised to 20 kW. When the temperature of the gas falls below 673 K (400°C) the switch opens and only 16 kW is supplied to the furnace. It is known that the power input to the furnace is related to the gas temperature at X by the transfer function:

$$G(s) = \dfrac{8}{(1+s)(1+s/2)(1+s/3)}$$

The transmitter and thermocouple have a combined steady-state gain of 0.5 units and negligible time constants. Assuming the solenoid switch to act as a standard on–off element determine the limit of the disturbance in output gas temperature that the system can tolerate.

Solution

Apart from the non-linear element **N**:

the open-loop transfer function $= \dfrac{0.5 \times 8}{(1+s)(1+1/2s)(1+1/3s)}$

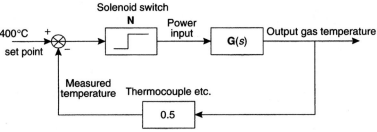

Figure 7s. Block diagram for Problem 7.20

In order to determine the maximum allowable value of **N**, it is necessary to determine the real part of the open-loop transfer function when the imaginary part is zero.

Thus, the open-loop transfer function:

$$\mathbf{G}_1(s) = \frac{4}{(1+s)(1+1/2s)(1+1/3s)}$$

Hence:
$$\mathbf{G}_1(i\omega) = \frac{24}{(1+i\omega)(2+i\omega)(3+i\omega)}$$

$$= \frac{24}{6(1-\omega^2) + i(11\omega - \omega^3)}$$

$$= \frac{24\{6(1-\omega^2) - i(11\omega - \omega^3)\}}{36(1-\omega^2)^2 + (11\omega - \omega^3)^2}$$

Thus:
$$\mathscr{R}e\{\mathbf{G}_1(i\omega)\} = \frac{24 \times 6(1-\omega^2)}{36(1-\omega^2)^2 + \omega^2(11-\omega^2)^2}$$

$$\mathscr{I}m\{\mathbf{G}_1(i\omega)\} = \frac{-24(11\omega - \omega^3)}{36(1-\omega^2)^2 + \omega^2(11-\omega^2)^2}$$

For $\mathscr{I}m = 0$: $\quad \omega^3 = 11\omega \quad \text{and} \quad \omega = 0 \quad \text{or} \quad \omega \approx \sqrt{11}$

For $\omega = \sqrt{11}$, $\quad \mathscr{R}e = -\left(\dfrac{24 \times 6 \times 10}{36 \times 100}\right)$

$$= -0.4.$$

The maximum $\mathbf{N} \times \mathscr{R}e = -1.0$ for stability by the Nyquist criterion,

Thus: $\mathbf{N} \leq 2.5$ for stability.

But, for an on–off element:

$$\mathbf{N} = \frac{4Y_0}{\pi x_0} = \frac{4 \times 2}{\pi x_0} \quad (Y_0 = \pm 2 \text{ deg K})$$

Thus:
$$\underline{\underline{x_0 = \frac{8}{2.5\pi} \approx 1 \text{ deg K}}}$$

PROBLEM 7.21

A gas-phase exothermic reaction takes place in a tubular fixed-bed catalytic reactor which is cooled by passing water through a coil placed in the bed. The composition of the gaseous product stream is regulated by adjustment of the flow of cooling water and, hence, the temperature in the reactor. The exit gases are sampled at a point X downstream from the reactor and the sample passed to a chromatograph for analysis. The chromatograph produces a measured value signal 600 s (10 min) after the reactor outlet stream has been sampled at which time a further sample is taken. The measured value signal from the chromatograph is fed via a zero-order hold element to a proportional controller having a proportional gain K_C. The steady-state gain between the product sampling point at X and the output from the hold element is 0.2 units. The output from the controller J is used to adjust the flow Q of cooling water to the reactor. It is known that the time constant of the control valve is negligible and that the steady-state gain between J and Q is 2 units. It has been determined by experimental testing that Q and the gas composition at X are related by the transfer function:

$$\mathbf{G}_{xq}(s) = \frac{0.5}{s(10s+1)} \quad \text{(where the time constant is in minutes)}$$

What is the maximum value of K_C that you would recommend for this control system?

Solution

Block diagrams for this problem are given in Figure 7t.

$$\text{Transform for zero-order hold element} = \frac{1 - e^{-Ts}}{s}$$

$$\text{Hence, total open loop } \mathbf{G}(s) = 0.2(1 - e^{-Ts})\left(\frac{K_c}{s^2(10s)}\right)$$

$$= \mathbf{G}_1(s)\mathbf{G}_2(s)$$

where: $\quad \mathbf{G}_1(s) = 1 - e^{-Ts} \quad \text{and} \quad \mathbf{G}_2(s) = \dfrac{0.2K_c}{s^2(10s+1)}$

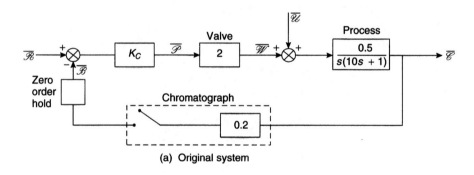

(a) Original system

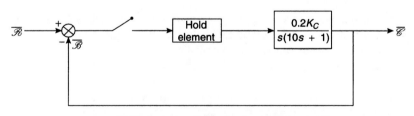

(b) Equivalent unity feedback system for stability

Figure 7t. Block diagrams for Problem 7.21

Corresponding z-transform $G_1(z) = \dfrac{z-1}{z}$

For $G_2(z)$:
$$G_2(s) = 0.2K_c\left(\dfrac{A}{s^2} + \dfrac{B}{s} + \dfrac{C}{10s+1}\right)$$

where:
$$A(10s+1) + Bs(10s+1) + Cs^2 = 1$$

putting $s = 0$, $A = 1$
Coefficients of s: $10A + B = 0$, $B = -10$
Coefficients of s^2: $10B + C = 0$, $C = 100$

Thus:
$$G_2(s) = 0.2K_c\left(\dfrac{1}{s^2} - \dfrac{10}{s} + \dfrac{10}{s+1/10}\right)$$

From the Table of z-transforms in Volume 3:

$$G_2(z) = 0.2K_c\left(\dfrac{Tz}{(z-1)^2} - \dfrac{10z}{z-1} + \dfrac{10z}{z-e^{-0.1T}}\right)$$

$$G(z) = G_1(z)G_2(z) = \dfrac{z-1}{z}\cdot G_2(z)$$

$$= 0.2K_c\left\{\dfrac{T}{z-1} - 10 + \dfrac{10(z-1)}{z-e^{-0.1T}}\right\}$$

The sampling time, $T = 10$ min (600 s)

Thus:
$$G(z) = 0.2K_c\left[\dfrac{10}{z-1} - 10 + \dfrac{10(z-1)}{z-e^{-1}}\right]$$

$$= \dfrac{2K_c(1 - 2e^{-1} + ze^{-1})}{(z-1)(z-e^{-1})}$$

The characteristic equation is $1 + G(z) = 0$

or: $(z-1)(z-e^{-1}) + 2K_c(1 - 2e^{-1} + ze^{-1}) = 0$

In order to apply Routh's criterion,

$$z = \dfrac{\lambda+1}{\lambda-1}$$

Thus: $\left(\dfrac{\lambda+1}{\lambda-1} - 1\right)\left(\dfrac{\lambda+1}{\lambda-1} - 0.368\right) + 2K_c\left[0.264 + 0.368\left(\dfrac{\lambda+1}{\lambda-1}\right)\right] = 0$

$$0.632\lambda + 1.368 + 0.632\,K_c\lambda^2 - 0.528\,K_c\lambda - 0.104\,K_c^2 = 0$$

$$0.632\,K_c\lambda^2 + (0.632 - 0.528\,K_c)\lambda + (1.368 - 0.104\,K_c) = 0$$

For the system to be stable, the coefficients must be positive. Hence, the system will be unstable if:

$$0.104 K_c \geq 1.368$$

That is:
$$K_c \geq 13.2$$

or:
$$0.528 K_c \geq 0.632$$
$$K_c \geq 1.2$$

Hence $K_c < 1.2$ for a stable system.

The recommended value of K_c is 0.8 in order to allow a reasonable gain margin.

PROBLEM 7.22

A unity feedback control loop consists of a non-linear element **N** and a number of linear elements in series which together approximate to the transfer function:

$$\mathbf{G}(s) = \frac{3}{s(s+1)(s+2)}$$

Determine the range of values of the amplitude x_0 of an input disturbance for which the system is stable where (a) **N** represents a dead-zone element for which the gradient k of the linear part is 4; (b) **N** is a saturating element for which $k = 5$; and (c) **N** is an on–off device for which the total change in signal level is 20 units.

Solution

A block diagram for this problem is shown in Figure 7u.

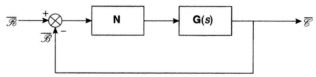

Figure 7u. Block diagram for Problem 7.22

The closed loop transfer function for a set-point change is given by:

$$\frac{\overline{\mathcal{E}}}{\overline{\mathcal{R}}} = \frac{N\mathbf{G}(s)}{1 + N\mathbf{G}(s)}, \quad \text{where } \mathbf{G}(s) = \frac{3}{s(s+1)(s+2)}$$

(a) When **N** represents a dead-zone element, then:

$$\mathbf{N} = \frac{k}{\pi}(\pi - 2\beta - \sin 2\beta) \qquad \text{(equation 7.201)}$$

where $\beta = \sin^{-1} \delta/x_0$ and the dead-zone is 2δ as shown in Figure 7.82.

338

Using the substitution rule:

$$G(i\omega) = \frac{3}{i\omega(i\omega+1)(i\omega+2)}$$

$$= \frac{-3}{3\omega^2 + i(\omega^3 - 2\omega)} = \frac{-3[3\omega^2 - i(\omega^3 - 2\omega)]}{9\omega^4 + (\omega^3 - 2\omega)^2}$$

$$= \frac{-9\omega^2}{9\omega^4 + (\omega^3 - 2\omega)^2} + i \cdot \frac{\omega^3 - 2\omega}{9\omega^4 + (\omega^3 - 2\omega)^2}$$

From equation 7.204 in Section 7.16.2, Volume 3, the conditionally stable conditions are given when:

$$G(i\omega) = -\frac{1}{N} \quad \text{as shown in Figure 7.87 in Volume 3}$$

$-\frac{1}{N}$ lies on the real axis, that is where $\mathscr{I}_m[G(i\omega)] = 0$

or where: $\qquad \omega^3 = 2\omega$

and: $\qquad \omega = 0 \quad \text{or} \quad \omega = \sqrt{2}.$

At this point: $\qquad G(i\omega) = \mathscr{R}_e[G(i\omega)] = \dfrac{-9\omega^2}{9\omega^4 + (\omega^3 - 2\omega)^2}$

When $\omega = \sqrt{2}$, then:

$$G(i\omega) = \frac{-18}{36 + \sqrt{2}(2-2)^2} = -0.5$$

That is: $\qquad -\dfrac{1}{N} = -0.5 \quad \text{and} \quad N = 2$

When $k = 4$, $N/k = 0.5$ and, from Figure 7.83 in Volume 3:

$$x_0/\delta = 2.5$$

Hence, for $k = 4$, the system will be unstable if:

$$\underline{\underline{x_0 > 2.5\delta}}$$

(b) For a saturating non-linear element with $k = 5$,

$$N/k = 2/5 = 0.4 \quad \text{(as shown in Figure 7.84.)}$$

From Figure 7.84(b), the corresponding value of x_0/δ is approximately 3.1.
Hence, in this case, the system will be unstable if:

$$\underline{\underline{x_0 > 3.1\delta}}$$

(c) For an on–off device having a total change in signal level of 20 units, $Y_0 = 10$ units as shown in Figure 7.80, Volume 3.

Thus: $\quad\quad\quad N = 4Y_0/\pi x_0 \quad\quad\quad\quad\quad\quad$ (equation 7.198)

Hence: $\quad\quad\quad x_0 = 4Y_0/N\pi = 40/2\pi = 6.4$

N is as in part (a) of the solution.

Thus, in this case, the system will be unstable if:

$$\underline{\underline{x_0 > 6.4 \text{ units}}}$$

CPSIA information can be obtained
at www.ICGtesting.com
Printed in the USA
LVOW09s1800040517
533268LV00004B/209/P